SPATIOTEMPORAL ENVIRONMENTAL HEALTH MODELLING:
A Tractatus Stochasticus

SPATIOTEMPORAL ENVIRONMENTAL HEALTH MODELLING:
A Tractatus Stochasticus

by

George Christakos

and

Dionissios T. Hristopulos

School of Public Health
Department of Environmental Sciences and Engineering
The University of North Carolina at Chapel Hill

KLUWER ACADEMIC PUBLISHERS
Boston / Dordrecht / London

Distributors for North, Central and South America:
Kluwer Academic Publishers
101 Philip Drive
Assinippi Park
Norwell, Massachusetts 02061 USA

Distributors for all other countries:
Kluwer Academic Publishers
Distribution Centre
Post Office Box 322
3300 AH Dordrecht, THE NETHERLANDS

Library of Congress Cataloging-in-Publication Data

Christakos, George.
Spatiotemporal environmental health modelling : a tractatus stochasticus / by George Christakos and Dionissios T. Hristopulos.
p. cm.
Includes bibliographical references and index.
ISBN 0-7923-8211-0 (alk. paper)
1. Environmental health--Mathematical models. 2. Stochastic processes. I. Hristopulos, Dionissios T. II. Title.
RA566.C58 1998
616.9'8'015118--dc21 98-24443
CIP

Printed on acid-free paper.

Printed in the United States of America

Dedicated to the memory of my father Panayiotis,
who gave and cared so much,
and to my mother Mary who continues to do so.

George Christakos

To my parents, Theodoros and Anna, and my wife, Lisa,
for their love and support;
To Greece, for all that she has given me.

Dionissios Hristopulos

Forward

"La pensée n'est qu'un éclair au milieu de la nuit.
Mais c'est cet éclair qui est tout"."

H. Poincaré

Protecting and promoting health at the population level requires a rigorous interdisciplinary scientific base. Health is the product of an array of complex physiological, environmental, and behavioral processes. Success in understanding and addressing these processes require the capacities of many different scientific disciplines. The science of public health has made remarkable advances in recent years as its practitioners have explored new ways of linking and applying knowledge from core disciplines in the mathematical, natural, and behavioral sciences. Environmental health is an essential element of public health science, and its concepts and methods epitomize the interdisciplinary nature of public health.

The successful integration of concepts and methods from diverse fields of study requires both creativity and depth of knowledge. At its best, such integration creates a scientific synergism that enables new insight regarding the problems and issues under study. In this book, George Christakos and Dionissios Hristopulos create just such a synergism for the practice of environmental health science. The authors explore a new and powerful approach to environmental health science --one which brings together principles from modern stochastic theory with more traditional methods of examining environmental processes and their associate health effects. This approach offers fresh views of common environmental health issues, and new insight regarding methods for assessment and control.

Public health scientists, decision-makers, and students alike have much to gain from the concepts and methods described in this book. By advancing our understanding of the natural and physiological processes which determine health, the authors make a substantial contribution to the field of public health and its mission of assuring population health. In doing so, the authors underscore the importance of interdisciplinary scientific approaches to public health issues.

William L. Roper, MD, MPH
Dean, School of Public Health
The University of North Carolina at Chapel Hill

Preface

"We haven't the money, so we've got to think."
E. Rutherford

This book is the culmination of an investigation into the applicability of modern stochastics to problems of environmental health science. The result is a stochastic interpretation of processes (physical, chemical, biological, epidemic, etc.) that account for changes in the state of the environment and human health. This interpretation exposes the inherent uncertainty in such processes, and advocates prediction in terms of spatiotemporal probability laws and statistical statements. The deterministic approach is based on the hypothesis that a complete characterization of the system is available. Deterministic theories provide inadequate representations for most environmental health processes, due to the incomplete knowledge of the underlying mechanisms and the partial characterization of model parameters. In the stochastic viewpoint the focus is shifted from the analysis of a single, perfectly determined system, to an ensemble of possible systems, which represent all the potentialities permitted by the existing knowledge of the system's properties. Thus, instead of concentrating on single system properties, the stochastic approach calculates ensemble properties. Finally, conclusions about the former are derived on the basis of the latter. Stochastic analysis can be viewed as a mathematical generalization of conventional deterministic analysis. This generalization is more flexible in practice and can handle problems in environmental health science more accurately and efficiently than the often-inadequate deterministic methods. In recent years a new synthesis has been emerging in the study of natural phenomena that combines physical models with stochastic methods, which account for uncertainties and variabilities. The new synthetic view, called modern stochastics, and its applications in environmental health science will be the focus of this book. The answers that modern stochastics supplies are not always obvious to the deterministically minded, and in certain cases deep-rooted prejudices may need adjustment.

As its title dictates, this book uses stochastic concepts and methods to build links between models and techniques of environmental sciences, on the one hand, and health sciences on the other. To our knowledge, this is a task that is not shared by any other book currently available in the literature. It is, however, a natural development, because the environment and human health are closely related and interdependent in many ways.

Rigorous quantitative analysis of the associations between, e.g., subsurface contaminant transport models and human exposure, or between space-time ozone distribution and disease incidence rates, are of great significance in health management. Such associations are crucial for site selection and human exposure assessment in epidemiologic studies. Another example is carcinogenesis, which is a random multistep process controlled both by cancer genes and environmental conditions. Separated, environmental and health models rarely lend each other assistance; only when their associations and interactions have been clearly established, they can illuminate and enrich one another.

In light of the above considerations, one may distinguish two main objectives of the book. The first is to provide a coherent and unified account of modern spatiotemporal stochastics for the environmental health scientist. At this level the book can function as a textbook as well as a reference volume, since it contains a number of novel methods and results in stochastic environmental health modelling. The second goal is to apply modern stochastics in environmental health science problems in a way that emphasizes their close relations and strong interdependence. Of course, since environmental health science is progressing very fast and in various directions, no text of any reasonable size can exhaust all the possibilities. Therefore, in this book we have chosen to focus on a number of carefully selected applications. However, modern stochastics has applications in phenomena that cover a considerably wider range than what is presented here, and practically includes every aspect of environmental health science.

Although this is definitely an application-oriented book, it contains several philosophical theses and comments on methodological issues. The reason is that, in our view, modern stochastics is not just a collection of abstract equations and techniques for problem solving. Instead, it constitutes a scientific paradigm based on a more flexible conceptualization of reality than its deterministic counterpart. Efficient use of the stochastic paradigm presupposes an appropriate conceptual framework, within which the significance and emerging applications of notions such as uncertainty, variability, heterogeneity and risk can be debated in the scientific forums. In addition, the new paradigm can help scientists develop an ability to argue coherently and convincingly on a wide range of issues, a skill which is extremely important in communicating their efforts to broader audiences in the public and political arenas.

Throughout the book our approach is conceptual, having a strong pedagogic orientation with many examples presented for edification and assimilation of the conceptual principles and the mathematical techniques. The variety of applications also provides a framework for evaluation, and a realization that there can be a systematic organization for many concepts, tools and perspectives. Furthermore, the aim of this approach has been to find quantitative solutions to environmental health problems that are serviceable to research scientists and practitioners who work in these fields.

Based on the fact that the roots of many useful stochastic concepts and tools have multiple branches that penetrate different scientific disciplines, the book has been written with a diverse audience in mind. It should be of interest to scholars and researchers in the environmental sciences, epidemiology, health sciences, statistics, risk analysis, mathematics, and decision making. Furthermore, practitioners in the above areas will find a variety of theoretical and computational tools that should be of immediate practical use. The book is also intended as a textbook for graduate-level modelling courses in environmental modelling, health risk assessment, geostatistics, descriptive epidemiology, biostatistics, and medical geography.

In writing this book our approach has been guided by the following dictum: "True education is what survives when a person has forgotten all that he/she has been taught." The aim of this book goes beyond providing a compendium of technical recipes, black-box computer algorithms and simplified remedies for certain problems. Instead, this book aims to give aspiring scientists some motivation and tools for addressing environmental health issues with a critical mind. The book should be read in the same spirit, and we encourage the reader to question the arguments that we use, and to think of counter-examples. The book is intended to stimulate thought, not to be an alternative to it. If readers examine the material presented critically, they will discover areas where they may agree or disagree with the authors. It is our hope that in either case they will clarify their own beliefs and, perhaps, prejudices in the process.

We are grateful to several colleagues for stimulating discussions, especially to C.T. Miller, D. Crawford-Brown, P. Bogaert and G. Cassiani. Valuable contributions were made by the first author's former and current students, V.M. Vyas, B.R. Killam, J-J. Lai, X. Li, D.L. Oliver, G. Thesing, A. Kolovos, M. Serre and K-M. Choi. Finally, we greatfully acknowledge funding from the National Institute of Environmental Health Sciences (Grant no. P42 ES05948-02), the Department of Energy (Grant no. DE-FC09-93SR18262), and the Army Research Office (Grant no. DAAL03-92-G-0111). Without their support, this book would have never been written, for better or for worse.

George Christakos
Dionissios Hristopulos
Chapel Hill, N.C.

Note: The following notation is used throughout the book:

- §5 denotes the 5th section in the same Chapter.
- §II.3 denotes the 3rd section of Chapter II.
- Eq. (4.2) denotes the 2nd equation of section 4 in the same Chapter.
- Eq. (IV.9.5) denotes the 5th equation of section 9 in Chapter IV. This notation is used when we refer to an equation from within a different Chapter.

Table of Contents

Chapter I: FUNDAMENTAL PRINCIPLES OF STOCHASTIC ENVIRONMENTAL HEALTH MODELLING

"If science is not to degenerate into a medley of ad hoc hypotheses, it must become philosophical and must enter upon a thorough criticism of its own foundations".
A. N. Whitehead

1. ON THE METHOD OF ENVIRONMENTAL HEALTH SCIENCE

Environmental health science is concerned with the study of connections between two distinct kinds of processes: (a) natural processes that produce pollutants which can affect the state of human health, and (b) health processes describing this state. Generally, a *process* is a coordinated series of changes in the complexion of reality, a group of events linked to each other either causally or functionally. The term *natural process*, in particular, refers to changes in the physical, chemical and biological properties of the environment (e.g., it is used to denote a soil parameter, hydrogeologic variable, exposure indicator, protein production, chemical agent concentration, climatic and atmospheric parameter). Of particular importance are natural processes which contribute (directly or indirectly) to the production and distribution of environmental pollutants in space-time. The term *health process*, on the other hand, is used to denote the variables describing changes in the health state of a receptor (human body, organ, skin, etc.) or a group of receptors (population). Health state includes diseases such as cancer or asthma and population events such as epidemics.

EXAMPLE 1: Radon concentration, precipitation, moisture, net radiation, wind velocity, ozone concentration, and cell growth rate are all specific examples of natural processes. Carcinogenesis, mesothelioma mortality, coronary heart disease incidence, breast cancer incidence, and regional counts of AIDS disease are all examples of health processes. □

Scholium 1: *In this book we will use the term environmental health processes to refer to either natural or health processes.*

Environmental health processes are emplaced in the coordinate order of space and time (space-time continuum, §3), within which they establish connections and integrations, and emphasize structured change governed by laws (physical, biological, etc.). A successful mathematical representation of these characteristics is accomplished in terms of the *field* concept (§4 and Chapter IIff).

Environmental health science is by definition based on an integration of sciences of the environment with sciences of the human body. Such an integrated study of natural phenomena and health effects may require a shift away from the *mechanistic* model of the world. According to the mechanistic model, nature is viewed as an aggregate of separately existent parts which influence each other across their separateness. In many cases this model is both necessary and appropriate in order that complicated phenomena be reduced to manageable components. A water resources project, e.g., may be divided into separate sub-projects, each one of which requires tools from a different discipline: surface hydrology, porous media hydrodynamics, meteorology, systems analysis, and soil mechanics. However, the study of environmental health effects requires a different, multidisciplinary approach introduced by the following model.

The holistic model of environmental health science: *Natural processes and the related health effects should be considered as an integrated whole that has a reality greater than the sum of its parts.*

The holistic model of environmental health sciences involves the integrated study of the following components and their interactions

$$\textit{Natural processes} \leftrightarrow \textit{Environmental pollutants} \leftrightarrow \textit{Health effects.} \tag{1}$$

The holistic model advocates that environmental health science is more complex than the sum of its components --environmental sciences and sciences of the body. Understanding an environmental health situation, e.g., may require besides the knowledge of environmental pollutant distribution and human physiology an understanding of how human physiological functions react to environmental exposure.

The basic holistic idea, therefore, is to view the world as a unified macroprocess consisting of several duly coordinated subordinate microprocesses. The main advantage of the holistic model is that it enables us to synthesize and understand the cognitive phenomena that confront us throughout the study of the environment and its effects on humans. Certain environmental health studies use existing exposure data in order to assess its health effects. Such studies are necessary to determine whether, e.g., increased exposure to ultraviolet light results in a detectable increase in melanoma mortality (Smans *et al.*, 1992); or, whether exposure to high levels of ozone concentration leads to an increase of the population ratio experiencing respiratory problems (Whitfield *et al.*, 1995). Other studies may use health damage information in search for approaches to control the pollutant distribution causing the damage.

EXAMPLE 2: Assessing the health risk of a population exposed to a certain pollutant according to the holistic model involves an estimate of exposure (perhaps based on a pollutant concentration map), and a measure of the accuracy of the estimate; determining the accuracy involves making additional assumptions or evaluating based on models properties of the natural processes that produce the pollutant distribution. □

REMARK 1: In fact, the reader may find it amusing that there exists an interesting linguistic connection between the words "health" and "holistic". The word "health" is based on the Aglo-Saxon word "hale" that means "whole". The word "science", on the other hand, is an anglicized version of the Latin "scientia", which means knowledge.

Karl Pearson, in his Grammar of Science (1951) reasoned that the unity of all science consists alone in its *method*, not in its material. The method of environmental health science is concerned with the study of observations and the formulation of logical inferences. These two components of the scientific method are closely related to each other. Progress in environmental health science is based on this relationship, which involves both inductive (from the specific to the general) and deductive (from the general to the specific) inferences.

Since environmental health science deals with real problems that concern people and their everyday environment, it should be considered an applied science. This does not mean, however, that environmental health scientists and engineers are not often confronted with very difficult problems, the solution of which requires progress in basic research.

The great task of environmental health science is to paint a coherent and understandable picture of the world that can lead to improvements in the human health. The cornerstones in this effort are:

(a) A wonderful combination of speculative creativity and openness to experience which leads to a powerful *method of thinking.*

(b) A quantification approach that allows the reduction of qualitative arguments to quantitative statements. It is this approach that makes it an *exact* science.

(c) A large box of *tools*, i.e., theories, models, and mathematical techniques that may have originated in the fundamental sciences or they have been developed for the specific problem.

The scientific method of thinking aims at reaching as high a level of coherence as possible, while avoiding confusion and contradictions. This requires that the concepts and tools be thoughtfully examined and understood, before they are employed to study real world problems. Tools may simply provide a collection of recipes for everyday use, or they may offer the means for improving one's method of thinking. Without underestimating the usefulness of the former in providing quick and efficient solutions to the everyday needs of environmental health scientists, the latter is of far greater importance for the advancement of science and the future benefit of the society.

Environmental health scientists should evaluate their work critically in the light of well-established scientific methods and sound philosophical arguments. Critical thinking combining scientific methodology and philosophical argument is absolutely essential for scientific evolution and the replacement of older models and ideas with newer, improved tools. Indeed, by analyzing the arguments for and against a research approach in light of scientific method, and by relating these arguments with well-studied theses, researchers may either build confidence in the approach, or refute it and replace it with a better one. Also, critical methodico-philosophical thinking helps scientists clarify precisely their thoughts, prejudices and beliefs. In the process, it develops an ability to argue coherently on a wide range of issues - a useful transferable skill. Such a skill is extremely important, e.g., for environmental health scientists in their efforts to communicate with the public, politicians, etc..

While critical thinking is absolutely necessary for research scientists, one may argue that it is not required by science practitioners. It can, nevertheless, have a positive influence on how practitioners understand their work. In addition, even if they do not question the soundness of the formulas and techniques they use in their everyday routine, critical thinking is necessary to stay abreast of new developments. After all in today's dynamic world, formulas and technical procedures may change many times in the course of a scientist's career. Hence, a scientist may find it more profitable in the long-run to develop the ability to think critically, rather than to rely on collecting recipes and techniques which may be soon become obsolete.

Einstein was once asked how the West had come to the idea of scientific discovery. His answer was that (Boorstin, 1994) "Development in Western Science is based on two great achievements, the invention of the formal *logical system* (in Euclidean geometry) by the Greek philosophers, and the discovery of the possibility to find out causal relationship by systematic *experiment* (Renaissance). In my opinion one has not to be astonished that the Chinese sages have not made these steps. The astonishing thing is that these discoveries were made at all." Einstein's comments emphasize the synergistic power of theoretical and experimental investigations. These statements are true today as well, since the essence of scientific method has not changed significantly. Computational investigations have gained wide acceptance in recent years, but the scientific method still involves both inductive and deductive inferences. These inferences establishes relationships between theoretical, experimental and computational investigations. In Fig. 1 an illustration is provided by

The triangle concept: *Environmental health science rests on three pillars forming a powerful triangle: theoretical, experimental, and computational investigations.*

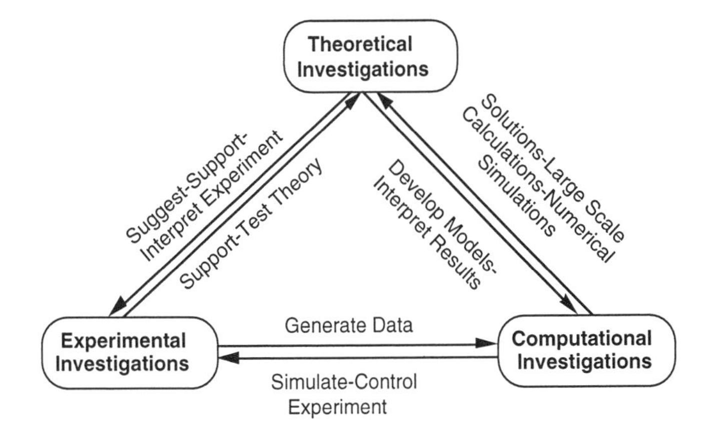

FIGURE 1: The three pillars of environmental health science.

Experimental investigations in the laboratory and in the field are primarily concerned with the accumulation of factual data and information. Theoretical investigations use mathematical tools and physical concepts, and are mainly directed towards ordering factual data and information into logically coherent patterns and laws.

REMARK 2: Theoretical and experimental investigations rely on closed interactions and mutual checks to direct them. Whether theory or experiment comes first has been the subject of a long-running debate among the methodologists of science (Poincaré, 1929; Popper, 1934; Polanyi, 1958). This seems like "the egg and the chicken question". In practice, one starts with a conceptual viewpoint and an observational framework that lead to the formulation of a model. This model suggests a set of experiments, the outcome of which provides intuition that can lead to modification of the original model. The new model predictions must be tested by improved experimental procedures. The outcomes of the new experiments can suggest additional model improvements or even lead to a completely new model that can adequately explain the observations, and so on.

Computational investigations employ numerical simulation techniques, and aim at overcoming difficulties associated with both the theoretical and experimental investigations. Computational investigations are becoming increasingly important: (i) in situations where theory is confronted with complex nonlinearities, lack of symmetries and large number of degrees of freedom; or, (ii) in cases where direct observation is hindered because experiments are either too difficult or impossible to perform, certain conditions are not easily controlled or simulated in the laboratory, or the financial cost is prohibitively high (especially in very large or very small space-time scales). Computer experiments can be used to perform detailed sensitivity analysis in a perfectly controlled environment, thus allowing evaluation of the relative importance of the numerous parameters of a complex system, before expensive technology is used to built an experimental set-up in the laboratory or in-situ. Computer simulations can also be used to study complex mathematical problems and complicated patterns by means of modern visualization techniques, and to provide valuable guidance regarding future theoretical investigations.

The three kinds of scientific investigations are closely related and complement each other. A combination of all three allows scientists to gain an understanding of the world and to obtain useful results. An impressive demonstration of the interplay between scientific methods is provided by Wegener's hypothesis for the motion of continental plates (e.g., Ziman, 1978), which is discussed in the following

EXAMPLE 3: Based on the excellent fit of the continental margins, Alfred Wegener put forward in 1912 the hypothesis of continental drift, which he argued, was due to tidal forces from the Sun and moon. Despite the excellent fit demonstrated by experimental investigations, his model led to a series of incorrect conclusions (such as the fixity of the main continental land masses) and fantasies (e.g., “land bridges” that eventually sank without trace into the deep waters of the South Atlantic) which survived for many years. Later, it was shown by mathematical calculations (not by new geological theories or new experimental evidence) that the tidal forces suggested by Wegener were insufficient to bring about such large effects, and Wegener's theory was abandoned (later, the continental drift idea was revived as data obtained with new geophysical techniques became available and became an accepted geophysical model, but the new theory was based on the concept of plate tectonics). □

This example demonstrates the interdisciplinary of science, expressed in the following

Scholium 2: *The combination of theoretical analysis, experimental investigations, and computational techniques is a much more efficient means for advancing scientific knowledge than independent, disciplinary investigations.*

This book is concerned with the quantitative analysis and modelling of environmental health processes. We will, therefore, discuss mathematical models and methods that provide the necessary vehicles for advancing from a qualitative understanding of the underlying processes to quantitative analysis and prediction that are extremely useful in health risk assessment and the development of effective public health policies.

An adequate quantitative analysis and modelling of environmental health processes should be based on three building blocks: the stochastic mode of thinking about uncertainty (§2); the concept of a space-time continuum (§3); and the notion of a field (§4). The above provide the necessary background for interpreting and processing various knowledge bases (§5), for choosing the appropriate scale of analysis (§6), and for deriving meaningful quantitative results at every stage of the holistic environmental health model (§7).

2. FUNDAMENTAL PRINCIPLES OF THE STOCHASTIC MODE OF THINKING

The main message that this book is trying to convey is the importance of modern stochastics in environmental health science. The stochastic mode of thinking, in particular, is very useful in the quantitative analysis of uncertainty. Environmental health scientists, therefore, can greatly benefit by learning the stochastic techniques and trying to apply them to the fields of their expertise.

Many classical scientific theories --like Mendel's theory of inheritance-- are explicitly non-deterministic. Several supposedly deterministic models actually involve experimental parameters which are described only in terms of probabilities (e.g., due to imperfect measuring devices and experimental errors). Postclassical mathematics and science --e.g. Gödel's logic and Bohr's quantum physics-- have demonstrated the fundamental incompleteness of deterministic knowledge. Deterministic thinking leads to insufficient representations of many phenomena, due to incomplete knowledge of the underlying mechanisms, partial characterization of model parameters and inherent uncertainty.

Most scientists and philosophers today accept that scientific reasoning is essentially reasoning in accordance with the formal principles of the *stochastic mode of thinking*. The latter is the fundamental process of combining (a) intellectual creativity with (b) physical knowledge (usually incomplete) and (c) mathematical techniques, in order to evaluate the uncertainties and adequately represent (or idealize) the mechanisms underlying a phenomenon and to make predictions. This process is faced with questions of concept as well as questions of fact. An obvious difference between the stochastic mode of thinking and the traditional, deterministic mode of thinking is that the former explicitly includes the concept of *uncertainty* in the development and implementation of methods. But, as we shall see below, there is more than that.

The mathematical language of the stochastic mode of thinking is based on the methods of *modern stochastics.* The latter provides a quantitative representation of reality, which improves in accord with advances in the three components of the stochastic mode of thinking above (theoretical understanding, factual knowledge, and mathematical tools). Hence, stochastic techniques are accurate only to the extent that the biological models and physical theories underlying these techniques are in reasonable agreement with reality. *Ab igne ignem*, you can not expect modern stochastics or any other method to produce accurate results when the input is wrong or inadequate.

We will discuss next a few principles which form the basis for the approach presented in this book. These principles are essentially *epistemic ideals*, i.e., qualities that we would like our scientific reasoning to possess. Of fundamental importance in the foundation and development of modern stochastics is the

Complementarity principle: *A multiplicity of potentialities (realizations) which are in agreement with factual knowledge should be considered in order to achieve a complete understanding of an environmental health process.*

In light of the complementarity principle, the stochastic mode of thinking is truly liberating and its deep philosophical roots can be clearly seen and appreciated. In ancient times Epicurus argued that if several potentialities are consistent with an observed phenomenon one should study them all, for the level of understanding achieved by multiple potentialities is usually sufficient for *human happiness*, and because it would be unscientific to prefer one potentiality to another when both are equally in agreement with what is known about the phenomenon (Bailey, 1928). In modern times the above principle is closely related to the *complementary* mode of description of quantum phenomena (Bohr, 1958) as well as the concept of *general economy* (Bataille, 1988-1990).

The type of reasoning introduced by complementarity is especially useful in risk analysis where alternative courses of action can lead to significant improvement or catastrophic results. It is then important to engage our imagination to simulate various possible situations and consider the outcomes predicted on the basis of different scenarios. In addition, complementarity offers an interesting interpretation of the cause-effect concept: An event E_1 may be considered as causing an event E_2 if both E_1 and E_2 occur in the observed realization, but in the vast majority of the other realizations in which E_1 does not occur, E_2 does not occur either.

REMARK 1: For a poetic reference to the complementarity principle, consider the words of Odysseus Elytis (1979 Nobel prize for literature): "*...We are the negative of the dream: that is why we see black and white, and we experience decay over a minimal reality...*"

Modern stochastics is instrumental in the practical implementation of scientific reasoning. The latter is generally concerned with (a) *knowledge bases* $\mathcal{K}$ established from experiments, observations, evidence, and previous experience, and (b) *hypotheses* H regarding a physical phenomenon, a biological process, etc. (see, also, the relation between theoretical and experimental investigations in Fig. 1 above). Relations between $\mathcal{K}$ and H may be expressed by means of deterministic or stochastic arguments, depending on the situation. While a deterministic argument can be stated as

$\mathcal{K}$ *is available, therefore* H *is valid,* (1)

a stochastic argument has the form

$\mathcal{K}$ *is available, therefore probably* H *is valid.* (2)

In other words, as "therefore" is the characteristic mark of a deterministic argument, so "therefore probably" is the characteristic mark of a stochastic argument. The term "probably" expresses the *uncertainty* about H given $\mathcal{K}$, or, *alio intuitu,* it expresses the support that the knowledge bases $\mathcal{K}$ gives to H. The assessment of this uncertainty should lead to some quantitative statement that expresses the degree of support that $\mathcal{K}$ provides about the validity of H. Such a quantification may be achieved, e.g., by establishing some kind of scale, say from a to b, where a is a real number associated with knowledge bases that completely oppose hypothesis H, and b is a number associated with experimental results which are completely in favor of H; any other possible experimental outcome is properly assigned a real number p which lies between a and b. i.e., $p \in [a,b]$. In stochastic analysis, these numbers are called probabilities of occurrence of H given knowledge $\mathcal{K}$, and are conveniently assigned the values a=0, b=1 and $p \in [0,1]$. The statement (2) may be now replaced with

$\mathcal{K}$ *is available, therefore* H *is valid with probability* $p \in [0,1]$. (3)

In the concise language of stochastic mathematics, statement (3) is expressed by the probability function

$$P[H|\mathcal{K}] = p \in [0,1], \qquad (4)$$

which means that "the probability of H given $\mathcal{K}$ is p". In other words, the probability function quantifies the support that knowledge $\mathcal{K}$ gives to hypothesis H. The preceding analysis is thus summarized by the

Knowledge support principle: *Quantification of the support that a knowledge basis* $\mathcal{K}$ *gives to a hypothesis* H *is made in terms of a probability function.*

Another way of expressing the meaning of this principle is by saying that probability judgments about a hypothesis H are relative to knowledge $\mathcal{K}$. Indeed, statements like (3) assert logical relations between bodies of knowledge $\mathcal{K}$ and hypotheses H regarding an environmental health process. The knowledge support principle assigns different support (probability) to the various realizations considered by the complementarity principle above.

REMARK 2: Other knowledge support principles have been also proposed, e.g., by means of fuzzy functions (Kosko, 1993). Certain of these proposals have been criticized for lack of originality, and for adding nothing significant to the existing theories of logic (Haack, 1996). The knowledge support principle above is based on a well-developed logical probability theory and is fully satisfactory for the environmental health applications considered in this book.

The knowledge $\mathcal{K}$ may involve various *knowledge bases* (general knowledge, case-specific or specificatory evidence, etc.; see, §5 below). A fundamental rule of logic is that one should use all available knowledge in deciding whether a proposition is true or false, probable or improbable, and in determining the probability associated to a hypothesis or theory (Burks, 1977). This point is of vital importance and deserves to be made explicitly by the

Total knowledge principle: *For the stochastic approach to be accurate and useful, it should fully incorporate the current state of knowledge $\mathcal{K}$ regarding the environmental health processes of interest.*

The above principle is an epistemic ideal of considerable importance from a modelling point of view. A significant amount of knowledge $\mathcal{K}$ is available in the form of models. Modelling involves considerably more than curve fitting; it offers theoretical justification to data analysis, as well as guidance on how to avoid the many pitfalls that confront those who have to interpret collections of numerical data. As we discussed earlier, the accuracy of stochastic solutions is limited by the models that we use to describe the state of our knowledge (physical laws, data, etc.). To illustrate this, we discuss briefly the history of black-body radiation (the term "black-body denotes bodies that absorb and emit electromagnetic radiation of all frequencies). In a first-order approximation, the Sun may be considered as a black body radiating at 5,800 K.

EXAMPLE 1: An illuminating demonstration of the crucial role of modelling is provided by the black-body radiation problem studied by Max Planck (e.g., Morrison, 1990). By examining various sets of experimental results covering a wide range of cavity temperatures T and electromagnetic radiation frequencies n, and by applying a healthy dose of physical intuition, Planck deduced the following empirical formula for the radiation energy density $\rho(v,T)$

$$\rho(v,T) = av^3/[\exp(bv/T) - 1], \tag{5}$$

where a and b were empirical coefficients. Eq. (5) provided an excellent fit to the experimental data. However, Planck was not satisfied with the excellent fit that Eq. (5)

provided to his experimental data. In hindsight, considering the Wegener example (where analysis based on the excellent fit of experimental data led to crucial fallacies), Planck's skepticism was justified. Searching for a rigorous theoretical justification of Eq. (1), Planck developed some novel and radical concepts and hypotheses, by means of which he derived the now famous radical law

$$\rho(\nu,T) = 8\pi h\nu^3 / \{c^3[\exp(h\nu/k_B T) - 1]\}; \quad (6)$$

where c is the speed of light, and h and k_B are the so-called Planck's and Boltzmann's constants, respectively. By comparing the empirical Eq. (5) with the theoretical Eq. (6) Planck derived expressions for the fitting coefficients a and b that appeared in his empirical form and, thus, he obtained an improved physical understanding of the phenomenon. More importantly, the theoretical investigations that led to Planck's model (6) paved the way for the development of quantum physics, which is responsible for some of the greatest scientific and technological advances of the last century. □

REMARK 3: The knowledge basis $\mathcal{K}$ may include all kinds of valid knowledge that are available at a given moment and can be obtained by the competent environmental health scientist using effectively a meaningful scientific procedure. In this sense, the availability of knowledge is objective. Subjective bias enters when the scientist fails to use the appropriate procedures leading to the valid knowledge, even when they are available.

Note that when $p = 1$, the stochastic statement (3) reduces to the deterministic statement (1). Hence, the deterministic viewpoint is a special case of the stochastic viewpoint. Indeed, in theory modern stochastics is a generalization of the conventional deterministic analysis. In practice, modern stochastics is more flexible and can handle problems in applied environmental health science more accurately and efficiently than the often-inadequate deterministic methods.

An illustration of the stochastic concept is given in Fig. 2: In the vast majority of practical situations our knowledge about an environmental health process is not "perfect", that is, we cannot predict with certainty its values at unmeasured points in space-time. In stochastic terms, the situation is represented by means of: (a) a set of possible values $\{\chi_1,\ldots,\chi_k,\ldots,\chi_m\}$ of the environmental health process of interest; and (b) a function $f_x(\chi)$ expressing the probability of their occurrence. The width L of the χ-domain of the probability function $f_x(\chi)$ quantifies our incomplete knowledge, uncertainties, etc.. Therefore, it will be called the *level of uncertainty* L. As more data and information become available, our knowledge about the specific process is improved. As a consequence, the uncertainty level L reduces and reaches its limiting zero value ($L \to 0$) in the ideal but rather unrealistic situation of perfect knowledge. In this case, the range of possible values $\{\chi_1, ..., \chi_k, ..., \chi_m\}$ is replaced by a unique value χ_k and the process becomes deterministic. This consistency requirement is the essence of the following fundamental principle.

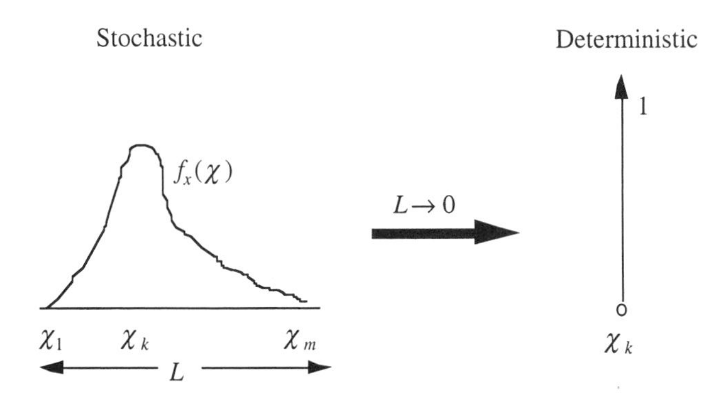

FIGURE 2: An illustration of the stochastic concept; L expresses one's incomplete knowledge, uncertainties, etc.

Correspondence principle: *In the zero uncertainty limit (i.e., when all uncertainties are practically negligible), the laws of modern stochastics must reduce to their deterministic counterparts.*

In light of the correspondence principle, modern stochastics do not render deterministic analysis invalid. Instead, deterministic analysis simply is overtaken and incorporated into modern stochastics. In fact, we could use modern stochastics, even when perfect knowledge is available. The correct deterministic result will always be obtained in the uncertainty limit, in which stochastic laws reduce to deterministic laws. This, however, does not work the other way around, i.e., if we use a deterministic approach in case of incomplete and uncertain knowledge, we may obtain incorrect results, the magnitude of the error increasing sharply with our uncertainty level L.

Mutatis mutandis, an interesting analogy with basic science is revealed: The stochastic correspondence principle can be compared with the correspondence principle which establishes the consistency of quantum mechanics with classical mechanics at the classical limit. The epistemic concept of the "level of uncertainty L" -which restricts the applicability of the deterministic approach- operates similarly with the concept of the "physical scale ", which defines the domain of applicability of classical mechanics. As one moves from the macrocosmos of everyday experience into the atomic microcosmos the physical scale becomes increasingly smaller. There is a limit beyond which the laws of classical mechanics break down and totally new laws, those of quantum mechanics, apply. Quantum mechanics, based on the *stochastic* concepts of probability and uncertainty, has led to cutting-edge technological advances and will continue to do so in the future.

EXAMPLE 2: In the quantum world a completely deterministic picture of physical reality is impossible, and all phenomena are governed by stochastic laws. Consider Fig. 2 above in the context of the wave-particle concept of quantum mechanics. A quantum particle (e.g., a photon or an electron) has a dual nature, i.e., it is also a wave. The amplitude of the wave determines the probability that the particle will be observed at a particular position. Assuming that the curve at the left hand side of Fig. 2 represents the amplitude of the wave and $\{\chi_1,...,\chi_k,...,\chi_m\}$ represent possible locations of the particle, the wave amplitude at χ_k represents the probability that the particle is observed at χ_k. Since the wave has a finite spread, the position of a particle at any given moment is not fixed. Thus, perfect knowledge (determinism) as shown at the right hand side of Fig. 2 is never attained in the microscopic quantum world. □

REMARK 4: *Ad summam*, one can distinguish between two major areas of application of modern stochastics: the macroscopic and the microscopic worlds. In the macroscopic world, the use of modern stochastics is necessary in cases of insufficient information. While there are certain situations where practically perfect knowledge allows us to use the deterministic method, in the majority of real world problems only limited knowledge is available. In the microscopic world, on the other hand, the deterministic method is rather useless and the only applicable method is its generalization, the stochastic method. One can no longer meaningfully consider, e.g., the interaction between single particles, but only between multiple complementary ensembles.

Modern stochastics provides a rational approach for conducting scientific inferences, supplying the technical means and the consistent rules which will improve the existing state of knowledge as soon as new data become available, and quantifying and conveying physical knowledge all the way to practical design and implementation. An important role in this respect is played by the

Prior-posterior principle: *A scientific method should balance two requirements: high expected prior information about the environmental health process of interest given general knowledge and high posterior probability for the final outcomes given case-specific data.*

The above is yet another important epistemic ideal. The probability at issue at the prior stage is probability with respect to general knowledge. The probability we seek to maximize at the posterior stage is probability on case-specific evidence. As we shall see in Chapter V, this principle has important applications in space-time mapping.

EXAMPLE 3: By expressing prior information in terms of Shannon's information measure (Shannon, 1948), the above principle leads to the maximization of an entropy function (which can incorporate various kinds of general knowledge in a logical manner) and the derivation of a posterior probability from incomplete case-specific knowledge. Entropy maximization implies that among all possible prior probability laws one should choose the one that maximizes the expected prior information and satisfies all prior knowledge. Bayes theorem updates the prior probability in light of additional information and experimental data, thus leading to the posterior probability law (Chapter V). □

The above principle introduces a fruitful combination of the entropic and the Bayesian concepts that can resolve certain paradoxes of probability theory. One of them is described in the following

EXAMPLE 4: One may consider applying recursively Bayes' theorem to update an arbitrary prior probability p_1, then use the resulting posterior probability p_2 as a new prior in a second application of Bayes' theorem, and so on until the procedure converges to some stable posterior. This approach is incorrect and produces misleading results and logical inconsistencies. The prior-posterior principle above can help us to avoid such pitfalls by stating that the correct prior is the one that maximizes the expected prior information, while it satisfies at the same time the given knowledge and data. □

The last principle of the stochastic mode of thinking is the

Relativity principle: *There can be no absolute truth or validity of a scientific theory, that is unequivocally established once and for all. What counts is the relative superiority of a scientific theory to its rivals, this superiority being an objective feature.*

This principle simply implies that a scientific theory, model or method are never conclusively verifiable. As Karl Popper (1934) has stated in his falsification theory, "every scientific theory is to the mercy of a counter example". Hence, no theory can be ever proven as the true one. A theory can be only established as superior to other theories existing at the given time.

EXAMPLE 5: According to the relativity principle, questions like "which model is the true one?" have no meaning. What is of real value in scientific investigations is answers to questions like "which model performs best among the various ones currently available?" □

Observational evidence is not evidence in the sense that a theory can be inferred from it, but rather in the sense that it can constitute a genuine reason for preferring one theory or model over others. Application of the relativity principle in practice requires the development of objective criteria by means of which the superiority of a model to its competitors can be established. This leads to the introduction of a sense of superiority, such as "more accurate" or "more informative", and tests demonstrating this kind of superiority of one method to its competitors.

EXAMPLE 6: In Chapter VI we show that the Bayesian maximum entropy mapping method is more accurate than kriging techniques, in the sense that it is capable of incorporating additional sources of information and offering more accurate predictions. □

These are the fundamental principles of modern stochastics, the rigorous mathematical development and application of which in environmental health science is the task of the subsequent Chapters of this book. When appropriately used, these principles give deeper insights and allow the solution of difficult problems. Although intuition and common sense are certainly very important in this endeavor, sophisticated mathematics leads to useful and sometimes unexpected results. After all, it is the duty of the scientist to look for the truth no matter how difficult the road may be. Sophisticated analysis is not unnecessary or “too theoretical” when it is the only means for understanding and predicting the outcome of processes that have a significant impact on society.

3. THE NOTION OF SPATIOTEMPORAL CONTINUITY

Another fundamental concept of environmental health science is the concept of a *space-time continuum*. The main implication of this concept is that environmental health processes do not evolve separately in space and in time, but in space-time. We first discuss some philosophical aspects of space and time and then introduce the operational concept of the space-time continuum to be considered in this book.

3.1 Philosophical Aspects of Space and Time

Being the most fundamental entities of the world we live in, space and time have excited the human intellect since the ancient times. Space has always been considered an intrinsic property of the universe we live in, which does not require the presence of a conscious observer. Space is the medium that contains the universe and all forms of life. The space that we experience in our every day life is three-dimensional (superstring theories propose that we live in a compactified ten-dimensional space with nine spatial dimensions plus time; however, the compactified six spatial dimensions can not be probed except at very high energies, which are impossible to attain with today's accelerator technology). Motion is possible in all directions in space, both forward and backward with respect to a reference direction. Time on the other hand is a more elusive concept. Augustine said of time that "we all know what it is until we are asked to define it". Time has provided a very useful bookkeeping device for following the order of events and measuring the rate of motion in space. However, it has always been an issue of philosophical debate whether time represents an inherent property of the natural world or merely an *ad hoc* human invention, an artificial yardstick introduced by the ancient astronomers in order to study the periodicity of planetary orbits and the cyclic return of the seasons (Smart, 1964; Hinkfuss, 1975; Langran, 1992).

The question of the nature of space and time has been fervently debated over the centuries. Several interpretations have been suggested, which distinguish between conceptual, perceptual, physical, and abstract space-time. The reader is referred to Lucas (1973) and Weyl (1952, 1987) for detailed discussions on the deep philosophical aspects of space and time. Most philosophers agree that while space and time share common properties, they are also fundamentally different in certain other ways. One of the outstanding differences between space and time is the irreversibility of time in the macroscopic world. Even though the microscopic dynamic equations of motion are symmetric with respect to time inversion, in the world of our everyday experience it is not possible to move backward in time. This asymmetry is usually interpreted by means of the second law of thermodynamics, which postulates that the entropy of a closed system increases with time (e.g., Prigogine, 1980). Since the entropy of closed systems increases monotonically, macroscopic events are irreversible. Thus, time machines that permit travel in the past, the holy grail for science fiction writers of the sixties, are prohibited by the second law of thermodynamics.

Nevertheless, it seemed natural in the past to distinguish between space and time, since after all everyday experience suggests that there are other differences besides the irreversibility of time. Hence, until the beginnings of this century time and space were treated as two independent entities. It was believed that time advances linearly and continuously from the past toward the future. Space, on the other hand, was obviously curved due to the shape of the Earth and geomorphologic features of the Earth's surface. This classical (Newtonian) view has been superseded following the discovery of the special and general theories of relativity by Einstein. In the modern world view, space and time form a unified geometric medium which is curved in the vicinity of gravitational sources. The curved nature of space-time is a significant factor in the study of physical phenomena that involve the motions of stars and galaxies, but it barely affects phenomena that occur at the scale of the solar system (the precession of the perihelion of Mercury being the most noticeable manifestation of space-time curvature effects in the solar system; Weinberg, 1972). Therefore, the relativistic space-time structure does not affect phenomena that occur at the scale of the Earth, with which this book is concerned.

3.2 The Space-time Continuum Concept

In a similar vein, the concept of the space-time continuum, being fundamental in the scientific description of the world, has been assumed several interpretations (e.g., Zwart, 1976; and Jeans, 1981). Putting philosophical considerations aside, the operational importance of the space-time continuum concept is its bookkeeping efficiency that permits an ordered recording of physical measurements and establishing relations between measurements by means of physical theories and mathematical expressions.

More specifically, in the classical space-time continuum (i.e., a three-dimensional continuum that evolves along a separate time dimension) the principle of simultaneity applies: all events that are simultaneous in one reference frame are also simultaneous in all other frames, stationary or moving with respect to the first. This principle is not valid in relativistic (Einsteinian) space-time frames, in which time is a fourth dimension interacting with space (space and time are unified in a four-dimensional continuum). However, relativistic effects become evident only at very high speeds or near large gravitational sources (e.g., stars). In environmental health science, observational data and processes commonly occur within a non-relativistic frame and, thus, the following space-time continuum idea will suffice for the purposes of our analysis.

Space-time Continuum: *Space-time is viewed as a unified physical entity defined in terms of the Cartesian product of space (which determines the location of environmental health processes) and time (which order the sequence of events).*

The importance of the classical continuum concept is paramount in representing the evolution of environmental health processes which assume values at any point in space-time, thus requiring continuously varying spatiotemporal coordinates. In such an integrated but non-relativistic space-time continuum, space-time cross effects play an important role. We are not concerned, e.g., merely about the distance between two geographical locations s and s', but rather about the distance between a location s at time t and a location s' at time t'. In the same context, the following scholium points out a significant feature of the space-time continuum.

Scholium 1: *Spatial arrangements in the space-time continuum are combined with the temporal order of events in a way that reflects relationships determined by physical, chemical and biological laws.*

3.3 Natural vs. Health Processes

Most natural processes evolve simultaneously in space and time. Natural variability is usually characterized by complicated spatiotemporal trends and erratic fluctuations. Restricting the analysis of natural processes purely in the spatial or temporal domain may lead to considerable loss of information and large prediction errors (Christakos, 1992; Christakos and Vyas, 1998a). In addition, arbitrary combinations of mathematical models that are valid separately in space and time are not, in general, valid in the space-time domain. The laws of nature lead to causal connections in space-time. Hence, valid stochastic models should incorporate natural laws in an integrated space-time domain. The inclusion, thus, of these spatiotemporal laws into stochastic analysis cannot be carried out if only the purely temporal or the purely spatial domains are considered. For these reasons, a *composite* space-time analysis is not just an improvement over more traditional statistical methods that treat space and time separately, but a qualitative step ahead.

EXAMPLE 1: Spatiotemporal data (i.e., that involve variations in both space and time) include precipitation measurements consisting of long time series at various locations in space, water-vapor concentrations, pollutant distributions in environmental media, earth-surface temperature data, morphogenetic processes that shape and organize developing organisms and stabilize the forms of adult organisms. □

Applications of the spatiotemporal continuum may be found in environmental health science, particularly in the study of microscopic patterns of energy absorption in receptors (human organs, tissues, etc.) exposed to ionizing radiation. There are also applications involving the analysis of probabilistic health indicators (e.g., disease incidence, mortality) that are not, in general, continuously measured in space. Instead, for modelling purposes all of the cases in a space-time unit Λ are considered concentrated at one point $\boldsymbol{p} \in \Lambda$, which is selected on the basis of statistical, administrative, or other criteria [to denote that the unit is centered around $\boldsymbol{p}$ we will sometimes use the notation $\Lambda(\boldsymbol{p})$]. Then, an incidence contour map can be produced so that disease rates are represented in terms of a geographically continuous process. This is a common modelling decision in medical geography and epidemiology (e.g., Christakos and Lai, 1997).

EXAMPLE 2: A regional measurement (e.g., rate per resident) is localized at the center of gravity of Λ, which is obtained by weighting according to population density. □

The choice of Λ depends on certain factors such as (Esteve *et al.*, 1994): (a) Administrative and political constraints. (b) Statistical stability of the disease rate. An appropriate ratio must be maintained between incidence or mortality from one geographical unit to the next. This may imply, e.g., that smaller, sparsely populated units need to be grouped together in Λ. (c) Criteria related to disease etiology. If the goal of the health process is to illustrate an exposure-disease relationship, the choice of Λ could be affected by the observation scale of exposure (this issue is revisited in the following Chapters).

By way of a summary, the notion of a spatiotemporal continuum is fundamental in the interpretation and analysis of numerous processes (natural, anthropogenic, environmental, epidemiologic, disease). It is, therefore, no surprise that it will play a central role in the present work.

3.4 Space-Time Coordinates

Analysis in the spatiotemporal domain involves spatial coordinates $\boldsymbol{s} = (s_1, \ldots, s_n)$ considered in the Euclidean space R^n (i.e., $\boldsymbol{s} \in S \subset R^n$), and a temporal coordinate t along the time axis $T \subset R^1$. For many applications it is sufficient to investigate the temporal evolution after an initial time instant in the not-too-distant past. Then, the initial time is set equal to zero and $T \subset [0,\infty)$. Exceptions are processes that have long-range correlations, such as the fractional Brownian motions. The space-time coordinates $\boldsymbol{p} = (\boldsymbol{s},t)$ are defined on the Cartesian product $S \times T$. Spatiotemporal continuity implies an integration of space with time and is a fundamental property of the mathematical formalism of environmental health processes. There are more than one ways to define a point in a space-time domain:

First Approach: *Consider a geometrical point in space with coordinates* $\boldsymbol{s}_i \in S$ *and a time instant* $t_j \in T$, *and then define a geometrical point* P_{ij} *in space-time as*

$$(\boldsymbol{s}_i, t_j) = \boldsymbol{p}_{ij} \to P_{ij}. \tag{1}$$

Second Approach: *Consider a geometrical point* P_i *in the space-time domain, and then derive its projections on the spatial and temporal axes, thus leading to the mapping*

$$P_i \to (\boldsymbol{s}_i, t_i) = (\boldsymbol{s}, t)_i = \boldsymbol{p}_i. \tag{2}$$

In the first approach the space-time point is denoted by a pair (i,j) of space and time labels. In the second approach a unified space-time label (i) is used. Approach (1) is more convenient when explicit reference is made to the times and the spatial locations. Approach (2) is more efficient in other cases, because it provides a simpler notation.

Scholium 2: *The reader must keep in mind that both approaches will be used interchangeably throughout the book.*

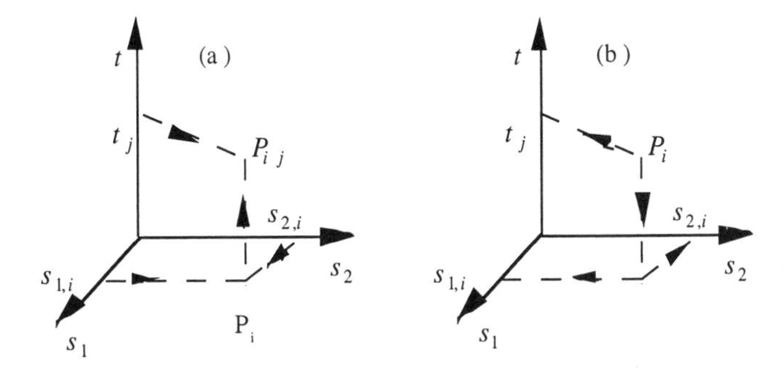

FIGURE 3: (a) The first approach; and (b) the second approach to define a point in the space-time domain $R^2 \times T$.

EXAMPLE 1: In Fig. 3 an illustration is given of the two approaches of defining a point in the space-time $R^2 \times T$ domain. □

In many health processes --such as disease incidence or mortality-- two time variables may be used: one denoting the time of the measurement (e.g., date), and the other denoting the age of the population considered in the investigation.

EXAMPLE 2: Coronary heart disease incidence is a function of the geographical location of the population of interest, as well as the interval of time and the age during which the incidence occurs. The age density function of a population is a function of the calendar time as well as age. □

4. THE FIELD CONCEPT IN ENVIRONMENTAL HEALTH SCIENCE

Another fundamental idea of environmental health science is the idea of a *field*. A field attributes mathematical quantities --scalars, vectors, or tensors-- to points of the space-time continuum. For the field to obtain physical meaning these entities must represent values of some natural or health process (e.g., soil moisture, fluid velocity, radon concentration and disease rate). These thoughts lead to the following description of

The field concept: *A field generally presupposes a continuum of space-time points and then attributes values of environmental health processes to these points. Specification of the values at all points in the space-time continuum specifies a realization of the field.*

The concept of the field is an important component of the holistic philosophy of environmental health science. A field may be considered as a model, i.e. a mathematical construction for representing the distribution of processes in space and time. As models, fields have certain interesting features. They represent interactions through which natural and health processes influence each other even at a distance. Field-based representations of reality may involve a hierarchy of fields. There is not a unique field representing every aspect of reality. Instead, one field describes one feature of reality and another field some other feature. Among the many kinds of fields, some represent material variables (e.g., the space-time distribution of moisture in the soil, or the concentration of a contaminant in water), and others express interactions (e.g., the earth's gravitational field, or the electromagnetic field). Fields in modern physics are considered to be more fundamental than matter.

In light of the above considerations, a field may be generally denoted as a function of the space-time coordinates, i.e.,

$$X(\boldsymbol{p}), \tag{1}$$

where $\boldsymbol{p} = (\boldsymbol{s}, t) \in S \times T$. In addition to the coordinates, certain environmental health processes are functions of an event e related to a health condition (e.g., a specific disease or death), as well as other factors (age, sex, race and marital status, etc.) denoted by the vector $\boldsymbol{f} = [f_1, \ldots, f_m]^T$. Therefore, in certain situations it may be more appropriate to denote the field as

$$X(\boldsymbol{p}, e, \boldsymbol{f}). \tag{2}$$

If the field (2) describes a biomarker (radon intake dose, ozone burden, etc.) without any reference to health effects, $e = \text{null}$ and the $\boldsymbol{f}$ may describe human features of the individual or the population.

EXAMPLE 1: At the population scale, the field may represent disease incidence e that varies geographically/temporally as denoted by $\boldsymbol{p}$, as well as with the human population characteristics $\boldsymbol{f}$, such as f_1 =age, f_2 =dietary habits, f_3 =individual's interaction with the environment, etc.. At the individual receptor scale, a field may represent the concentration of a pollutant in the organs or tissues of the human body. The variations of fields representing biomarkers are usually divided into two groups: *intersubject* variation in exposure between individuals (each individual being associated with different space-time coordinates $\boldsymbol{p}$), and *intrasubject* variation in exposure for an individual (associated with the individual's characteristics $\boldsymbol{f}$). While intersubject variation is mainly due to the geographical and temporal state of the environment, intrasubject variation results from the human aspects of the population. □

If the field (2) represents exposure to a pollutant, $e = \text{null}$ and $\boldsymbol{f}$ includes the specific experimental conditions and physical laws governing the natural processes involved in exposure as well as the impact pathway factors (contact surface, route, medium, exposure intensity and duration).

EXAMPLE 2: Variations of permeability distribution are associated with different space-time coordinates $\boldsymbol{p}$ (natural variation), as well as the specific experimental setup and assumptions f (e.g., one-dimensional flow along a soil column in the laboratory, homogeneous specimen). □

For simplicity, notation (1) is often preferred to (2) in theoretical analysis or applications. When this happens, the e and the f are explicitly stated in the description of the problem under consideration.

Most environmental health processes have rather complicated dependence on space and time (e.g., ozone concentrations, soil moisture distribution, or coronary heart disease incidence). In very simple cases, it is possible to distinguish clear patterns, e.g., seasonal periodicities and spatial symmetries. In these simple cases the dynamics of the processes are controlled by few dominant degrees of freedom. However, often many degrees of freedom cooperate to generate complex spatiotemporal patterns that seem to have erratic components. In these cases, the variability of the environmental health process in space and time can not be represented exactly, because it is impossible to describe the dynamics of all the contributing degrees of freedom. As a consequence, the process is not sharply defined but, instead, it has an uncertain or indeterminate structure. The need to account for this uncertainty gave rise to the complementarity principle of §2: a multiple parallel processing of field realizations should be considered; these realizations are all different and yet necessary for a complete understanding of the process. This leads to the so called *random field* model, in which randomness manifests itself as an ensemble of possible realizations $\chi_i(\boldsymbol{p},e,f)$, $i=1,2,\ldots,m$, regarding the environmental health process under consideration; i.e.,

$$X(\boldsymbol{p},e,f) \rightarrow \begin{cases} \chi_1(\boldsymbol{p},e,f) \\ \chi_2(\boldsymbol{p},e,f) \\ \vdots \\ \chi_m(\boldsymbol{p},e,f) \end{cases} . \tag{3}$$

Thus, the random field is a combination of the concepts of space-time continuum, field and complementarity: A random field $X(\boldsymbol{p},e,f)$ is a collection of complementary field-realizations $\chi_i(\boldsymbol{p},e,f)$ associated with the values of an environmental health process at points $\boldsymbol{p}=(s,t)$ belonging to a space-time continuum $S \times T$. The random field is a central topic of this book (Chapter IIIff).

Given the large number of uncertainties involved in environmental health investigations, identifying and assessing the role of uncertainties is the most important, and yet many times the most neglected aspect of environmental health risk assessment (e.g., Morris, 1990). There exist several sources of uncertainty regarding the behavior of environmental health processes in space-time. Uncertainty sources may be classified in several broad categories. Certain kinds of uncertainty may be considered as *inherent properties* of natural processes. *Insufficient knowledge* of the underlying physical or biological mechanisms is another important source of uncertainty. Yet another source of uncertainty is related to *model inadequacy*.

EXAMPLE 3: Many physical and biological processes take place at the molecular or atomic scale, but are measured with macroscopic devices (inherent uncertainty). The space-time distribution of next week's rainfall cannot be exactly forecasted due to insufficient information regarding a series of atmospheric phenomena; in cancer studies sufficient exposure data for specific populations are not always available (uncertainty due to insufficient knowledge). Experimental results show that many models of multiphase flow in porous formations provide inadequate descriptions of reality; models of human exposure do not adequately capture differences among individuals (uncertainty due to modelling inadequacies).

Uncertainty has important consequences for environmental health studies: the task of evaluating how much we don't know is often more difficult and more important than evaluating how much we do know. Uncertainty characterization should aim to accomplish a number of *goals*, such as: (a) to provide information about the major uncertainty sources and, thus, to determine whether further research is needed to reduce them; (b) to prevent large financial and social costs due to incorrect decisions that are based on inaccurate estimates; (c) to provide the decision makers with risk estimates *and* uncertainties associated with these estimates; (d) to recognize the limits of our knowledge based on the available information. Risk assessment that does not address these issues is a disservice to environmental health risk management.

EXAMPLE 4: Propagation of initial uncertainties by physical laws describing the evolution of pollutants can be crucial. A small initial uncertainty in one of the parameters of the law can be so rapidly magnified by the differential equation governing the phenomenon that in the end no practical predictability is left (this behavior is related to the well-documented phenomenon of *chaos*). Realistic health risk analysis and management, on the other hand, strongly depend on the various uncertainties of the natural and health processes involved. Ignoring these uncertainties may have serious social and economic implications. □

5. KNOWLEDGE BASES IN THE SPACE-TIME DOMAIN

In spatiotemporal analysis of environmental health processes it is desired to use the best available (most complete) *knowledge bases* in the spatiotemporal domain, denoted by the symbol $\mathcal{K}$. Here we distinguish between two types of knowledge: General knowledge and case-specific knowledge.

General knowledge includes general natural laws, structured patterns and assumptions, justified beliefs regarding the investigated situation, and previous experience independent of any case-specific observations. The knowledge thus obtained is considered "general" in the sense that it characterizes a large class of fields or situations. More specifically, general knowledge can be divided into two main groups:

(i) *Analytic* statements involve logical relations (e.g., if A or B and not B, then A) between general quantities (abstract symbols, concepts, etc.); these quantities may or may not have physical counterparts, but the validity of the statements rests solely on definition and the rules of logic (e.g., every bachelor is unmarried).

(ii) *Synthetic* statements involve factual statements, natural laws, and theories. A fact is, e.g., "the breast cancer incidence rate for the year 1992 in Wake county, North Carolina, was 132.4 per 100,000 cases". A law formalizes a relation that has been observed consistently in the past and is expected to occur consistently in the future (e.g., Darcy's law of groundwater flow); in this sense, a law is a summary representation of numerous facts. A theory is an intellectual construct that has explanatory power. Laws may be derived from theories (e.g., under certain conditions Fick's law of fluid diffusion may be deduced from first principles). Theories may also include non-observable quantities. A law that is based solely on observation but can not be derived from a theory is called empirical.

Case-specific knowledge includes measurements and perceptual evidence, empirical propositions, expertise with the specific situation, incomplete evidence, etc.. Case-specific knowledge can be divided into two main groups:

(i) *Hard data,* which are measurements of the environmental health processes obtained from real-time observation devices (monitoring stations, etc.). The vector, e.g.,

$$\boldsymbol{\chi}_{\text{hard}} = [\chi_1\ \chi_2 \cdots \chi_{m_h}]^T \quad (1)$$

denotes a set of measured values χ_i at the space-time points $\boldsymbol{p}_i$ ($i = 1,\ldots,m_h$).

(ii) *Soft data,* such as interval values, probability data and functional relationships. Some examples of soft data are given in Table 1, including intervals [Eq. (2)], probability functions [Eqs. (3) and (4)], empirical formulas or engineering charts [Eq. (5)].

TABLE 1: Soft Data Examples (from Christakos, 1998a)

Interval data : Unmeasured values χ_i of the environmental health process at points $\boldsymbol{p}_i$ lie within the known intervals I_i ($i = 1,\ldots,m$), i.e.,

$$\boldsymbol{\chi}_{\text{soft}} = \{[\chi_{m_h+1}\ \chi_{m_h+2} \cdots \chi_m]^T : \chi_i \in I_i = [l_i, u_i],\ i = m_h + 1,\ldots,m\}. \quad (2)$$

Probability data : The probability law of the unmeasured χ_i is known, i.e.,

$$\boldsymbol{\chi}_{\text{soft}} = \{[\chi_{m_h+1}\ \chi_{m_h+2} \cdots \chi_m]^T : P_x[\boldsymbol{x}_{\text{soft}} \le \boldsymbol{\zeta}] = F_x(\boldsymbol{\zeta}),\ \boldsymbol{\zeta} = [\zeta_{m_h+1}\ \zeta_{m_h+2} \cdots \zeta_m]^T\}; \quad (3)$$

or certain interval probabilities are known, i.e.,

$$\boldsymbol{\chi}_{\text{soft}} = \{[\chi_{m_h+1}\ \chi_{m_h+2} \cdots \chi_m]^T : P_x[\boldsymbol{\chi}_{\text{soft}} \in \boldsymbol{I}] = p_x(\boldsymbol{I}),\ \boldsymbol{I} = [I_{m_h+1}\ I_{m_h+2} \cdots I_m]^T,$$
$$I_i = [l_i, u_i]\}. \quad (4)$$

Functional data : An empirical formula, graph or chart are available, i.e.,

$$\boldsymbol{\chi}_{\text{soft}} = \{[\chi_{m_h+1}\ \chi_{m_h+2} \cdots \chi_m]^T : W(\boldsymbol{\chi}_{\text{soft}}, \boldsymbol{\psi}) = 0,\ \boldsymbol{\psi} = [\psi_{m_h+1}\ \psi_{m_h+2} \cdots \psi_m]^T\}, \quad (5)$$

where the functional form of $W_i(\cdot)$ is known and ψ_i are values of a measurable environmental health process $Y(\boldsymbol{p})$.

---

6. SPATIOTEMPORAL SCALES

In addition to using the best available knowledge bases, spatiotemporal analysis of environmental health processes should provide meaningful concepts and models for operating on these bases. The concept of *scale* is of primary importance.

In fact, spatiotemporal analysis involves a variety of scales. There are considerable differences in choosing the appropriate scale for natural vs. health processes. These differences are related to the peculiarities of the processes such as the medium involved (i.e., natural environment vs. human body), as well as the goal of the analysis. On the other hand the fact that natural and health scales are considered in a holistic framework implies that the analysis must ensure an internal consistency of scales.

6.1 Spatiotemporal Scales for Natural Processes

Below, we define and discuss several scales that are useful in the description of physical systems, spatiotemporal data analysis, estimation and mapping. We define scale concepts based on intuition rather than mathematical rigor. This approach will give us the flexibility to handle various situations. Depending on the specific application, the term scale may refer to distance, time intervals, surface, volume, or a four-dimensional element of space-time.

Physical scale: *It characterizes the mechanisms involved in natural processes. It is determined by the intrinsic properties of the natural process and by the structure of the medium within which it takes place.*

EXAMPLE 1: Consider dispersion of an injected tracer in a heterogeneous porous medium with a resident fluid: The mixing length is a physical scale that determines the extent of the zone within which the two fluids are mixed. The correlations of the heterogeneous medium are important in determining the mixing length. If the medium has correlations, i.e. structures and patterns at different scales, more than one length scales may be relevant. □

Measurement scale: *It denotes the scale of the instrument and the measurement process.*

Most measurements involve some averaging of the measured quantity over an instrument bandwidth. Averaging leads to a loss of small scale information, and hence limits the resolution of the instrument. The modeller must decide whether the measurements can be used as point values of the spatiotemporal random field they purport to represent, or whether the coarse graining effect must be taken into account. In this respect, the effect of the instrument or measurement process on the physical law must also be considered.

EXAMPLE 2: If the data are sufficiently homogenized by the measurement process, non-local relations may lead to local empirical expressions. The measurement scale is determined by the available technological resources and the resourcefulness of the experimentalists. □

Sampling scale: *It refers to the average spacing between measurements.*

The sampling scale, thus, decreases with increasing sampling density. A large sampling density is desirable for accurate estimation. On the other hand, in practice economic considerations and time constraints impose severe constraints on sampling.

EXAMPLE 3: A network of stations for monitoring changes in the global environment, should be properly designed so that a certain number of stations are placed in boundaries between ecoclimatic regions. With the highest degree of instability of the ecosystems and the greatest sensitivity of their components to various forms of pressure occurring there, these boundaries are good places for detecting environmental changes (Bailey, 1996). □

Modelling scale: *It is the scale where a specific physical model is valid.*

EXAMPLE 4: Consider fluid flow within a porous medium. The internal motion of fluid particles at the atomic scale is determined by the laws of quantum mechanics; at the molecular level the collection of fluid molecules is organized according to the laws of statistical mechanics; at a higher scale, the molecular fluctuations in density are averaged out and the fluid behaves as a continuum; at the pore scale fluid flow is governed by the Navier-Stokes equations; for relatively large blocks of porous media it becomes too expensive to solve the Navier-Stokes equations, which are hence replaced by coarse grained partial differential equations (PDEs), e.g., Darcy's law or some generalization; finally, at field scales the local permeability fluctuations are homogenized leading, under certain conditions, to an effective flow equation. □

Support or observation scale: *It refers to the scale of the space-time domain used in a specific application for the estimation of point values or functions of environmental-health variables.*

EXAMPLE 5: The support scale may be expressed as the number of neighboring samples in the estimation of a natural process at an unmeasured point. This seems similar to the sampling scale, but the latter is global, while the support may vary locally. □

Within the support, estimates of the natural process can be obtained using the same model. The observation scale is not known *a priori*, but it is determined from the natural variability of the data using some optimization criterion (see, also, following Chapters). In the case of theoretical investigations, the support scale is identified with the domain scale.

Mapping scale: *It defines the size of the domain within which estimates are sought (estimation domain).*

If the physical laws are not known or can not be solved, estimation (mapping) schemes are used, which are in essence educated interpolation techniques (Chapters V and VI).

Discretization scale: *It defines the resolution of the numerical grid, that is, the spacing of the lattice on which the numerical solution is obtained.*

In the case that maps of a natural process are generated, the discretization scale determines the size of the smallest features that can be resolved on the map.

The choice of scales depends on many factors including availability and quality of data, physical reality, computational resources (e.g., memory space and computer time), reliable model inferences, and mapping accuracy. The observation scale is intimately related with space-time variability assessment and with the accuracy of predictions. If a soil parameter varies smoothly relative to the observation scale, estimates of a phenomenon at this scale are very accurate, and, thus, highly reliable.

The mapping resolution and the sampling scale are also crucial in determining and representing the variability: High sampling density is desired in order to resolve space-time patterns, and a sufficiently fine numerical grid should be used for representing this variability. The importance of physical scales can not be overemphasized. In many situations interactions of phenomena between scales lead to great difficulty, especially when it is not possible to decouple effects occurring at very different scales.

EXAMPLE 6: Consider the competition between global and local features in determining the movement of pollution distributions. Global terrain morphology influences the large scale meteorological processes that are important factors in the fate of airborne pollutants; on the other hand, local terrain features (hills, etc.) can affect the wind field locally and lead to turbulence, which then influences pollutant concentrations in their neighborhood. □

6.2 Spatiotemporal Scales for Health Processes

In the case of health processes, determining appropriate scales of analysis as well as understanding the causes for differences in the apparent disease distributions across scales are important research issues (Cleek, 1979; Christakos and Lai, 1997).

In toxicology, the scale usually refers to *individual* receptors (e.g., the skin area or the volume of the target organ). Certain biological studies may focus on atomic or molecular scales.

In epidemiological studies, on the other hand, *groups* of people rather than individuals constitute the basic unit. Usually, we distinguish three main kinds of scales, as follows.

Population scale: *It determines the population under study, usually a political unit (city, county, state or nation).*

Observation scale: *It is the temporal-geographical unit on which population measurements are made.*

Discretization scale: *It defines the resolution of the numerical grid, i.e., the lattice spacing on which estimates of the health process are obtained using a mapping technique.*

The choice of the appropriate health scale depends on the goals of the analysis. If the goal are to derive maps that fulfill a purely descriptive need, the choice of scales should account for the effects of the study population and the level of data aggregation, and also satisfy desired visualization goals. If the focus of the analysis is the demonstration of an exposure-disease relationship, the choice of scales should account for socioeconomic factors, exposure distribution in space-time (e.g., scales should match those of the exposure distribution), and the statistical stability of disease rate. Scale issues related to health processes are also discussed in Chapter II.

REMARK 1: Health processes may involve physical scales, as well. Microdosimetry is concerned with the quantification of the spatiotemporal distribution of absorbed energy in irradiated receptors (organs, tissues, etc.). Microdosimetric analysis is particularly interested about the physical scale in the irradiated receptor (usually represented by the dimensions of a spherical site) in which the specific energy differs considerably from its average value.

6.3. The Hierarchy of Scales

In the natural sciences, great progress has been made in the interpretation of macroscopic material properties in terms of the microscopic (atomic, molecular, etc.) structure of matter. We believe that future progress in environmental health science depends on the analysis and understanding of natural and health processes at various scales, and on establishing suitable quantitative connections between the results obtained at each of these scales. This *hierarchy of scales* includes microscopic scales (molecular, particle and cell) and a sequence of macroscopic scales (representative elementary volume, target organ, field, and population).

EXAMPLE 7: The practical importance of the concept of hierarchy of scales is illustrated below in a rather simple manner by means of the five data sets of Fig. 4. Assume that the spatial data set P corresponds to a human exposure pattern at a microscopic scale (e.g., molecular or cellular scale radiation). The fundamental biological models apply at this scale where, though, experiments are not usually possible (when possible, the detection of small amounts of a pollutant at the microscopic scale within a target organ or tissue of the human body can increase considerably the effectiveness of intervention strategies). At the macroscopic scale, on the other hand, the existing measuring devices perform some kind of averaging of local values (spatial averaging within a representative elementary volume, a cell, etc.). One such averaging may lead to picture P_1, in which each macroscopic value is the result of averaging two neighboring microscopic P-cells in the horizontal direction. Similarly, the other three pictures (P_2, P_3 and P_4) were constructed by applying different types of averaging (e.g., performed by different measuring devices) and, therefore, they provided very different images of reality than P and P_1 (each macroscopic value in P_2 is obtained by averaging two neighboring microscopic P-cells in the vertical direction; each cell value in P_3 is the average of two neighboring cells of P_2 in the horizontal direction; finally, each value in P_4 is the result of averaging nine neighboring microscopic P-cells). The implications of these different pictures in human exposure assessment can be crucial. The exposure statistics indicated were obtained using only values above a critical exposure threshold level $\zeta = 10$ units. The thick lines represent 10-unit contour lines. Notice the different extension and orientation of the contour lines corresponding to different spatial variability exposure patterns. Indeed, analysis at the microscopic level P reveals $n = 15$ cells where exposure exceeds the critical threshold value; analysis at the macroscopic levels P_1, P_2, P_3 and P_4 predict $n = 8, 4, 1$ and 0 cells being exposed, respectively. □

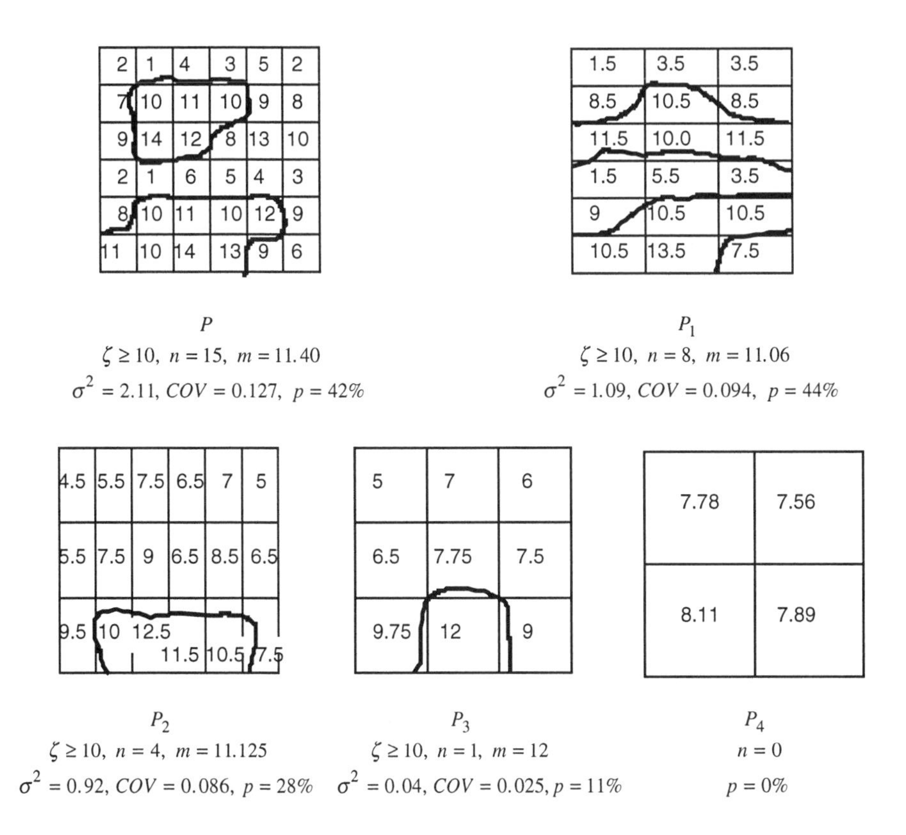

FIGURE 4: An illustration of the implications of the hierarchy of scales in exposure assessment; n is the number of elementary P-cells considered at the specific scale, m is the exposure mean and σ^2 is its variance; p is the % of areas where the pollutant exceeds the critical threshold level ζ.

This simple example illustrates the fact that considerable differences may exist between human exposure variation and statistics at different scales. The appropriate scale choice should, therefore, lead to a model which is sufficiently detailed to reproduce precisely all the important parameters for the specific application, and yet not so detailed as to make computations impractical. Exposure differences between scales can obviously have important consequences in environmental health analysis (health risk assessment, dose-response relationships, in-situ remediation, etc.).

In conclusion, by looking at the same phenomenon through different data windows (associated with different observation scales), we get different images of reality with different physical behavior and statistics. Then, questions such as the following arise naturally: (a) How to relate one scale to another? (b) How to transfer knowledge and statistics between scales? (c) Which is the most appropriate window for the case at hand? From an epistemological point of view, what we discussed above is a quantitative restatement of long-standing philosophical ideas, such as the one expressed in the following statement by the eminent methodologist of science, F. Bacon (1561-1626): "There is a hierarchy of explanatory laws to be discovered in physical sciences, and the scientific investigator should expect to make a gradual ascent to more and more comprehensible laws, each law leading to new experiments strictly within certain domain and scale of validity."

7. AN OVERVIEW OF THE ENVIRONMENTAL HEALTH PARADIGM

The two important features of the environmental health paradigm considered in this book are summarized by the words *holistic* and *stochastic*. Fig. 5 presents an overview of an environmental health study involving six basic stages, which will be discussed to considerable extent in subsequent Chapters. The justification for a holistic approach stems from the fact that the these six stages form an integrated whole that is more important than each one of them individually. In this whole, the six stages are closely linked and influence each other considerably (e.g., in the health risk assessment stage information from the exposure assessment stage may be used in order to estimate health effects; or, in the exposure assessment stage information may be used from the health risk assessment stage in order to suggest ways to control the exposure distribution causing the health damage). The stochastic approach is necessary, as a result of the various uncertainties involved at every stage (e.g., for the vast majority of pollutants there is no sufficient knowledge that allows a unique deterministic characterization of the physical and biological processes involved in environmental health assessment).

Hazard identification is the initial stage of environmental health studies, and it is concerned with evaluating the potential of a pollutant to cause damage to humans (Chapter II). It may include various methods such as epidemiological studies, animal data, short-term tests and general evidence (e.g., Smith and Fingleton, 1982; Nabholz, 1991). Hazard identification determines the specifics of the human risk to exposure which is studied in more detail in the following stages of the environmental health study.

Exposure assessment is based on both monitoring (measurements) and spatiotemporal modelling (e.g., transport and dispersion models, space-time mapping; Chapter VII). Exposure assessment can involve one or more environmental media and compartments, and

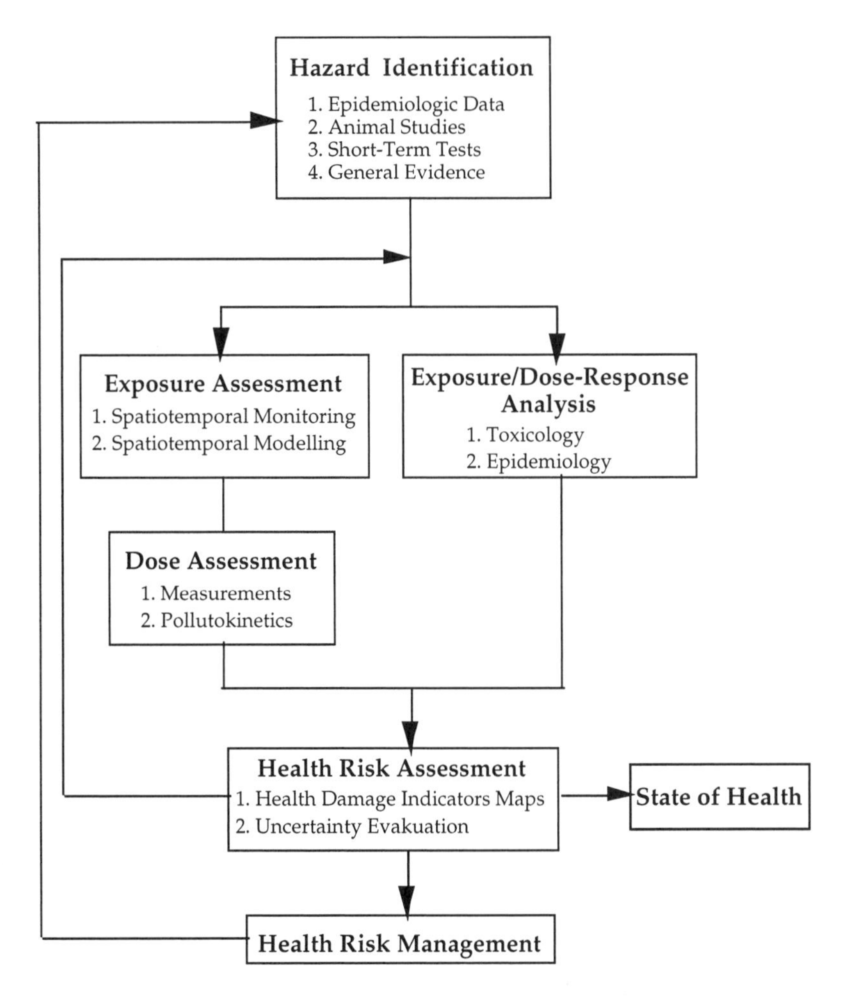

FIGURE 5: A holistic overview of the environmental health paradigm.

it is concerned with pollutant sources, environmental evolution mechanisms (e.g., transport, transfer and transformation), and population exposure (Chapters II-VII and Chapter IX). As a result of the considerable uncertainties involved in every stage, exposure assessment is fundamentally stochastic in nature. The movement of a pollutant from a source to the environment and ultimately to the human population involves a number of natural processes, each of which is characterized by different kinds of variabilities, heterogeneities and uncertainties (e.g.,Covello and Merkhoffer, 1993; Neely, 1994).

Environmental transport models have not, generally, been developed with the aim to determine human exposure, but rather to estimate the distribution of natural processes and pollutants in space-time. Determination of human exposure is the task of *dose assessment* studies, which focus on the transfer and transformation processes that occur in the body following exposure to a pollutant (Chapters II and VIII). This also includes intake pathways by means of which the pollutant is delivered to the target organ, methods for translating exposure to dose at biologically important sites and measures of the interaction between the pollutant and the target organs.

Exposure and *dose-response analysis* is concerned with the impact of a specific exposure level to human health, and it can be carried out at the individual or the population level (Chapters II and VIII). Animal toxicology provides data that, under certain conditions, can be extrapolated to humans (Loomis, 1978). Epidemiology provides methods for deriving probabilistic dose-response relationships (Esteve *et al.*, 1994). In comparison with animal toxicology, epidemiology has the advantage of studying directly human subjects. Due to the complicated biological mechanisms involved, exposure and dose-response analysis is usually the component with the largest uncertainty in environmental health studies.

Health risk assessment combines information from all the above stages in order to evaluate the state of health of a human population. On the basis of this evaluation, recommendations can be made to decision makers that include health risk estimates and the associated measures of estimate uncertainty. *Health management* is the final stage of a public health study, during which appropriate measures for protecting the health state of the population are decided (e.g., if the probability of a future exposure leading to an adverse health effect is considerable, action will be initiated to prevent health damage). The health management process involves political, social and economic issues, which will not be addressed in this book (though appropriate references will be given, when necessary).

Chapter II: ENVIRONMENTAL EXPOSURE FIELDS AND THEIR HEALTH EFFECTS

"The whole is more than the sum of the parts".
Aristotle, Metaphysica 1045a, 10f

1. THE HOLISTIC ENVIRONMENTAL EXPOSURE -HEALTH EFFECT PERSPECTIVE

Environmental science is concerned with the study of natural cycles and systems and their component natural processes. The state of the environment can be described in terms of *fields* (§I.4) that represent spatiotemporal distributions of natural processes. It is well-documented that the state of the environment affects the state of human health in the following way.

The environment as a whole and humans in particular are homeostatic systems characterized by processes which take values within certain ranges. The biological state, which includes processes such as blood pressure and heart rate, is influenced by the environmental state. Hence, biological organisms in general and humans in particular, function adequately only within certain ranges of environmental variables (e.g., temperature or pressure). Diversity and variability in these ranges exist to a certain extent, and they can be enforced by the evolutionary process. Inhabitants of high plateaus, e.g., breathe more comfortably in conditions of low atmospheric pressure than visitors from lower altitudes. Over long time scales adaptation may allow humans to live in a significantly different environment. However, for the short term humans can be considered as biological machines with strict environmental specifications.

Environmental pollution and human health should be considered in the same framework. If health is defined as the homeostatic state of an individual or population (i.e., the state in which all physiological variables are in the "normal" range), environmental pollutants can be viewed as substances that tend to disrupt homeostasis.

For many years, the combined actions of natural fields within a specific environmental medium or compartment have been generating a variety of pollutants, which were defined as toxic or otherwise harmful substances present in measurable concentrations (e.g., pesticide concentrations in the soil, or radioactivity concentrations in the air).

Environmental pollution involves natural fields that occur in a variety of spatial and temporal scales. Fig. 1, e.g., shows several overlapping spatial and temporal scales in chemical contamination.

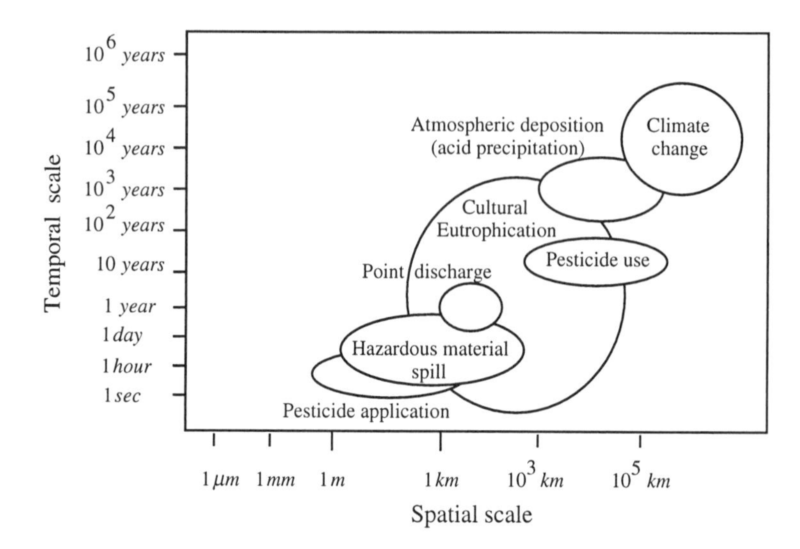

FIGURE 1: The overlap of spatial and temporal scales in chemical contamination (reconstructed from Landis and Yu, 1995).

Human populations are exposed to pollutants generated by two major sources: natural phenomena (e.g., volcanic eruptions) and human activities (industry, automobiles, cigarette smoking, oil spills, etc.). Due to their adverse health effects, the space-time distributions of these pollutants are of particular interest in environmental health studies (Lippmann and Schlesinger, 1979; Allaby, 1996).

From the health effect perspective, a most common categorization of environmental pollutants includes: (i) substances that lead to deterioration of human health by means of direct interaction with human receptors (e.g., airborne particulates, toxic and carcinogenic substances in food and water); (ii) substances that lead to secondary health effects by causing changes in local or global environmental conditions (e.g., increase in skin cancer incidence due to depletion of the ozone layer); and (iii) substances that generate environmental trends which can have significant socioeconomic implications (global warming, etc.). A single substance may belong to more than one of the above categories. In addition, the health role of the same substance may vary depending on its location in the environmental system. Ground-level ozone, e.g., is an environmental pollutant that causes respiratory problems and even chronic illness in humans, while stratospheric ozone forms a protective layer that shields the surface of the Earth from the harmful UV-B radiation. Based on the above, the following description of environmental pollution seems adequate:

Environmental pollution: *It refers to the occurrence of a disturbance in the natural environment that is caused by the introduction of substances or production of harmful conditions which can lead to primary or secondary health effects, or trigger natural events with significant socioeconomic impact.*

While environmental sciences have developed a plethora of physical models of environmental transport and fate that provide estimates of pollutant concentrations at any point in space-time, these models do not evaluate the actual effects of these pollutants on human health. This is, in fact, the task of *human exposure* analysis.

Human exposure analysis studies the distribution of pollutants within the human body via different pathways (e.g., respiratory system and blood circulation) and their potential for causing adverse health effects. Pollutants that are known to have significant impact on human health include several organic compounds (VOCs and PCBs, pesticides), toxic metals (mercury, lead), inorganic chemicals (particulates, ozone, NO_x and sulfur oxides), acids (nitric, hydrochloric, and sulfuric), radioactive materials, and biological contaminants (pathogenic microorganisms). Other types of environmental contaminants such as CFC compounds have secondary effects on human health by causing harmful changes in the environmental conditions. Exposure varies in space and time in response to changes in environmental conditions. The effects of exposure variation on human health also vary as a consequence.

This Chapter is concerned with quantitative aspects of human exposure analysis. Our approach will adopt a distinctly *holistic* perspective, already discussed in brief in §I.1. The term "holistic" has certain important implications as emphasized by the following model.

The holistic model: *The quantitative analysis of environmental exposure and health state should be considered as an integrated system, in which exposure modelling has direct effects on the human health state.*

REMARK 1: The holistic model of environmental health science may be seen as a spider web. When one strand of the web is altered, other strands are affected through the various linkages constituting the web.

The holistic model, therefore, gives primacy to the whole unit --environmental health science-- rather than to its parts. Natural processes (physical, biological, etc.) are of interest in the holistic framework if they are related (directly or indirectly) to a health effect. In this framework, human exposure should combine knowledge about the state of the natural environment surrounding the receptors with information regarding receptor-environment interactions that affect human health. This viewpoint has several important consequences. Decisions made regarding exposure assessment (scales , etc.) should take into consideration the modelling characteristics of the health effect, and vice versa. Another important consequence of the holistic model becomes clear in the context of environmental health policy: *Tradeoff* analysis shows that policies aiming to reduce a specific target risk may lead to an increase in countervailing risks (Graham and Wiener, 1997).

EXAMPLE 1: The 1977 Clean Air Act required that coal-burning power plants install scrubbers to remove sulfur dioxide from smokestacks. This requirement has produced tons of sulfur sludge that needs to be disposed. In addition, scrubbers reduce the energy efficiency of power plants, leading to an increase in fuel consumption per unit of generated electrical output, and consequently to increased emissions of other pollutants such as the greenhouse gas carbon dioxide. □

In addition, the environmental health processes involved in holistic human exposure analysis are represented as *spatiotemporal random fields* (§I.4 and Chapter IIIff), reflecting the fact that both exposure and its health effects vary from place to place and from time to time, and their variations are characterized by considerable uncertainties.

As we saw above as well as in §I.6, the issue of scale is an important factor in environmental health studies. For the purpose of human exposure analysis we will focus on two fundamental spatiotemporal scales defined with respect to the individual or the group of individuals considered.

Scales of human exposure analysis: *(a) The individual receptor scale, in which case the health effect is expressed in terms of measurable health state transitions for the individual receptor of interest. (b) The population scale, in which case health effects are expressed in terms of the number of receptors within the population of interest affected by the exposure.*

While in some disciplines (e.g., toxicology) environmental health analysis focuses on the individual receptor scale, in other disciplines (e.g., epidemiology) the population scale is of primary concern. When both scales are involved in a human exposure study, physical and mathematical requirements make it necessary that these scales are internally consistent.

The theoretical framework of the holistic exposure-health effect approach considered in this Chapter is based on the following steps:

(a) The distribution of pollutants within environmental media varies in space and time. Thus, it is represented by means of a *spatiotemporal distribution*.

(b) In the case of pollutants with primary health effects, *exposure* occurs when the pollutant comes in contact with the receptor.

(c) The process by means of which the pollutant enters into the receptor's body is referred to as *intake* (the term also denotes the amount of pollutant intake).

(d) The fraction of the intake that is actually retained in the body is called *uptake*.

(e) Uptake represents a *burden* on a target organ, tissue or cell.

(f) The *biologically significant burden* is the fraction of the burden that becomes biologically active.

(g) The biologically active form of the pollutant interacts with target organs, tissues or cells; the frequency of the interaction is measured by means of the *dose rate*.

(h) The dose rate can initiate a *transition* process, which changes the receptor's state of health.

REMARK 2: In the case of secondary health effects the pollutant remains external to the receptor; hence, steps (c)-(f) above may be not necessary in exposure analysis.

In the following sections we will be concerned with the quantitative formulation of the holistic exposure-health effect model. In particular, item (a) is the subject of §2, which describes the main concepts underlying the rigorous characterization of environmental fields; items (b) through (d) are discussed in §3, which focuses on exposure concepts; and items (e) through (h) are the subject of §4, which investigates the relation of exposure concepts to health effects. In this Chapter, therefore, the reader will become familiar with a series of environmental health processes that are important in the quantitative description of exposure-health effect relationships. In the following Chapters, these processes will be modelled stochastically by means of spatiotemporal random fields. In this respect, Chapter

II has a certain continuation in Chapter IX which contains several random field-based exposure indicators at the individual as well as at the population scales.

2. STUDYING ENVIRONMENTAL EXPOSURE FIELDS

Exposure assessment is an important stage of environmental health risk assessment (§I.7), the latter been driven by the former. Risk assessment is often initiated by public concerns regarding the effects (documented or potential) of specific exposures on human health. Exposure assessment focuses primarily on modelling the exposures, for it is generally concerned with effects of pollution sources at distant locations, future estimates of exposure, or estimates at geographical locations where data are not available (spatiotemporal mapping, Chapters V and VI). Exposure assessment can be used to identify populations at risk, demonstrate paths of exposure to human populations and provide rigorous quantification of the exposure. It also serves as input to the analysis of the adverse health effects associated with an environmental hazard.

Exposure assessment is based on both monitoring (measurement) and modelling. It includes pollutant source determination, environmental evolution mechanisms and population exposure, and involves natural fields and pollutants occurring in one or more environmental media and compartments. Studying environmental exposure encompasses a broad range of scientific disciplines (chemists, hydrologists, meteorologists, biologists, physicists, engineers, geographers, statisticians, etc.) and, also, demands a considerable variety of skills.

2.1 Pollution Sources

Identification of the pollutant source plays an important role in the determination of the physical and chemical forms of the pollutant, the medium to which it has been released, the pathways it will follow, the exposure routes, etc.. Release assessment studies the extent to which a risk source releases or otherwise introduces pollutants into the environment (air, soil and water).

There exist *routine* emission sources (e.g., dioxin emissions from a specific plant), as well as *occasional* releases (e.g., ships polluting the sea while cleaning their oil tanks) and *accidental* releases (e.g., methyl-isocyanate release in Bhopal, India) of pollutants into the atmosphere (Morris, 1990; Covello and Merkhofer, 1993).

Risk assessment methods for these sources include *monitoring* the status of a risk source while it is functioning in its normal way (collecting data on the integrity of hazardous waste at a disposal site, statistical data analysis), *performance testing* (collecting data about a risk system under conditions that are extreme or otherwise of special concern), *accident investigation* (reconstruction of the accident on the basis of post-accident information, simulation of possible accident scenarios), and *modelling* (the source is decomposed into a set of individual systems, each of which are studied separately to identify which elements of the source contribute the most to the final release).

Finally, there are several situations in which one is not concerned with sources but simply with pollutant distributions in space-time (e.g., many studies focus on the distribution of a pollutant in the air rather than how it got there).

2.2 Environmental Pollutants

There is a great variety of environmental pollutants that cause primary health effects, but they can be conveniently grouped into three major categories as follows:

(i) *Physical* pollutants are substances that act primarily by virtue of their physical properties rather than by means of chemical or biological reactions. Hence, their impact is due to the transfer of mass or energy --in various forms. Examples of such pollutants include particles and fibers, radiation and radioactivity, heat, and noise.

(ii) *Chemical* pollutants are substances that act primarily by means of chemical reactions in the body. Most known pollutants belong in this category which includes agricultural pesticides, industrial waste products, and most atmospheric and groundwater pollutants.

(iii) *Biological* pollutants are microorganisms that cause health effects by means of biological reactions that they initiate in the receptor. Examples include viruses, bacteria, protozoa, and fungi.

EXAMPLE 1: A plot of the average human body burden of some toxic heavy metals is shown in Fig. 2. □

Even though human activities are usually blamed for the production of pollutants, it should be noted that certain environmental pollutants occur naturally. Common examples include volcanic ash released during eruptions, and pathogenic microorganisms that abound in certain ecosystems. The modelling and monitoring of large scale natural events that have long range effects on the environment, such as volcanic activity, is desirable and has significant health implications. However, in most cases it is the man-made pollution that presents immediate health risks for the human ecosystem. This is due to a number of factors that include the proximity of pollution sources to population centers and distribution systems, high toxicity and long residence times of certain pollutants in the environment, and continuous operation of the sources.

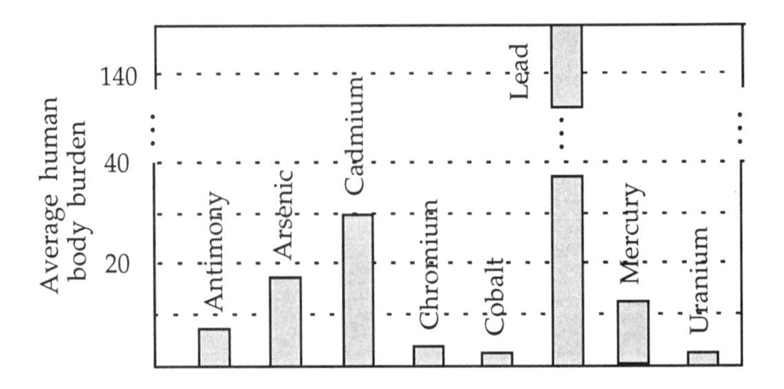

FIGURE 2: Average human body burden of toxic heavy metals -- *mg* in a 70 *kg* person (Botkin and Keller, 1998).

REMARK 1: Implicit in the description of environmental pollution is the notion of natural scales (§I.6). Environmental trends depend on the temporal and spatial scales of the observation windows, and the health implications vary with geographical/temporal scale. Obviously, the average distribution of a pollutant over a large geographical area is less informative than a detailed distribution profile near densely populated areas.

2.3 Characterization of the Environment

Exposure assessment requires an adequate characterization of the environmental conditions, which can be accomplished only by means of a multidisciplinary approach that involves various natural sciences. A few examples of exposure situations that require a combination of different types of information are presented below.

EXAMPLE 2: Assessment of ozone exposure requires information about the spatiotemporal distribution of ozone into the air, and the pathways by means of which it enters into the human body. □

EXAMPLE 3: Exposure assessment of animal waste spills requires an understanding of the hydrologic principles governing the transport of waste materials as runoff into the water supply system, characterization and measurement of the bacteria and inorganic pollutant concentrations contained in the waste, as well as estimation of the potential health risks due to the spills. □

EXAMPLE 4: Exposure to pesticides (such as DichloroDiphenylTrichloroethane-DDT or organophosphates) is a result of significant spatiotemporal pesticide distributions in the environment. The use of DDT has been banned in the U.S. because of its long residence time, adverse effects on animal populations, and ability to bioaccumulate in animal fat. Organophosphates affect the environment via different pathways. They are easily absorbed through the skin which can cause pesticide applicators to receive high exposures. In addition, they are water soluble and thus easily transported with agricultural runoff, which eventually percolates through the soil to the groundwater table. Consumers are also exposed to pesticides in the form of food residuals. □

EXAMPLE 5: Asbestos is a natural fibrous material that has been widely used as an insulator for its heat resistance properties. Exposure to high levels of asbestos can lead to chronic illnesses including lung cancer. The main pathway for asbestos accumulation in humans is via the respiratory system. Hence, assessment of health risks require modelling the transport of asbestos fibers in the air and their distribution in the lung tree. □

EXAMPLE 6: Radon gas is an indoor pollutant that is generated primarily via the radioactive decay of uranium-238 in uranium rich soils, rocks, or water. According to the Surgeon General, Radon is the second leading cause of lung cancer in U.S. after smoking. Human exposure studies require knowledge of the spatiotemporal source distribution, modelling of the movement of Radon gas in the subsurface and its escape routes into the ambient air. □

These examples show that the study of exposure distribution in space-time is a necessary step for assessing the health effects of environmental pollution. For a systematic characterization of environmental exposure it is helpful to identify certain components of the global environment. These components are known as environmental compartments and media. Their physical properties and the mechanisms by means of which they interact with each other and with humans are important.

2.4 Environmental Compartments

For the purposes of environmental health science the environment is generally divided into four major compartments, as follows:

(i) *Atmosphere* is the mantle of gas and vapor that surrounds the Earth. It consists of the following zones, in order of increasing distance from the Earth's surface: troposphere, tropopause, stratosphere, stratopause, mesosphere (upper atmosphere), mesopause and thermosphere. Atmospheric pollutants can be divided into two different categories based on the scale at which their impact is manifested: macro effects occur at the global scale (e.g., global warming, depletion of stratospheric ozone), and micro effects occur locally (toxic concentrations of VOCs in the air, photochemical pollution in urban areas, etc.).

(ii) *Lithosphere* comprises the outer shell of the Earth's crust which extends 50 to 100 *km* from the surface and consists mainly of soil and rocks. The soil is divided into two zones (root and non-root) with respect to vegetation: closer to the surface is the root zone, via which pollutants are absorbed by plants. With respect to water content the lithosphere can be divided as follows: Near the Earth's surface extends the aeration zone, in which the soil's pores are partly filled with water and partly with air. The upper part of the aeration zone is called moisture belt; water that enters in this belt is either used by plants or evaporates in the atmosphere. Below the aeration zone lies the saturation zone, in which all pores are filled with groundwater. The boundary between the aeration and the saturation zones is called water table. A capillary fringe forms at the water table as water rises above the saturation zone due to the capillary forces.

(iii) *Hydrosphere* includes all the water in the Earth's system. It consists of oceans (97.3% of the total water), groundwater (0.61%), surface water in land areas such as rivers and lakes (0.0171%), soil and atmospheric moisture (0.006%), glaciers and icecaps (2.14%). For groundwater contamination micro effects are most important, since contamination incidents, which occur when pollutant concentrations exceed regulatory standards, are usually localized. Pollutants are transported in the form of suspended particles, colloidal particles or dissolved materials (which are the most difficult to control).

(iv) *Biosphere* includes all living organisms, i.e., flora, fauna and microbes. The following levels of organization are present in the biosphere: individual, population or community, ecosystem, landscape and biome (i.e., regional community of plants and animals). There are continuous interactions in the biosphere as well as interactions of biospheric components with the other three major compartments.

2.5 Environmental Media

Environmental compartments are complicated systems that consist of smaller units, and different phases and materials. *Environmental media,* on the other hand, refer to specific substances through which the movement and interaction of pollutants takes place.

Environmental media can consist of a single phase (e.g., solid, liquid, and gas), or of multiphase mixtures (e.g., as in vapors which are liquid-gas mixtures).

EXAMPLE 7: Aerosols are suspensions of liquid droplets and solid particles in a gaseous phase. Aerosols play a central role in several industrial processes and air pollution. They are released into the atmosphere by natural events such as volcanic eruptions, forest fires, soil and rock erosion, ocean spray, the metabolic cycle of plants and animals, and by human activities, e.g., fuel combustion. As environmental pollutants they appear in dust, smoke, mist, fume, and fog. The size of particulate matter that can be suspended and transported in the air varies between 0.1 and 50 micrometers. Particles less than 10 micrometers in diameter (PM_{10}) are inhalable and can cause serious health problems. EPA has recently revised the standards for particulate matter exposure under the national ambient air quality standards (NAAQS). The new annual exposure standards are 50 $\mu g / m^3$ for (PM_{10}), and 15 $\mu g / m^3$ for ($PM_{2.5}$), and the corresponding 24-hour standards are 150 $\mu g / m^3$ for (PM_{10}), and 65 $\mu g / m^3$ for ($PM_{2.5}$). □

The same medium can be found in more than one environmental compartments. Air, e.g., is found in the atmosphere, but also in the pores of the aeration zone in soil formations. The hydrologic cycle, to be discussed in Example 11 below, presents a characteristic example of the movement of one medium (water) between different environmental compartments (atmosphere, lithosphere and hydrosphere). Finally, an environmental compartment may contain more than one media.

EXAMPLE 8: The atmosphere contains a gaseous medium which is a mixture of nitrogen (78% by volume), oxygen (21%), carbon dioxide, argon, neon, helium and methane, as well as a medium of water vapor. □

To the basic media above that represent different phases of matter and mixtures thereof, one may add *biological* materials, particularly in the form of food.

2.6 Natural Fields

Characterization of an environmental medium or compartment involves the quantitative description of the natural fields that are present in the medium or the compartment and affect environmental health. The natural fields represent measurable properties which may vary in space and time (some fundamental properties are summarized in Table 1).

TABLE 1: Fundamental Measurable Pollutant Properties

- The *mass* measures the quantity of a material or medium.
- The *density* represents the mass per unit volume of the medium.
- The *concentration* measures the quantity of pollutant (measured in units of mass or volume) per unit volume or mass of the medium.
- The *surface density* is the quantity of pollutant per unit surface area.
- The *pressure* measures the force per unit area.
- The *flux* is equal to the quantity of fluid crossing a surface per unit area and time.
- The *fluence* represents the quantity of pollutant crossing a surface per unit area.
- The *spatial gradient* measures the rate of change of a quantity in space.
- The *rate* represents the temporal rate of change.
- The *velocity* measures the rate of spatial displacement with respect to time.
- The *mass flow rate* represents the transported total mass per unit time.
- The *temperature* is a macroscopic measure of the average mechanical energy carried per particle in the system.
- The *heat* is a measure of the energy exchange that occurs as a result of a temperature difference when no mechanical work is produced.
- The *attenuation or extinction* determines the rate of energy loss of a wave (e.g., electromagnetic radiation) traveling through medium.
- The *pH* is a logarithmic measure of the concentration of free hydrogen ions in a solution. A value of *pH* larger than 7 denotes an alkaline solution, while a value less than 7 denotes an acidic solution.

Spatiotemporal variability is of significant importance and leads to the following notion of the natural field (also, §I.3).

Natural field: *It is the spatiotemporal distribution of the values of a natural process or a pollutant property.*

EXAMPLE 9: The data sets shown in Fig. 3 represent rainfall levels at two different

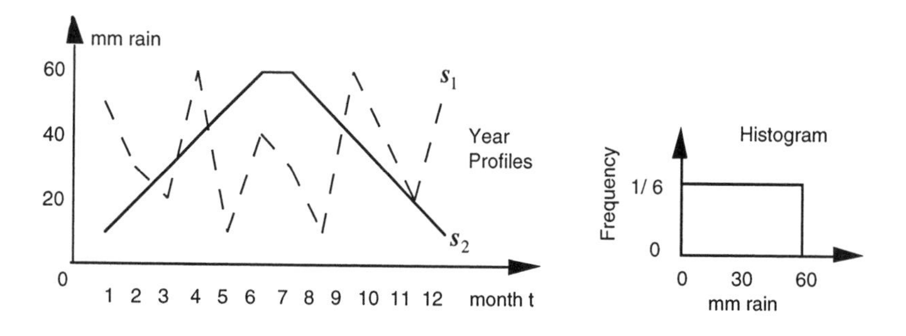

FIGURE 3: Monthly rainfall level measured at locations s_1 and s_2 sharing the same statistics.

monitoring stations s_1 and s_2. The data sets have the same classical statistics (mean=35 mm, variance=292 mm^2 and histogram), but very different physical behavior in space and time. In particular, at s_1 the rainfall height shows significant variability in time t. On the contrary, the data at s_2 imply a systematic temporal behavior. It seems that these data sets contain more information than what is reflected by their classical statistics. Indeed, as it turns out the key parameter in an adequate description of these data sets is the spatiotemporal variability of the rainfall field. □

An efficient way for visualizing spatiotemporal variability of natural fields is by means of spatiotemporal *maps* (Chapters V and VI). On the basis of a limited number of observations, spatiotemporal mapping generates estimates at numerous unobservable points in space-time, i.e., it pertains to the goal of capturing significant *generalizations* that go beyond the small data set available.

EXAMPLE 10: A map of ozone concentrations over Eastern U.S. is shown in Fig V.1a. Maps of wetting and non-wetting phase pressure contours for two-phase flow in a heterogeneous medium are shown in Fig V.2. □

REMARK 2: In environmental health science a significant part of our knowledge originates from observations (§I.5). These observations must be, however, organized in an efficient way to become part of our knowledge processing mechanisms. An organization may not cause any change in the information content of the data (e.g., the data arrangement in Fig 3). Other organizations however, such as the maps mentioned in Example 10 above, involve a change in the information content of the data by means of a generalization from the little we actually observe to a much larger environmental health process --a part of which is the observed data set .

2.7 Evolution Mechanisms

We begin our discussion of evolution mechanisms by presenting a principle that is of great importance in environmental health science.

Principle of nature: *Everything comes from other things and gives rise to other things. Nothing is lost in the process.*

This general principle finds a more formal expression in the form *of conservation laws.* The first explicit statement of the principle of nature is due to the Roman philosopher Lucretius (1st century BC). Since then, it has been restated under various forms. To a large extent, the study of natural systems focuses on the investigation of the laws that govern different manifestations of the principle of nature. In general, it is possible to distinguish three groups of laws:

(A) *Deterministic* laws. These laws reflect the macroscopic space-time regularity of natural fields that follows from deterministic conservation laws. The ideas of determinism and necessity date back to Aristotle (4th century BC) who claimed: "Nature does not act without a goal". The laws of celestial mechanics discovered by Newton and Kepler were the crown jewel of deterministic thought. Determinism was thought to provide an adequate description of nature until the advent of statistical mechanics.

(B) Laws of *chance.* These laws govern the microscopic, random variations of natural fields and focus on a quantitative description of the erratic fluctuations. The central limit theorem, e.g., describes the behavior of the sum of a large number of random variables, independently of the physical quantities that these variables represent. From the philosophical standpoint, laws of chance reflect the Democritean idea that "both chance and necessity are essential characteristics of physical phenomena".

(C) Laws *relating* (A) and (B). Laws of this type are associated with dynamic connections in space-time. They implement relationships between distinct components of physical models and across different scales. Constitutive relations --e.g., pressure-saturation equations-- determine the interdependence of physical variables. The Langevin equation extends the deterministic Newtonian law of motion to cases, in which the force includes a random component due to microscopic fluctuations.

The essence of (C) laws is ideally captured in the following ancient Sufi teaching: "You think that because you understand *one* you also understand *two*, because one and one make two; but you must also understand *and.*" This perspective in the analysis of physical phenomena helps to reveal relations that transcend determinism and classical statistics.

EXAMPLE 11: An interesting application of the principle of nature is provided by the *global hydrologic cycle* (Gupta, 1997). Water on the Earth exists in the hydrosphere which extends about 15 *km* into the atmosphere and about 1 *km* below the surface into the lithosphere. The continuous transformations and circulation of water form the global hydrologic cycle, which involves various natural fields (such as precipitation, evaporation, infiltration and groundwater flow). These fields are distributed in space-time in a way that involves a combination of the laws (A), (B) and (C). Precipitation, e.g., is a hydrologic field which depends on the causal laws of atmospheric physics, on the random distributions of clouds and wind directions, as well as on the space-time interdependencies and correlations. The principle of nature ensures that water is not lost during these transformations. Hence, the amount of water that falls on the Earth as precipitation is counterbalanced by runoff, evaporation and storage. □

Modelling of the physical processes involved in the global hydrologic cycle based purely on concepts and techniques from classical statistics may account for the laws of chance, but it cannot account for the (A) or the (C) laws of the principle of nature.

EXAMPLE 12: In Example 9 above we saw that rainfall data with the same classical statistics may correspond to very different physical behavior. This behavior in space-time is rigorously expressed in terms of a spatiotemporal random field (Chapter IIIff). □

REMARK 3: We should add at this point that certain natural phenomena obey deterministic laws, and yet their evolution in space-time appears to be erratic and unpredictable. Such phenomena exhibit quite different behavior than random systems. This behavior is called *chaotic*, and it is the subject of the theory of dynamical systems. Chaotic behavior results when small perturbations are compounded during the space-time evolution of the system. In the case of chaotic behavior even a small uncertainty regarding the initial state severely limits the predictability of the system in the future. This lies at the root of the well-known difficulties associated with weather prediction.

The fundamental principle of nature ensures the conservation of matter in the environment. However, the environment is in a state of continuous change which is represented by the space-time evolution of natural fields. Change is realized by means of the three evolution mechanisms, *transport*, *transfer* and *transformation* to be discussed in more detail below. Much of environmental science is concerned with the development of theories and mathematical models that aim to explain these mechanisms and predict their effects. Transport provides information about the distribution and the movement of environmental pollutants, transformation describes the changes between different physical phases and chemical compositions, and transfer provides information about the flow of pollutants between different media or compartments.

More specifically, transport processes focus on movements of pollutants within the same environmental medium and compartment. Transport processes include the following: *Diffusion* (random movement of particles due to kinetic energy), *advection* (the transport of particles within a flow field), *sedimentation* (gravitationally driven accumulation of particles obtained by deposition processes and chemical precipitation), *carriage* (movement of pollutants by means of currents), *inertial transport* (due to applied mechanical forces), and *electromagnetic transport* (movement of polar pollutants driven by electromagnetic fields). Transport processes are mathematically described by PDEs (e.g., models of atmospheric dispersion, or solute transport in porous soil formations; Chapter VII).

EXAMPLE 13: A plume released into the subsurface disperses in all directions, partially because of diffusion; the exact spatial distribution of the plume depends on soil properties. Bacteria released into rivers eventually settle on sediments on river floors. Following rainstorms, pesticides absorbed in soils are carried by surface runoff. Earthquakes supply tidal waves with inertial energy. Aerosol particles often carry electric charge which allows their manipulation by means of electric fields. □

Transformation is the process by means of which matter or energy change form within an environmental medium or compartment. It occurs by means of chemical reactions that change the chemical composition of the reacting substances, by physical mechanisms such as radioactive decay, radiation absorption and emission, which modify the physical state of particles or by biological mechanisms that lead to new forms of biological organization and involve the mediation of living organisms. Transformation is described mathematically by differential equations (e.g., 1st-order kinetics; Chapter VIII).

EXAMPLE 14: Chemical transformation is involved in the reaction of sulfur dioxide with moisture in the atmosphere, which generates sulfuric acid that later precipitates as acid rain. Physical transformation is exemplified by means of the solar radiation cycle: a fraction of the solar radiation entering the atmosphere is reflected back into space by the cloud cover, while the remaining fraction either is absorbed by the atmosphere or travels to the surface of the Earth; part of the radiation that reaches the surface is reflected back into the atmosphere, and the remaining part is absorbed by the land and the hydrosphere. A useful environmental application of biological transformation is the biodegradation of organic contaminants. This process --which was used successfully in the cleanup of the *Exxon Valdez* oil spill-- involves microorganisms that transform the hazardous waste into nutrients degrading it at the same time. Biodegradation is also used in waste ponds on animal farms for the treatment of animal waste. □

Finally, transfer processes move a pollutant between different environmental media or compartments. Transfer processes are described mathematically by systems of equations that enforce the conservation laws. Transport and transfer processes are related, since transfer of a pollutant between media involves a sequence of transport processes.

EXAMPLE 15: Radon gas is transferred between drinking water, indoor and outdoor air, and the human body. □

3. HUMAN EXPOSURE CONCEPTS

3.1 Basic Notions

The study of human exposure to environmental pollutants requires the introduction of three fundamental concepts: *pollutant concentration*, *exposure* and *biomarker*. All these concepts may involve various physical and biological fields in space-time. Biomarkers are most critical, for they assess the effects of exposure on human health.

Definitions of human exposure to a pollutant may refer to a critical site of an individual receptor at the space-time point $\boldsymbol{p} = (\boldsymbol{s}, t)$ or a group of receptors (population) within a space-time domain $\Lambda(\boldsymbol{p})$ around $\boldsymbol{p}$. We first formulate the mathematical framework at the level of an individual receptor. Later, we extend our exposure analysis to populations by taking into consideration the population density distribution in space-time.

In the following, the term "pollutant" will denote a toxic substance that has adverse health effects on humans. Generally, the specification of a pollutant must account of what is possible with the analytical and sampling techniques available. One of the most fundamental characteristics of a pollutant is its space-time concentration.

Definition 1: *The concentration* $C(\boldsymbol{p})$ *of the pollutant at the space-time point* $\boldsymbol{p} = (\boldsymbol{s}, t)$ *is defined as*

$$C(\boldsymbol{p}) = \lim_{\delta \upsilon(\boldsymbol{p}) \to 0} \delta m(\boldsymbol{p}) / \delta \upsilon(\boldsymbol{p}), \tag{1}$$

where $\delta\upsilon(\boldsymbol{p})$ *is an elementary space-time domain centered around point* $\boldsymbol{p}$ *and* $\delta m(\boldsymbol{p})$ *is the mass of the pollutant contained in this domain.*

EXAMPLE 1: If we consider a pollutant in a flow field with velocity $\boldsymbol{v}(\boldsymbol{p})$ and a cross-sectional area A perpendicular to the flow direction and centered around $\boldsymbol{s}$, the mass of the pollutant flowing through the cross-section in time δt is given by $\delta m(\boldsymbol{p}) = A\, v(\boldsymbol{p})\, \delta t\, C(\boldsymbol{p})$. Hence, the pollutant concentration is $C(\boldsymbol{p}) = \lim_{\delta t, A \to 0} \delta m(\boldsymbol{p}) / A\, v(\boldsymbol{p})\, \delta t$. □

Since the point concentration $C(\boldsymbol{p})$ is a natural field, Definition 1 presupposes the existence of a space-time continuum (§I.3). Concentration values can be estimated at any point of the continuum using the mapping techniques of natural fields (see Chapters V and VI). The $C(\boldsymbol{p})$ has units of mass per unit volume ($mg\,/\,m^3$) if we are interested in mass concentrations. Other measures of $C(\boldsymbol{p})$ besides the mass concentration defined in Eq. (1) above are, also, possible: *Number concentration* is equal to the number of particles per unit volume; e.g., Radon concentrations are measured in number of atoms/unit volume. *Volume concentration* is defined in terms of the total volume of pollutant/unit volume. *Surface concentration* is defined as the number or total mass of particles on a surface/unit area. Finally, the concentration of pollutants in a medium is often measured in parts per million (ppm) by volume or mass.

Exposure generally refers to the contact of a receptor with an environmental pollutant (e.g., Georgopoulos and Lioy, 1994; Zartarian *et al.*, 1997). In many cases, exposure is assumed to be identical with the concentration of the toxic substance. Since concentration values can be estimated at any point in space-time, exposure values are also available for receptors located at the same points.

Exposure is primarily a function of the pollutant concentration and the *exposure pathway* $\mathcal{J}_E$, which describes how the receptor is connected to the environment. The $\mathcal{J}_E$ involves the following components:

a. Receptor: This is the biological unit (e.g., person, human population) that receives the impact of exposure to the pollutant.

b. Medium: It refers to the environmental medium that carries the pollutant (e.g., air, water, soil or food).

c. Exposure (or contact) boundary: It represents the part of the receptor through which the pollutant gains entry into the body.

The exposure boundary can be the entire human body --in the case of radiation exposure-- or parts of it such as the skin or the respiratory tract epithelium --in the case of particle deposition. The toxicity of certain pollutants depends on the exposure pathway, but other substances are generally toxic. The following general notation of exposure accounts for the dependence on both the pollutant concentration and the exposure pathway:

$$E(\boldsymbol{p}, C, \mathcal{J}_E). \tag{2}$$

Expression (2) denotes the joint occurrence of two events: (a) the pollutant concentration $C = C(\boldsymbol{p})$ is present at the space-time point $\boldsymbol{p}$, and (b) the receptor is present at the same point and connected to the environment through $\mathcal{J}_E$. Exposure can be measured in terms of the *exposure rate*, which is the pollutant concentration that comes in contact with the receptor, and the *cumulative exposure*, which measures the total amount of pollutant within a specified space-time domain (rigorous definitions of exposure will be given later).

EXAMPLE 2: Consider a person in a room where the uniform concentration of nicotine in the air is $C = 45\,\mu g/m^3$. The $\mathcal{J}_E$ includes the receptor (person in the room), the medium (air), and the contact boundary (the whole body immersed in the nicotine-polluted air, the skin, the receptor's facial area, etc.). □

The requirement that the receptor and pollutant coexist in space-time presupposes an appropriate choice of scale. At finer resolution scales, toxic substances are distributed to different locations at different times before they reach the targets. An interaction then takes place between the receptor's contact boundary and the pollutant. For risk assessment purposes, information on the quantity of pollutant delivered to the receptor via the contact boundary is required. The transfer of the pollutant from the environmental medium into the receptor occurs via the *intake pathway* $\mathcal{J}_D$. Hence, another component is added to the exposure assessment framework as follows:

d. The *intake pathway* $\mathcal{J}_D$: It represents the route by means of which the pollutant is delivered to the receptor.

Typical intake pathways $\mathcal{J}_D$ involve ingestion, inhalation and skin absorption. Hence, intake occurs through the skin, mouth and nostrils, water and food, etc. A less common route involves injection of a chemical under the skin (hypodermic) or into the bloodstream (intravenous).

The quantitative description of the intake pathway $\mathcal{J}_D$ involves some quantitative measures, also known as *biological markers* or *biomarkers*.

Definition 2: *Biomarkers are measurable cellular, biochemical, or molecular alterations that occur in biological media, such as human tissues, cells or fluids.*

Biomarkers describe the application of physical, chemical, radiological and immunobiologic tests to human biologic samples, such as blood, urine and tissue (Perera, 1996). Three classes of biomarkers are considered: *intake*, *uptake* and *interaction* markers. In this section we will discuss the biological meaning of these biomarkers. Rigorous mathematical expressions of the biomarkers will be given in a following section.

Intake markers are used to quantify the total exposure to the pollutant. As with exposure, they include the *intake rate* I_r (i.e., the rate at which the pollutant is entering the body), and the *cumulative intake* or simply *intake* (i.e., the total amount of pollutant that has entered the body via a specified space-time domain).

Uptake biomarkers are used to specify the quantity of the pollutant that remains in the receptor after intake. They include the *cumulative uptake* or simply *uptake* U (i.e., the quantity of pollutant residing in the receptor immediately after intake), and the *uptake rate* U_r (i.e., the rate at which the pollutant is absorbed or deposited in the receptor). The *uptake* (or *absorption* or *deposition*) *fraction* f_u represents the fraction of the intake that remains in the receptor long enough to interact with *target organs* such as tissues and cells.

REMARK 1: While the term "uptake" is used for all pathways, absorption rate specifically refers to ingestion, injection and skin absorption, and deposition to inhaled particles that settle in the respiratory system.

The *retention half-time* or *half-life* $T_{1/2}$ is equal to the time required for half of the pollutant to exit an organ after uptake. The pollutant *burden* is equal to the fraction of the pollutant uptake that accumulates in target organs of the receptor. Examples of these markers are DDT and PCBs in serum and adipole tissue as a result of environmental contamination, plasma or salivary nicotine from cigarette smoke, urinary aflatoxin indicative of dietary exposure, etc. Most uptake markers have the advantage of being comparatively easy to measure, but they do not provide information about interactions with cellular targets.

Interaction biomarkers quantify the pollutant interactions with the receptor. An interaction biomarker may indicate, e.g., the amount of absorbed chemical that has interacted with critical subcellular targets, measured either in a target or surrogate tissue (Hulka *et al.*, 1990). The *biologically active form* of the pollutant is the fraction of the burden that transforms into a form that can interact with target organs. The *dose rate* D_r determines the interaction rate between the biologically active form and target organs in the receptor. The *dose* D is the cumulative quantity of the biologically active form delivered to the receptor. Therefore, the dose is the most important measure of the interaction between the pollutant and the target organs. DNA adducts are particularly interesting members of this category of biomarkers, because they are capable of initiating mutation events. Polycyclic aromatic hydrocarbons (PAH)-DNA adducts have been associated with workplace exposure to PAHs, with exposure to air pollution and cigarette smoke, as well as lung cancer (Sinopoli *et al.*, 1990; Crawford *et al.*, 1994).

REMARK 2: In this book we use the modern definition of dose as given above. However, a word of caution is necessary regarding the multiple uses of the term "dose" in the literature (Duan, *et al.*, 1989; Zartarian *et al.*, 1997). The term has been used to denote the amount of pollutant that enters the body through a contact boundary (which is, actually, the intake), or the amount of pollutant deposited inside the body during a specific time interval (which, in fact, refers to the uptake).

Pollutant-organ interaction does not necessarily imply that biological effects, which can possibly cause health impairment (e.g., a disease), will follow. When a response to a certain dose does indeed induce such effects, the latter are represented by the *effect biomarkers*. Biomarkers that fall in this category vary from DNA single strand breaks, to gene and chromosomal mutation, and alterations in target oncogenes or tumor suppressor genes. Effect biomarkers are sometimes classified as (Schulte and Perera, 1993): *early biologic effects* representing events correlated with, and possibly predictive of, health impairment; and *precursor biologic effects* (altered structure or function, DNA hyperploidy) which are even more closely related to a disease.

The above considerations may lead to the following notation for a biomarker, in general:

$$\mathcal{B}(\boldsymbol{p}, E, \mathcal{J}_D). \tag{3}$$

Expression (3) denotes the dependence of a biomarker on the exposure $E = E(\boldsymbol{p})$ present at the space-time point $\boldsymbol{p}$, as well as the variables characterizing the intake pathway $\mathcal{J}_D$. Biomarkers are affected, indeed, by both (i) *natural exposure variations* in space-time (the extent of the effect varies depending mainly on the biomarker's half-life), and (ii) *biological variability* linked to the individual receptor (body build, liver and kidney functions, pulmonary ventilation, etc.). In the following, the parameters E and $\mathcal{J}_D$ will usually be dropped and a biomarker will be simply denoted as $\mathcal{B}(\boldsymbol{p})$. Biomarkers are not usually directly measurable; instead they are estimated based on mathematical models that include exposure variables, the intake pathway, biological parameters, etc. (see, pollutokinetic modelling in Chapter VIII).

EXAMPLE 2 (cont.): In the dose notation (3) the exposure $E(\boldsymbol{p})$ is expressed as in Eq. (2), and the dose pathway $\mathcal{J}_D$ incorporates $\mathcal{J}_E$ and the fact that intake occurs (through the skin on which the nicotine settles, through the receptor's mouth and nostrils, or through food and water consumed by the receptor). □

The crux of the preceding discussion is the existence of a *continuum* of events between exposure and health effect (Fig. 4). Each biomarker represents an event in the continuum.

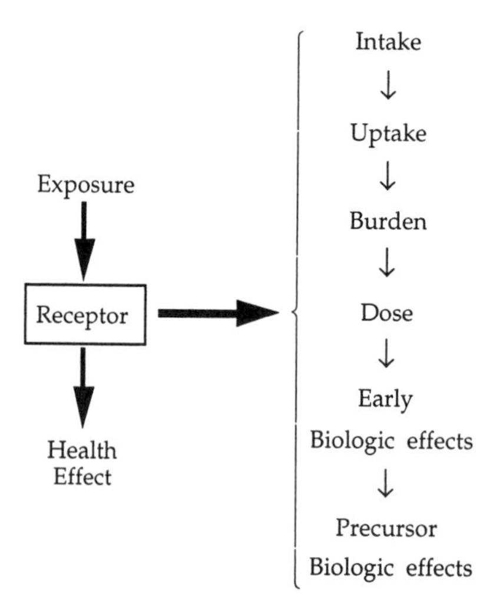

FIGURE 4: Biomarkers of the exposure-health effect continuum.

The relationships between biomarkers depend on genetic and other characteristics of the individual; in fact, biomarkers calculated on the basis of similarly exposed people may show significant variation. The study of the continuum biomarkers provides the means for improved environmental health analysis and management. For example, critical disease-related events may be detected earlier on a smaller scale, which makes it possible to focus on preclinical rather than clinical intervention approaches. Also, valuable hints may be obtained regarding the mechanisms relating exposure and health effect. To any component of the continuum *susceptibility* markers can be assigned, which are considered as indicators of increased or decreased risk for the particular component. For example, genetic and acquired factors (DNA repair, differences in metabolism, nutritional deficiencies, etc.) may lead to individual susceptibility to cancer.

REMARK 3: The application of biomarkers in studies of population health state raises a significant question: Does a biomarker imply the simultaneous occurrence of a health state (e.g., disease), can it be used as a predictor of the state, or is it correlated in some other way to the state? In fact, different options may be valid in different situations and, therefore, conclusions should be obtained on a case-by-case basis, before the markers are used in epidemiological investigations and health decision making.

Finally, when the effect biomarkers indeed imply the occurrence of a health effect, they are succeeded by *disease markers*. A disease marker is a measurable indicator of a biological or biochemical event that either represents a subclinical stage of disease or is a manifestation of the disease itself (Hulka *et al.*, 1990). This group includes (a) markers of *clinical disease* (tumor-associated antigen, serum alpha-fetoprotein, etc.), and (b) *health effect thresholds* (e.g., biomarker values above which the disease appears; §4 below) at the individual risk assessment level, as well as *population health indicators* at the group risk assessment level (see, also, stochastic health indicators in Chapter IX).

3.2 Mathematical Expressions for Exposure

Various quantitative expressions of exposure (2) and dose (3) exist in the literature. Below, we begin with a basic theoretical definition of the "space-time point exposure", which we then use to construct several other quantitative measures of exposure. A similar approach will be employed with biomarkers. For simplicity, the pathways $\mathcal{J}_E$ and $\mathcal{J}_D$ will not always be explicitly denoted. Nevertheless, these pathways are implicit in the exposure definitions and measurements.

Definition 3: *The space-time point exposure or exposure rate* $E(\boldsymbol{p})$ *is defined as the pollutant concentration* $C(\boldsymbol{p})$ *at the same space-time point* $\boldsymbol{p}$, *i.e.,*

$$E(p) = C(p). \tag{4}$$

The units of $E(\boldsymbol{p})$ are those of pollutant concentration. Notice that while $C(\boldsymbol{p})$ is a natural field with values at all space-time points, $E(\boldsymbol{p})$ represents the pollutant concentration only at points where a receptor is connected to the environment via $\mathcal{J}_E$.

EXAMPLE 2 (cont.): In this case the exposure rate to nicotine in the room is $E(\boldsymbol{p}) = C(\boldsymbol{p}) = 45\,\mu g\,/\,m^3$. □

Estimates of pollutant concentrations are based on ambient monitoring and mathematical modelling of natural fields. Exposure estimates are obtained from personal monitoring, as well as mathematical modelling of biological processes within the human body (Chapters V-VIII and IX).

Other definitions of exposure within the space-time framework involving averages over specific time intervals are also used. For example, let $\Lambda(\boldsymbol{p})$ denote a domain centered around $\boldsymbol{p}$, which is given by $\Lambda(\boldsymbol{p}) = V \times T$, where V denotes the volume or the area and T the temporal extent of the domain. Other notations have been also used (e.g., §V.2). Environmental standards for pollutant concentrations are often expressed in terms of time-averaged concentrations.

EXAMPLE 3: U.S. EPA has recently proposed an 8-hour based primary standard (0.008 ppm) for ground-level ozone. The 8-hour standard tends to be more stable than a 1-hour standard, because it averages fluctuations and smoothens short-lived peaks. In addition, it is a better predictor of health effects caused by prolonged exposure to moderate ozone concentrations. □

Since different standards (e.g., primary vs. secondary) require different measures of exposure, and since standards are subject to change, it is often desirable to have complete exposure rate profiles, i.e. continuous records of exposure at all points in space-time.

Definition 4: *The exposure rate profile over a space-time domain Ω is defined as the field of $E(\boldsymbol{p})$-values for all $\boldsymbol{p} \in \Omega$.*

EXAMPLE 4: The ozone concentration map over the Eastern U.S. shown in Fig. V.1 is a spatial profile $E(\boldsymbol{p})$ at a specified time period. □

Exposure rate profiles provide a complete characterization of exposure in space-time, but accurate estimation of profiles poses economic and computational challenges. In such cases, the development of *functional* or *summary* (or coarse-grained) measures of exposure may be more appropriate. Summary measures are defined as functionals of the exposure rate profile that involve averages over specific windows in space-time. Before we start our discussion of summary exposure measures, we emphasize that detailed maps of exposure rate are important in applications. For certain acute health effects, the space-time distribution of exposure is necessary in order to determine health-based policy regulations. A detailed profile provides information about incidents of extremely high and low exposures, which may be more important in predicting health effects and in planning health policy than information on summary exposure.

Definition 5: *The functional or summary exposure is defined as*

$$E_{\Psi}(\boldsymbol{p}) = \int_{\Lambda(\boldsymbol{p})} d\boldsymbol{p}' \, \Psi(\boldsymbol{p} - \boldsymbol{p}') E(\boldsymbol{p} - \boldsymbol{p}'), \tag{5}$$

where Ψ *is a space-time filter and* Λ *is the space-time integration domain centered around* $\boldsymbol{p}$.

The filter function Ψ incorporates the effects of summarizing. It may include the influence of the measurement device, non-local effects, population density inhomogeneities and other factors involved in determining the summary exposures.

Special cases of the functional exposure (5) are the *cumulative* exposures. In particular, if $\Psi(\boldsymbol{p} - \boldsymbol{p}') = 1$ and $\Lambda(\boldsymbol{p}) = V(\boldsymbol{p})$ is a volume centered around $\boldsymbol{p}$, then the

$$E_V^c(\boldsymbol{p}) = \int_{V(\boldsymbol{p})} ds' \, E(s - s', t) \tag{6}$$

denotes the *volume-cumulative* exposure; if $V(\boldsymbol{p})$ is a contact surface, Eq. (6) gives the *surface-cumulative* exposure. The *effective exposure time* around $\boldsymbol{p}$ is given by

$$T_e(\boldsymbol{p}) = \tau_e(\boldsymbol{p}) \, f_e(\boldsymbol{p}), \tag{7}$$

and is usually measured in $days$, where $\tau_e(\boldsymbol{p})$ is the exposure duration ($days$) and $f_e(\boldsymbol{p})$ is the exposure frequency ($\%$), i.e., the fraction of total exposure time $\tau_e(\boldsymbol{p})$ during which the person is actually exposed. For $\Lambda(\boldsymbol{p}) = \tau_e(\boldsymbol{p})$ and $\Psi(\boldsymbol{p}-\boldsymbol{p}') = f_e(\boldsymbol{s},t-t')$, Eq. (5) leads to the *time-cumulative* exposure

$$E_T^c(\boldsymbol{p}) = \int_{\tau_e(\boldsymbol{p})} dt'\, f_e(\boldsymbol{s},t-t')\, E(\boldsymbol{s},t-t'). \tag{8}$$

The *spatiotemporal-cumulative exposure* may be defined in a similar fashion, i.e.,

$$E_\Lambda^c(\boldsymbol{p}) = \int_{\Lambda(\boldsymbol{p})} d\boldsymbol{p}'\, f_e(\boldsymbol{p}-\boldsymbol{p}')\, E(\boldsymbol{p}-\boldsymbol{p}'), \tag{9}$$

where $\Lambda(\boldsymbol{p}) = \tau_e(\boldsymbol{p}) \times V(\boldsymbol{p})$.

Furthermore, special cases of Eq. (5) are spatial and temporal averages of the exposure rate profile over the receptor, as well as population averages. More specifically, the *spatially-averaged* exposure is given by

$$E_V(\boldsymbol{p}) = |V|^{-1} \int_{V(\boldsymbol{p})} d\boldsymbol{s}'\, E(\boldsymbol{s}-\boldsymbol{s}',t); \tag{10}$$

clearly, Eq. (10) is obtained from Eq. (5) for $\Lambda(\boldsymbol{p}) = V(\boldsymbol{p})$ and $\Psi(\boldsymbol{p}-\boldsymbol{p}') = |V|^{-1}$. If $V(\boldsymbol{p})$ is the receptor's volume, the above represents the *volume*-averaged exposure over $V(\boldsymbol{p})$. If $V(\boldsymbol{p})$ represents a contact surface, then (10) is the *surface*-averaged exposure. The volume-averaged exposure gives the amount of pollutant per unit volume (e.g., the pollutant per forced expiratory volume, FEV_1). Similarly, the contact surface-averaged exposure provides the amount of agent per unit area (e.g., skin surface). The *temporally-averaged* exposure is given by

$$E_T(\boldsymbol{p}) = |\tau_e(\boldsymbol{p})|^{-1} \int_{\tau_e(\boldsymbol{p})} dt'\, f_e(\boldsymbol{s},t-t')\, E(\boldsymbol{s},t-t'). \tag{11}$$

Note the obvious relations, $E_V^c(\boldsymbol{p}) = |V| E_V(\boldsymbol{p})$ and $E_T^c(\boldsymbol{p}) = |\tau_e(\boldsymbol{p})|\, E_T(\boldsymbol{p})$. Other useful exposure measures may also be defined by considering different averaging schemes.

EXAMPLE 5: Consider exposure to a uniform nicotine concentration $C(\boldsymbol{p}) = 45\, \mu g / m^3$ within a room of volume $|V| = 512\, m^3$ for $T = 9\, hrs$. The volume-cumulative and time-cumulative exposures are $E_V^c(\boldsymbol{p}) = 45 \times 512 = 23{,}040\, \mu g$ and $E_T^c(\boldsymbol{p}) = 45 \times 9 = 405$ $\mu g - hrs / m^3$ respectively. □

The *population-averaged* exposure $E_\theta(\boldsymbol{p})$ is defined using the general functional exposure Eq. (5) [Christakos and Vyas, 1998b]. The filter function in this case represents the population density at $\boldsymbol{p}'$, i.e. $\Psi(\boldsymbol{p}-\boldsymbol{p}') = \theta(\boldsymbol{p}-\boldsymbol{p}')$, and

$$E_\theta(\boldsymbol{p}) = \int_{\Lambda(\boldsymbol{p})} d\boldsymbol{p}'\, \theta(\boldsymbol{p}-\boldsymbol{p}') E(\boldsymbol{p}-\boldsymbol{p}'). \tag{12}$$

The population density $\theta(\boldsymbol{p}-\boldsymbol{p}')$ is defined as the average population per unit area/time period over a domain υ around $\boldsymbol{p}$.

EXAMPLE 6: Assuming that 5,000 *persons* in a region are exposed to the same amount of pollutant $E(\boldsymbol{p}) = 0.1\ mg/day$. The region consists of 4 counties, each one having an area of $25\ Km^2$ and population densities $\theta_1 = 40$, $\theta_2 = 20$, $\theta_3 = 100$ and $\theta_4 = 40$ $persons/Km^2$. At the population level the exposure is $E_\theta(\boldsymbol{p}) = (40+20+100+40) \times 25 \times 0.1 = 500\ person-mg/day$. Note that --due to the fact that all counties have the same area and a uniform exposure is assumed-- we could have obtained the same result by simply multiplying the exposure $0.1\ mg/day$ times the population of 5,000 *persons*. □

In some cases, the exposure $E_\theta(\boldsymbol{p})$ is also averaged over a domain υ around $\boldsymbol{p}$, leading to the population-weighted and spatially-averaged exposure $E_\theta^\upsilon(\boldsymbol{p}) = E_\theta(\boldsymbol{p})/|\upsilon|$.

EXAMPLE 7: In Fig. 5 the $E_\theta^\upsilon(\boldsymbol{p})$ map associated with the ozone exposure map over Eastern U.S. is shown (a more detailed discussion is given in Chapter IX). □

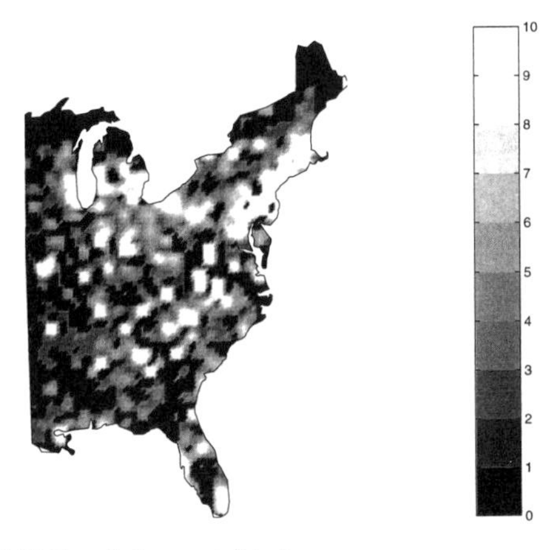

FIGURE 5: Population-weighted ozone exposure over eastern U.S.

The choice of the appropriate exposure measure depends on the specific goals of the study and previous knowledge regarding the health impact of the pollutants. An important modelling parameter is the relationship between exposure and health effect (see, also, §4 below). For certain effects (e.g., cancer), health policies are based on integrated exposure. For other effects (e.g., bronchitis and eye irritation induced by ozone), rate or other measures of functional exposure may be more important. Indeed, very high exposure rate or short-duration rate surges can produce significant health effects, even if the integrated exposure over a longer time period is low (Checkoway *et al.*, 1989).

REMARK 4: On the basis of the "generic" exposure concepts defined above, Chapter IX will introduce novel exposure indicators in a realistic stochastic context (including mean excess exposure above threshold levels, pollutant connectivity measures in space-time, and measures of level-crossing geometry). The stochastic framework accounts rigorously for exposure uncertainty, space-time variation and variability linked to the individual receptor.

3.3 Mathematical Expressions for Biomarkers

Between environmental exposure and the development of health effects, certain biomarkers $\mathcal{B}(\boldsymbol{p})$ were identified in §3.1 above, like intake rate, cumulative intake, uptake rate, cumulative uptake, burden and dose. Rigorous mathematical expressions for these markers can be derived in terms of empirical or mathematical expressions that involve exposure measures and intake pathway parameters.

As with exposure, two major groups of biomarkers may be considered: *point* (or *rate*) *biomarkers*, which refer to a specific space-time point $\boldsymbol{p}$; and *functional* or *summary biomarkers*, which refer to an average over a spatial and/or temporal domain $\Lambda(\boldsymbol{p})$. We start with some basic space-time point biomarkers. The *intake rate* is given by

$$I_r(\boldsymbol{p}) = E(\boldsymbol{p})\, I_{r,m}(\boldsymbol{p}) \tag{13}$$

in $\mu g\,/\,day$, where $I_{r,m}(\boldsymbol{p})$ is the medium intake rate (in $m^3\,/\,day$) which denotes the average inhalation rate of the receptor. If $f_u(\boldsymbol{p})$ denotes the *uptake fraction* of a receptor (experimental values are available; e.g., Crawford-Brown, 1997), the *uptake rate* is given as

$$U_r(\boldsymbol{p}) = f_u(\boldsymbol{p})\, I_r(\boldsymbol{p}) \tag{14}$$

in $\mu g / day$. The *space-time point dose* or *dose rate* $D_r(\boldsymbol{p})$ is the rate of interaction at the space-time point $\boldsymbol{p}$ between the biologically significant form of the pollutant and target organs. The $D_r(\boldsymbol{p})$ is measured in interactions per unit time. As with the exposure rate profile, the *dose rate profile* over a space-time domain Ω is defined as the field of space-time point doses within Ω.

Functional or *summary* biomarkers can be also defined on the basis of the space-time point biomarkers. A general definition is given first.

Definition 6: *The functional or summary biomarker is defined as*

$$\mathcal{B}_{\psi}(\boldsymbol{p}) = \int_{\Lambda(\boldsymbol{p})} d\boldsymbol{p}' \, \Psi(\boldsymbol{p}-\boldsymbol{p}') \mathcal{B}(\boldsymbol{p}-\boldsymbol{p}'), \tag{15}$$

where Ψ *is a space-time filter and* Λ *is the space-time integration domain centered around* $\boldsymbol{p}$.

Depending on the choice of Ψ and Λ, a variety of summary biomarkers can be defined. The *time-cumulative intake*, e.g., is given by

$$I(\boldsymbol{p}) = \int_{\tau_e(\boldsymbol{p})} dt' \, f_e(\boldsymbol{s},t-t') I_r(\boldsymbol{s},t-t'). \tag{16}$$

Similarly, the *time-cumulative uptake* is

$$U(\boldsymbol{p}) = \int_{\tau_e(\boldsymbol{p})} dt' \, f_e(\boldsymbol{s},t-t') U_r(\boldsymbol{s},t-t') = \int_{\tau_e(\boldsymbol{p})} dt' \, f_e(\boldsymbol{s},t-t') f_u(\boldsymbol{s},t-t') I_r(\boldsymbol{s},t-t'). \tag{17}$$

A few special cases illustrate the intuitive meaning of the above equations.

EXAMPLE 8: If $f_e = 100\%$ and the intake rate is assumed constant over the effective exposure time T_e, the cumulative intake and uptake for T_e are, respectively,

$$I(\boldsymbol{s}) = I_r(\boldsymbol{s}) \, T_e, \tag{18}$$

$$U(\boldsymbol{s}) = U_r(\boldsymbol{s}) \, T_e, \tag{19}$$

both in μg. □

As already mentioned, some biomarkers can only be determined by means of mathematical models. The pollutant *burden*, e.g., obeys the *pollutokinetics* equation (Chapter VIII)

$$dB(\boldsymbol{p})/dt = U_r(\boldsymbol{p}) - \lambda\, B(\boldsymbol{p}), \tag{20}$$

where λ is the constant transfer rate out of the organ. The solution of Eq. (20) is

$$B(\boldsymbol{p}) = U_r(\boldsymbol{p})[1 - \exp(-\lambda\, t)]/\lambda. \tag{21}$$

The *retention half-time* or *half-life* $T_{1/2}$ for a pollutant is the time length it takes for half of the pollutant to clear out of an organ after uptake, assuming 1st-order transfer or removal kinetics; i.e., $B(\boldsymbol{p}) = \frac{1}{2} B(\boldsymbol{s},0) = B(\boldsymbol{s},0)[1 - \exp(-\lambda\, T_{1/2})]$, $\boldsymbol{p} = (\boldsymbol{s},t)$, which gives

$$\lambda = 0.693/T_{1/2}. \tag{22}$$

($T_{1/2}$-values for various chemicals may be found in Droz, 1993). At *steady-state* (for very large t), Eq. (25) reduces to

$$B(\boldsymbol{p}) = U_r(\boldsymbol{p})/\lambda = U_r(\boldsymbol{p})\, T_{1/2}/0.693, \tag{23}$$

which is the pollutant burden in the lungs for the retention half-time $T_{1/2}$. As we shall see in Chapter VIII, in a stochastic framework Eqs. (21) and (23) express the *mean* burden 1st-order kinetics. For illustration let us discuss the following

EXAMPLE 9: Consider a receptor in the smoke-filled room (Example 2 above) where the exposure rate to nicotine is $E(\boldsymbol{p}) = C(\boldsymbol{p}) = 45\ \mu g\ /\ m^3$ at each space-time point $\boldsymbol{p}$. Let the medium intake rate $I_{r,m}(\boldsymbol{p}) = 22\ m^3\ /\ day$ denote the average inhalation rate of the receptor. The nicotine intake rate is $I_r(\boldsymbol{p}) = 45\ \mu g\ /\ m^3 \times 22\ m^3\ /\ day = 990\ \mu g\ /\ day$. Assuming an exposure duration $\tau_e = 100\ days$ and an exposure frequency $f_e = 60\ \%$, the time length of exposure is $T_e = 100\ days \times 0.6 = 60\ days$. The cumulative intake for $T_e = 60\ days$ is $I(\boldsymbol{p}) = 990\ \mu g\ /\ day \times 60\ days = 59{,}400\ \mu g$. Suppose that the characteristic value for the uptake fraction is $f_u = 0.35$ for a non-smoker receptor. Then, the uptake rate is $U_r(\boldsymbol{p}) = 990\ \mu g\ /\ day \times 0.35 = 346.5\ \mu g\ /\ day$, and the cumulative uptake ($60\ days$) is $U(\boldsymbol{p}) = 346.5\ \mu g\ /\ day \times 60\ days = 20{,}790\ \mu g$. Assume that the half-life $T_{1/2}$ for nicotine carriers in the lungs of a non-smoker is $1\ day$. The nicotine burden in the lungs under steady-state conditions (large t) for a 60% exposure frequency is $B(\boldsymbol{p}) = 346.5\ \mu g\ /\ day \times 1\ day \times 0.6/0.693 = 300\ \mu g$. □

Not all forms of burden $B(\boldsymbol{p})$ calculated using pollutokinetics constitute a health threat. In fact, it is the so-called *biologically significant burden* $B_s(\boldsymbol{p})$ that is capable of changing the state of health. Generally, $B_s(\boldsymbol{p})$ is found by multiplying $B(\boldsymbol{p})$ by a *transformation fraction* $f_t(\boldsymbol{p})$, i.e.,

$$B_s(\boldsymbol{p}) = B(\boldsymbol{p})\, f_t(\boldsymbol{p}). \tag{24}$$

In some cases, the $f_t(\boldsymbol{p})$ can be calculated experimentally; in several other cases it is unknown and, then, the biologically significant burden is assumed to be the same as the burden. As we already mentioned, while $B_s(\boldsymbol{p})$ implies the existence of the active form of the pollutant in a target organ, it does not necessarily mean that the pollutant actually interacts with the organ. It is the *dose rate* D_r which determines the interaction rate between the biologically active form and target organs in the receptor. Summary dose markers can be determined on the basis of Eq. (15) above. The *cumulative dose* or simply *dose* D is the cumulative quantity of the biologically active form delivered to the receptor, i.e.,

$$D(\boldsymbol{p}) = \int_{\tau_d(\boldsymbol{p})} dt'\, D_r(s,\ t - t'), \tag{25}$$

where $\tau_d(\boldsymbol{p})$ is the time period during which interactions between the biologically active form and target organs take place. The $\tau_d(\boldsymbol{p})$ may be larger than the exposure duration $\tau_e(\boldsymbol{p})$, for biologically active pollutants are retained in the body even after exposure is terminated. Unfortunately, in several cases the $\tau_d(\boldsymbol{p})$ is not known, which means that D_r can be found but not D.

EXAMPLE 10: An interesting concept of dose is found in radiation studies, in which the absorbed energy in cells provides a measure of the damage (injury or lethality). The dose D is expressed as the mean energy absorbed per unit mass of biological material. The dose units in this case are *Gy* (i.e., 1 joule/kilogram of material). The dose rate D_r is the rate of the mean energy absorbed per unit mass (e.g., *Gy* per minute). Note that these definitions are consistent with the general dose definition above in terms of interactions between the biologically active form and target organs in the receptor (i.e., the absorbed energy is proportional to the number of interactions between radiation and targets). □

The biomarkers above are concerned with biologic processes at the individual level. *Health indicators* assessing the state of health can be determined at the individual receptor and the population level. Certain of these indicators are discussed in §4 below. Furthermore, in Chapter IX the issue of biomarkers and health indicators is revisited in a stochastic context (e.g., cell-based health indicators, traditional population health indicators such as space-time distributions of incidence, prevalence and standardized mortality ratios are considered as spatiotemporal random fields taking into consideration uncertainty and space-time variations in exposure and biologic processes).

3.4 Exposure as a Practical Biomarker Estimator

It is evident from the definitions of the preceding sections that for health risk assessment biomarker information is usually more important than information about exposure. Certainly, exposure to a pollutant does not necessarily imply that health damage will ensue. The health damage is, in fact, determined by biomarkers (intake, uptake, and interaction biomarkers, particularly dose), which are more relevant than exposure for regulatory standards, epidemiological studies of health effects, or other health-related investigations.

However, exposure is considerably easier and cheaper to measure than most biomarkers, which is an important practical advantage. This is true at both the individual and the population levels. Many biomarkers, on the other hand, are not directly measurable for individual receptors, but instead they are estimated in terms of mathematical models (physiological pollutokinetic models, etc.; Chapter VIII). Also, it can be assumed with reasonable accuracy that people in a specific region are exposed to the measured pollutant concentrations. There is, however, considerable uncertainty regarding the biomarker values for the same individuals. This is due to the fact that individuals may vary considerably in the relationship between exposure and the biomarker (intersubject variability). Moreover, the presence of a biomarker does not by itself provide information about the exposure route --which can be a problem when the goal is to regulate exposure to a specific pollutant.

In epidemiological applications, while the spatiotemporal distribution of exposure can be measured by means of a network of monitoring devices and mapping techniques (Chapters V and VI), the distribution of the burden accumulated in the target organs of the numerous receptors constituting a population is more difficult to obtain. As a consequence, in practice exposure is often used as an estimator of burden (exposure and burden have even been used synonymously; Georgopoulos and Lioy, 1994). In order that this estimator make sense, however, certain conditions must be satisfied. For example, for exposure to be a meaningful estimator of burden the two must be proportionally related (a high exposure, e.g., should imply a high burden and vice versa). This is typically the case of air pollutants such as NO_x, SO_x, O_3 and particulates (Curtiss and Rabl, 1996).

4. HEALTH EFFECT CONCEPTS

4.1 The Health Impact Pathway and the Various Kinds of Health Effects

Human health is generally defined using the values of physiological properties that measure the well-being of an individual or a population. Environmental exposure to pollutants has in many cases an impact on human health. As we saw in the preceding sections, exposure of the receptor to environmental pollutants does not guarantee pollutant intake. Even if intake occurs, the amount may not be sufficient for measurable interactions with the target organs. If interactions actually occur, the concept of exposure is superseded by the concept of dose. Interaction *per se* does not necessarily imply a change in the *state* of the human health.

Changes of the health state occurring as responses to a certain dose can be described by means of the *health impact pathway* $\mathcal{J}_H$, which constitutes an extension of the dose pathway $\mathcal{J}_D$ and leads to the specific *health effect.* In mathematical terms, the relation between the three pathways can be represented as

$$\mathcal{J}_E \subset \mathcal{J}_D \subset \mathcal{J}_H. \tag{1}$$

Eq. (1) expresses the fact that pathway $\mathcal{J}_H$ is an extension of pathway $\mathcal{J}_D$, which is, in turn, an extension of pathway $\mathcal{J}_E$. The health impact pathway adds to the four components of the dose pathway $\mathcal{J}_D$, a fifth component as follows:

e. *Transition stage*: It is the stage during which transitions between different states of health occur due to dose interactions.

Based on the above, the following definition may be established [in the following, as usual, $\boldsymbol{p}$ is associated with an individual receptor located at the space time point $\boldsymbol{p} = (\boldsymbol{s}, t)$, or a population within a domain surrounding $\boldsymbol{p}$].

Definition 1: *A health effect* $H_I(\mathbf{p})$ *on a receptor* $\mathbf{p}$ *caused by exposure to a harmful pollutant is a transition from a state of health to another.*

Typical *qualitative gradations* of the state of human health are described as normal health, minor effect, moderate effect, and full effect. There are various kinds of health effects that can be produced by exposure to pollutants. In Table 2 a summary is given of the health effects recognized by the U.S. EPA (EPA, 1987; Covello and Merkhoffer, 1993).

TABLE 2: Human Health Effects Identified by the U.S. EPA

Noncancer:

- Cardiovascular (e.g., increased heart attacks)
- Developmental (e.g., birth defects)
- Hematopoietic (e.g., impaired heme -blood-synthesis)
- Immunological (e.g., increased infections)
- Kidney (e.g., dysfunction)
- Liver (e.g., Hepatitis A)
- Mutagenic (e.g., hereditary disorders)
- Neurotoxic Behavioral (e.g., retardation)
- Reproductive (e.g., increased spontaneous abortions)
- Respiratory (e.g., emphysema)
- Other (e.g., gastrointestinal)

Cancer:

- Lung
- Colon
- Breast
- Pancreas
- Prostate
- Stomach
- Leukemia
- Other

Epidemiological studies focus on the response of receptors at the population scale and permit cause-effect associations between diseases and substances that present health risks. Toxicological analysis, on the other hand, focuses on studies of individual receptors. Thus, it provides a detailed analysis of the cause-effect associations at the individual scale and elucidates the mechanisms of toxicity. Studies of individual receptor responses to varying levels of exposure provide crucial information on the nature of the responses and their underlying causes (Lippmann, 1989). Hence, toxicology investigates the transition stage (e) above. The final state in the transition stage depends on the substance and exposure details. Exposure to toxic chemicals, e.g., can result in residual disease and/or dysfunction, or it can cause death within a certain time period after exposure.

One can also distinguish, in general, between *acute* health effects that appear shortly after exposure and last for a short time period, and *chronic* health effects that occur over a long time period. Acute health effects often appear when the exposure rate $E(\mathbf{p})$ exceeds a threshold level $\zeta(\mathbf{p})$, and they cease when the rate recedes below the threshold $E(\mathbf{p}) < \zeta(\mathbf{p})$; the threshold approach is discussed in more detail below.

EXAMPLE 1: A central dogma in industrial hygiene is that exposure should not exceed certain thresholds, in order to prevent the occurrence of acute health effects. □

In the case of chronic health effects it is not always possible to identify a low exposure threshold below which the effects disappear. Chronic health effects usually persist for a period of time after exposure is terminated. This difference between chronic and acute health effects is reflected in their mathematical representation. The response in the case of acute health effects should involve a step function of the exposure (or dose) to model the threshold. On the other hand, the response in the case of chronic health effects can be a smooth function of exposure (or dose) possibly with hysteretic effects.

Adverse health effects have some important characteristics at the individual receptor or the population scale. These *health effect characteristics* include: (i) *Severity*, which measures the amount of health damage caused by a disease. (ii) *Temporal pattern of appearance* associated with each pollutant and each disease. (iii) *Frequency of the disease-related episodes* experienced by the individual or the group of individuals.

REMARK 1: In some cases a sixth component is added to the health effect pathway $\mathcal{J}_H$ representing the economic consequences of the health effects. While of considerable importance, this component will not be discussed further in this book.

4.2 Mathematical Models for Health Effect Indicators

In addition to the clinical disease biomarkers mentioned in the previous sections, there exist certain important health effect indicators which detect the appearance of a disease or, generally, the occurrence of a transition between the possible states of health.

A commonly used effect indicator is the *health effect threshold* indicator, usually measured in units of exposure rate or some other biomarker (e.g., dose rate or dose). When the exposure or biomarker value of a receptor exceeds the threshold, the health effect appears. This leads to the following definition.

Definition 2: *The health effect threshold indicator $\zeta(\boldsymbol{p})$ is defined as the exposure or biomarker value of a receptor $\boldsymbol{p}$ above which the health effect appears, i.e.,*

$$E(\boldsymbol{p}) \text{ or } \mathcal{B}(\boldsymbol{p}) > \zeta(\boldsymbol{p}) \Rightarrow H_I(\boldsymbol{p}), \tag{2}$$

where $E(\boldsymbol{p})$ or $\mathcal{B}(\boldsymbol{p})$ and $\zeta(\boldsymbol{p})$ have the same units.

EXAMPLE 2: If a receptor $\boldsymbol{p}$ is exposed to $E(\boldsymbol{p}) > \zeta(\boldsymbol{p}) = 2850$ ppm of carbon monoxide for 10 min, then $H_I(\boldsymbol{p})$ =death. □

The threshold $\zeta(\boldsymbol{p})$ for a receptor $\boldsymbol{p}$ is usually available from biological and toxicological data. For a variety of biophysical reasons, however, the threshold $\zeta(\boldsymbol{p})$ varies considerably among receptors. Some of these reasons are associated with differences among receptors with regard to the minimal damage that must be produced before the effect appears, the immune system, the burden at which a deactivation system for biotransformation saturates, the dose rate at which a repair system saturates or an unreliable "emergency" repair system begins to operate, etc. (Crawford-Brown, 1997). Thus, the threshold may be itself considered a random field $Z(\boldsymbol{p})$ with a probability density function (pdf) $f_z(\zeta; \boldsymbol{p})$ such that the probability

$$P[\zeta \leq Z \leq \zeta + d\zeta] = f_z(\zeta; \boldsymbol{p}) d\zeta \tag{3}$$

(a rigorous mathematical definition of the random field is given in Chapter IIIff). Then, the probability that a receptor $\boldsymbol{p}$ --exposed to $E(\boldsymbol{p}) = \varepsilon$ or showing a biomarker value $\mathcal{B}(\boldsymbol{p}) = \beta$-- is experiencing a specified health effect is written as

$$H_P(\boldsymbol{p}) = \mathrm{P}[Z(\boldsymbol{p}) < \varepsilon \text{ or } \beta]. \tag{4}$$

Assuming that the pdf $f_z(\zeta; \boldsymbol{p})$ is known (from previous toxicological studies, epidemiological investigations, etc.), the health effect probability can be written as

$$H_P(\boldsymbol{p}) = \int_0^{\varepsilon \text{ or } \beta} d\zeta \, f_z(\zeta; \boldsymbol{p}) = F_z(\varepsilon \text{ or } \beta), \tag{5}$$

i.e., the probability that a receptor $\boldsymbol{p}$ with $E(\boldsymbol{p}) = \varepsilon$ or $\mathcal{B}(\boldsymbol{p}) = \beta$ experiences the health effect is equal to the cumulative probability distribution of $Z(\boldsymbol{p})$, $F_z(\varepsilon \text{ or } \beta)$.

EXAMPLE 3: The dose rate for receptor $\boldsymbol{p}$ is $D_r(\boldsymbol{p}) = 0.5$ $units$. Assume a chi-square $(n = 1)$ pdf for the health effect threshold $Z(\boldsymbol{p})$ associated with a specific health effect. The probability of the receptor $\boldsymbol{p}$ experiencing the effect is $H_P(\boldsymbol{p}) = F_z(0.5) = 0.455$. □

4.3 Mathematical Models for Individual-Based and Population-Based Dose and Exposure-Response Curves

The concept of a *dose-response* (or *dose-effect*) curve is based on the premise that between the two extreme cases of (a) zero observable adverse effect and (b) maximum effect (usually defined as death), there exists a range within which dose leads to a graded response. Since the response at very low dose levels is difficult to measure, modelling assumptions are made. The simplest model assumes that the response varies *linearly* with dose. Hence this model predicts an adverse effect even for minute doses. The linear model is used for cancer risk assessment by the USA FDA. Certain European countries prefer nonlinear dose-response models that have safe-dose thresholds. Dose-response data on humans are rarely available, because of the risks involved. Hence, most of the testing is carried out using animal surrogates for humans (rats, mice, hamsters, rabbits, monkeys and dogs), and the resulting dose-response curves are extrapolated to humans. Both individual and population dose-response curves have been studied. A number of factors including sex, age, species, nutrition, and dose rate influence the shape of the dose-response curves.

Definition 3: *The individual dose-response (IDR) curve relates the dose* $D(\boldsymbol{p})$ *at a critical site* $\boldsymbol{p}$ *in the receptor with a health effect* $H_I(\boldsymbol{p})$, *i.e.,*

$$H_I(\boldsymbol{p}) = \mathcal{F}_{IDR}[D(\boldsymbol{p}), \mathcal{I}_H], \tag{6}$$

where the form of the functional $\mathcal{F}_{IDR}[\cdot]$ *is determined on the basis of experimental data and theoretical analysis.*

The health effect $H_I(\boldsymbol{p})$ on a receptor is expressed in terms of measurable health deterioration, disease, dysfunction, or death [e.g., $H_I(\boldsymbol{p})$ =Hepatitis A].

EXAMPLE 4: A typical IDR is plotted in Fig. 6. At the grey area no response can be observed. This may be due to the negligible effect of the low dose, regardless of the duration of the exposure; or it may be due to the fact that there are not data regarding the response of the individual to exposures for long time periods. Experimental determination of dose-response relationships is a goal of environmental toxicology. □

Recognizable health effects are usually expressed with respect to human populations and

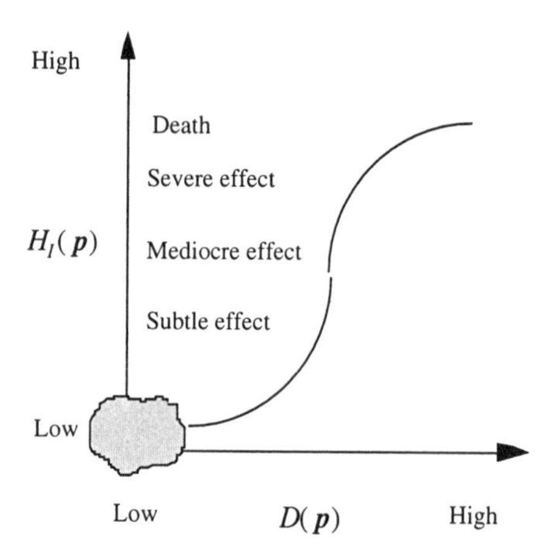

FIGURE 6: A typical IDR curve; the grey area corresponds to non-detectable effects.

are generally divided into two broad categories: *mortality* (i.e., number of deaths per unit of population), and *morbidity* (i.e., number of nonfatal cases of reportable disease per unit of population). Therefore, health effects are often expressed in terms of *percentages* or *probabilities*. This leads to the concept of the population dose-response curve.

Definition 4: *The population dose-response (PDR) curve relates the dose $D(\boldsymbol{p})$ received by a population around the space-time point $\boldsymbol{p}$ with the health effect $H_P(\boldsymbol{p})$ on the population; i.e.,*

$$H_P(\boldsymbol{p}) = \mathcal{F}_{PDR}[D(\boldsymbol{p}), \mathcal{J}_H], \tag{7}$$

where the form of the functional $\mathcal{F}_{PDR}[\cdot]$ is determined on the basis of experimental data and theoretical analysis.

Dose and response values at $\boldsymbol{p}$ always involve an average over some neighborhood around $\boldsymbol{p}$, because measuring devices and procedures have a finite bandwidth. The neighborhood size is presumed to be small compared to the size of the entire domain, but it varies depending on the scale of the study (§I.6). Clearly, the neighborhood size for individual dose-response studies is considerably less than for population studies.

EXAMPLE 5: Fig. 7 gives an illustration of the IDR and the PDR curves. The D vs. H_I

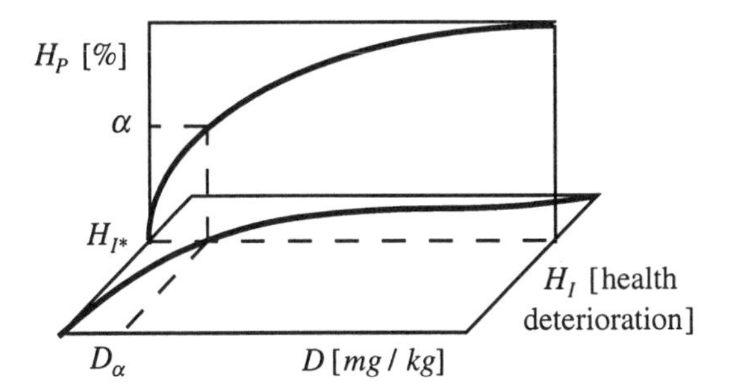

FIGURE 7: The IDR and PDR curves.

curve shown represents the IDR relation for one member of the population considered; for example, D represents the doses of dioxin (in $mg\,/\,kg$) causing various levels of health deterioration H_I to the receptor (similar curves exist for all other members of the population). Let D_α be the dose for which α % of the population shows a certain symptom $H_I = H_{I*}$ of bad health. The corresponding D vs. H_P curve represents the percentage of the population with the symptom $H_I = H_{I*}$ as a function of the dose D. □

In descriptive epidemiology and medical geography, the following modelling assumptions are often made (Esteve, *et al.*, 1994; Christakos and Vyas, 1998b): (a) A group of receptors (population) rather than the individual constitutes the basic unit of exposure-health effect analysis; inference for individual effects is based on group results. (b) The PDR curve is considered to be identical to the *population exposure-response* curve (*PER*), i.e.,

$$H_P(\boldsymbol{p}) = \mathcal{F}_{PER}[E(\boldsymbol{p})]. \tag{8}$$

An important feature of the population health effect (8) is that it can be generally expressed as the conditional probability of the individual health effect $H_I(\boldsymbol{p})$, i.e.,

$$H_P(\boldsymbol{p}) = \text{P}[H_I(\boldsymbol{p})|E, \mathcal{J}_H]. \tag{9}$$

Eq. (9) implies that the population health effect $H_P(\boldsymbol{p})$ is also a function of E, $\mathcal{J}_H$ and H_I. For simplicity, the variables E, $\mathcal{J}_H$ and H_I are usually dropped, and the population health effect is denoted by $H_P(\boldsymbol{p})$. When this happens, the specific exposure conditions are explicitly stated in the description of the problem under consideration.

Some typical examples of PER curves are shown in Fig. 8. The *linear* PER [curve (a)] is expressed mathematically as

$$H_P(\boldsymbol{p}) = \alpha(\boldsymbol{p})E(\boldsymbol{p}), \tag{10}$$

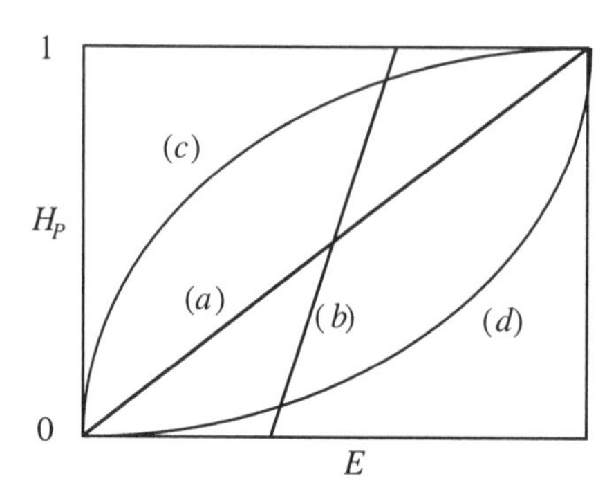

FIGURE 8: Typical PER curves: (a) linear, (b) polynomial, (c) supralinear, and (d) sublinear.

where $\alpha(\boldsymbol{p})$ is the slope of the PER curve. Several recent studies have demonstrated PER relationships of the form (10). Whitfield *et al.* (1995), e.g., derived a linear PER relationship (response rate vs. ozone concentration; FEV_1 decrement $\geq 10\%$, for 6.6 hours of exposure under moderate exercise). Similarly, Thurston *et al.* (1992) and Rombout and Schwarze (1995) reported a linear relationship between the ground-level ozone concentration and the frequency of daily hospital admissions.

The *threshold* or *polygonal* PER [curve (b)] can be written as

$$H_P(\boldsymbol{p}) = \begin{cases} 0 & \text{for } E(\boldsymbol{p}) < \zeta_1(\boldsymbol{p}) \\ \alpha(\boldsymbol{p})E(\boldsymbol{p}) & \text{for } \zeta_1(\boldsymbol{p}) < E(\boldsymbol{p}) < \zeta_2(\boldsymbol{p}) \\ 1 & \text{for } E(\boldsymbol{p}) > \zeta_2(\boldsymbol{p}) \end{cases}, \tag{11}$$

where $\zeta_1(\boldsymbol{p})$ and $\zeta_2(\boldsymbol{p})$ are threshold exposures.

The *curvilinear* PER [curves (c) and (d)] are represented by means of the expression

$$H_P(\boldsymbol{p}) = \alpha(\boldsymbol{p})E(\boldsymbol{p})^c. \tag{12}$$

An exponent $c > 1$ leads to a *supralinear* PER [curve (c)], while an exponent $c < 1$ leads to a *sublinear* PER [curve (d)]. When $c = 1$, Eq. (12) coincides with Eq. (10). In certain applications the PER models include a background health effect $H_0(\boldsymbol{p})$, even when the exposure $E(\boldsymbol{p})=0$. The background effect is due to confounding factors (e.g., other pollutants, age, pre-existing health conditions, etc.). In such cases the curvilinear PEE should be expressed as

$$H_P(\boldsymbol{p}) = H_0(\boldsymbol{p}) + \alpha(\boldsymbol{p})E(\boldsymbol{p})^c. \tag{13}$$

Other PER forms are also possible, depending on the existing human exposure conditions.

4.4 Spatiotemporal Coordinates in Human Exposure Analysis

Exposure and the health effect vary with geographical location and time, i.e., they are spatiotemporal fields. Therefore, the quantitative description of exposure and health processes must involve suitable space-time coordinates (§I.3). Spatiotemporal coordinates may enter the composite exposure-health effect analysis in two distinct ways:

Lagrangian approach: The receptor's response is measured as he/she is allowed to move freely in space and time. The spatial location is a function of time t and the specific individual i, i.e. $\boldsymbol{s}=\boldsymbol{s}(t,i)$. This approach is used in many medical studies, as well as in analytical epidemiology where health risks are studied in groups formed *a posteriori* from data collected at the individual level.

Eulerian approach: A space-time domain is defined and the response of groups of receptors within this domain is observed. The receptors are assigned fixed locations which are modeled by the nodes of a spatiotemporal grid associated with a set of Cartesian space-time coordinates $(\boldsymbol{s},t)\in R^n\times T$ (where usually $n=2$). Many descriptive epidemiological studies are based on an Eulerian description of groups of receptors (populations).

A complete study of the exposure-health effect problem usually requires a combined analysis of results obtained with both approaches. Both approaches are associated with a *global* domain Ω (e.g., Eastern USA). When the Lagrangian approach is used, the health condition of a moving individual and the surrounding environment must be monitored simultaneously.

EXAMPLE 6: The global domain Ω and individual paths for Lagrangian exposure-health effect description in $R^2\times T$ are shown in Fig. 9. The arrowhead lines denote the "world

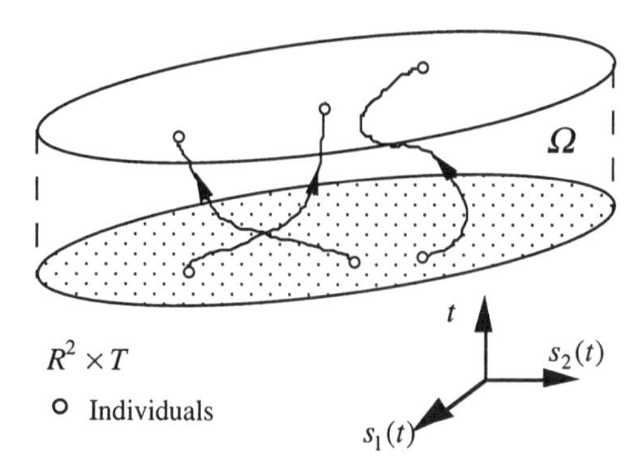

FIGURE 9: An example of Lagrangian exposure-health effect description in $R^2 \times T$.

lines" of each individual. □

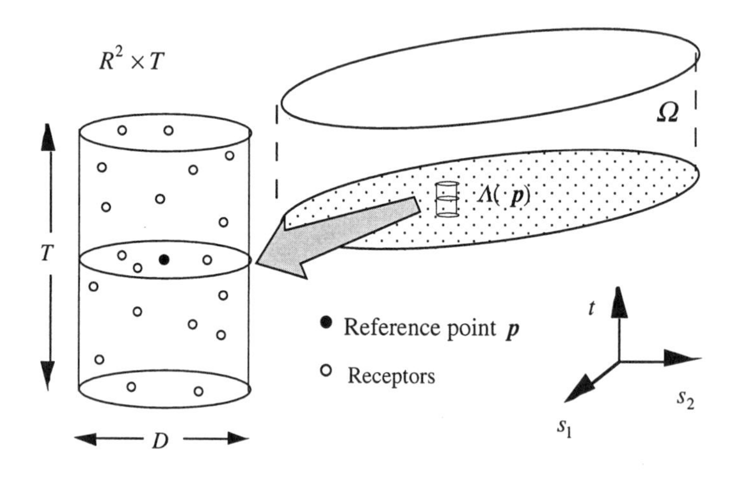

FIGURE 10: An example of Eulerian exposure-health effect description in $R^2 \times T$.

The Eulerian approach is concerned with the evolution of exposure-health effects over time for populations at specified geographical locations. In addition to exposure-response studies over a global domain Ω, the Eulerian approach can also handle *local* space-time domains (e.g., counties) with appropriate definitions for exposure and health outcome. Usually, the local space-time domain $\Lambda(\boldsymbol{p})$ within which the study population is located is around a reference point $\boldsymbol{p} = (\boldsymbol{s}, t)$. Naturally, $\Lambda(\boldsymbol{p}) \subset \Omega$. The space-time domains that we consider involve time and two or three spatial dimensions. Although unified space-time domains can be defined using the Cartesian product, this is purely a matter of convenience, and we can, whenever appropriate, consider separate neighborhoods in space or time.

EXAMPLE 7: Fig. 10 presents an Eulerian exposure-health effect description in $R^2 \times T$ with global domain Ω and local domain $\Lambda(\boldsymbol{p}) = D/T$ (this is a space-time cylinder of diameter D spatial length units and height T time units). □

Chapter III: SPATIOTEMPORAL RANDOM FIELDS IN EXPOSURE ANALYSIS AND ASSESSMENT

"Real knowledge is not possible; what matters is useful opinions".

Protagoras

1. A THEATRICAL INTRODUCTION TO THE RANDOM FIELD CONCEPT

What follows is a play which involves the philosophers St. Thomas Aquinas, Francis Bacon and August Comte (hence we will call it the ABC play). These philosophers promoted ideas that may help us understand important aspects of the random field concept. Aquinas (1225-1274) argued for a system of philosophy that is based on a combination of mystic Aristotelian faith and natural reason. Bacon (1561-1626) introduced the concept of scientific method that emphasizes experimentation and observations. Finally, Comte (1798-1857) promoted the view that one should focus on the connections between observable facts in order to make hypotheses and develop theories. In our theatrical play, Aquinas goes into an isolated room and generates a sequence of numbers. Our *prior (initial) knowledge* regarding the mechanism that Aquinas used to generate his data was that he tossed a die repeatedly (we assumed here that while God may not play dice, his Saints do!) and recorded the outcome, but the outcome itself is --of course-- unknown to us. Based on this limited knowledge we are asked to guess what the sequence looks like.

Let $\chi(t)$ represent the process developed by Aquinas, where t counts order in the tossing sequence; i.e., $t=1, 2, \ldots, T$. At each t, $\chi(t)$ is equal to an integer between one and six. Within the deterministic framework we lack the means to make any predictions, because the initial knowledge is incomplete. In the stochastic framework, however, it is possible to study the problem. Indeed, according to the stochastic method, the actual (but unknown) $\chi(t)$ is considered as one possible realization of the *mathematical model* $X(t)$ representing the process. Several possible realizations are plotted in Fig. 1, using the initial knowledge available. The various realizations of Fig. 1 give an idea of the inherent uncertainty in any attempt to model $\chi(t)$ and make predictions, given our prior status of knowledge. The total number of possible realizations is $N=6^T$ (if, e.g., $T=5$ there are $N=7776$ realizations).

To improve our modelling we ask Bacon to provide us some *experimental*

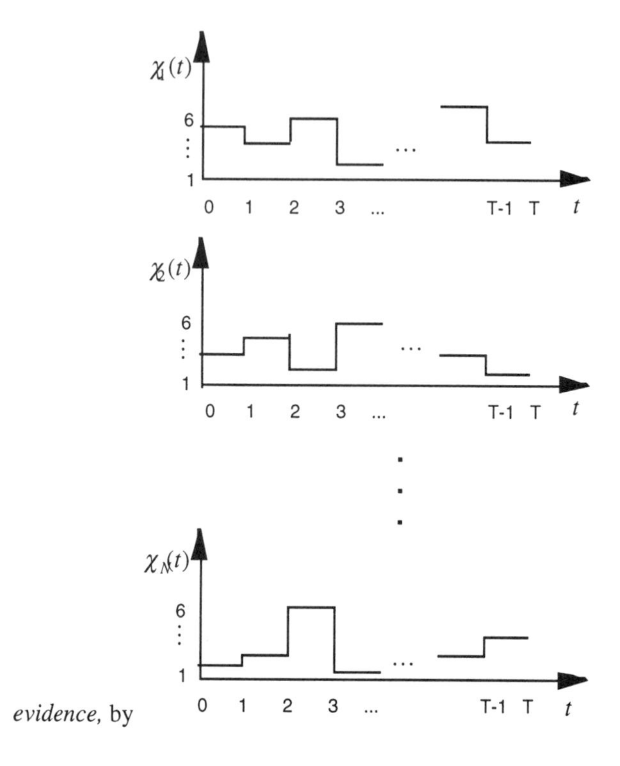

FIGURE 1: The ABC play.

evidence, by sneaking into the room and secretly having a look into Aquinas' records. Bacon later returns with the information that the values of $\chi(t)$ at the third and fifth toss respectively were $\chi_j(3)=5$ and $\chi_j(5)=1$. This evidence represents *hard data,* and it immediately excludes certain of the possibilities plotted in Fig. 1; in particular those realizations which do not honor the hard data. This leads to a reduced number of possible realizations and, thus, our uncertainty about $\chi(t)$ is also reduced (if, e.g., $T=5$ then the number of realizations is reduced from $N=7776$ to only $N'=6^3=216$ realizations). In practical terms, *data conditioning* of the process significantly reduces uncertainty about $\chi(t)$.

Still not satisfied with the degree of remaining uncertainty, we try to collect more information about $\chi(t)$. At this point Comte steps in and claims that, since Aquinas is a religious man, it is reasonable to assume that he will refuse to record the "evil" number six. This is a piece of *soft* knowledge which implies that every time the die turned up six, Aquinas refused to record the outcome and, instead, he tossed the die again until a different number turned up. Depending on how strongly we believe in Comte's conjecture, we may decide to use or discard this soft knowledge. Suppose that we trust Comte and decide to take his knowledge into account. This will eliminate an additional number of realizations in Fig. 1, namely all of these which include the number six at some time instant t. Our uncertainty was now represented by an even smaller number of $X(t)$ realizations (e.g., if $T=5$ the number of realizations is further reduced from $N'=216$ to $N''=5^3=125$ realizations). Based on these realizations, we can estimate the statistical (ensemble) characteristics of the stochastic model $X(t)$, and estimate the process recorded by Aquinas.

The rather simple situation described in the ABC play above contains some fundamental ideas of random field modelling, as described by the three theses of Table 1. These theses

TABLE 1: Three Theses Underlying the Random Field Concept.

Thesis 1 : The focus of the analysis is shifted from a single realization (traditional deterministic concept) to an ensemble of possible realizations that represent all the potentialities restricted only by our knowledge of the actual process' properties (including hard data, such as measurements, physical laws and soft data-§I.5).

Thesis 2: Instead of investigating properties of single realizations , the stochastic approach focuses on ensemble properties (probability laws, expected values, etc.).

Thesis 3 : Predictions about the behavior of the actual realization are derived on the basis of the ensemble of possible realizations .

are direct consequences of the Complementarity principle presented in §I.3. The rigorous mathematical representation of the theses will be the task of the following sections, with the hope of presenting a useful theory of spatiotemporal random fields. Before proceeding with the study of the following sections, the reader may wish to review §I.3 on spatiotemporal continuity.

2. THE SPATIOTEMPORAL RANDOM FIELD

A field generally presupposes a continuum of space-time points (§I.3) and represents values of an environmental health processes at these points. A set of values at all points in the space-time continuum specifies a realization of the field. Randomness manifests itself as an ensemble of possible realizations for the process under consideration. The random field offers a general framework for analyzing data distributed in space-time. This framework involves operations that enable us to investigate human exposure problems in a mathematically effective way, improve our insight into the physical mechanisms and, thus, to enhance predictive capabilities.

2.1 Basic Mathematical Formulation

A central element in the mathematical formalization of the theses of Table 1 is the spatiotemporal random field model (S/TRF). The idea of an S/TRF --which is the main tool of modern stochastics-- enters as the idea that values of environmental health processes are attributed to space-time points. Generally, a field may associate mathematical entities (such as a scalar, a vector, or a tensor) with space-time points. For the S/TRF to obtain physical meaning these entities must represent values of some environmental health process (e.g., soil moisture, fluid velocity, nicotine concentration or lung cancer mortality rates).

EXAMPLE 1: Let us imagine that space-time is represented by the nodes of a discrete lattice. Physical observables (e.g., temperatures) or health-related events (e.g., disease rates) are represented by means of numerical values at the nodes of the grid. The set of values on the entire grid defines an S/TRF realization. Since there is a random aspect in the observable, different realizations of the field are possible corresponding to potential states of the observable. The ensemble of all possible configurations constitutes the S/TRF. □

In the following, an S/TRF will be denoted either as $X(\boldsymbol{s},t)$, $\boldsymbol{s} \in S$ and $t \in T$, or as $X(\boldsymbol{p})$, $\boldsymbol{p} = (\boldsymbol{s},t) \in S \times T$. There exist various mathematical definitions of an S/TRF (Christakos, 1991; 1992). One of these definitions is presented below. It uses the concepts of the sample space Ω which includes all the possible outcomes (field realizations); the family of all events F, such that $F \subset \Omega$; and a probability $P(F) \in [0,1]$ associated with each event. In view of these definitions, it is clear that the sample space corresponds to the stochastic ensemble, and it may be infinite. On the other hand, the events constitute the observables of the S/TRF.

Definition 1: *A spatiotemporal random field (S/TRF)* $X(\boldsymbol{p})$ *over* $S \times T$ *is a mapping*

$$X\colon\ S \times T \to L_p(\Omega, F, P), \tag{1}$$

where $L_p(\Omega, F, P)$, $p \geq 1$, *denotes the* L_p*-norm on the probability space* (Ω, F, P).

Hence, an S/TRF $X(\boldsymbol{p})$ representing an environmental health process is a collection of realizations (possibilities, potentialities) for the distribution of the process values in space-time (Thesis 1 of Table 1 above). The S/TRF also can be viewed as a collection of correlated random variables at m points in space-time, i.e., $\boldsymbol{x} = [x_1,...,x_m]^T$ at the points $\boldsymbol{p}_1,...,\boldsymbol{p}_m$. The realization of an S/TRF at these points is denoted by the vector $\boldsymbol{\chi}_l^m = [\chi_1,...,\chi_m]^T$ or simply $\boldsymbol{\chi} = [\chi_1,...,\chi_m]^T$. We assume that the S/TRF takes values in the space of real numbers, since this assumption represents most applications in environmental health sciences. The probability that a realization $\boldsymbol{\chi}$ occurs is expressed in terms of the *multivariate probability law* of $X(\boldsymbol{p})$ --Thesis 2 of Table 1--

$$P_x[\boldsymbol{\chi}] = P_x[\chi_1 \leq x_1 \leq \chi_1 + d\chi_1,...,\chi_m \leq x_m \leq \chi_m + d\chi_m] = f_x(\boldsymbol{\chi})d\boldsymbol{\chi}, \tag{2}$$

where $f_x(\boldsymbol{\chi})$ is the multivariate probability density function (pdf). The pdf depends explicitly on the points $\boldsymbol{p}_1,...,\boldsymbol{p}_m$, but we do not explicitly show this for convenience. The probability that a S/TRF realization assumes at m points values less than or equal to $\boldsymbol{\chi}$ is given by the multivariate cumulative distribution function (cdf) $F_x(\boldsymbol{\chi})$ defined as

$$F_x(\boldsymbol{\chi}) = P_x[x_1 \leq \chi_1,....,x_m \leq \chi_m] = \int_{-\infty}^{\chi_1} d\chi_2 ... \int_{-\infty}^{\chi_m} d\chi_m \, f_x(\chi_1,\chi_2,...,\chi_m). \tag{3}$$

In many applications it is useful to consider the probabilities of different realizations at a single point in space-time. Because of the correlations, these probabilities are influenced by the state of neighboring points. This is accomplished by means of the marginal cdf $F_x(\chi_1)$ defined as follows

$$F_x(\chi_1) = P_x[x_1 \leq \chi_1] = \int_{-\infty}^{\infty} d\chi_2 ... \int_{-\infty}^{\infty} d\chi_m \, f_x(\chi_1,\chi_2,...,\chi_m). \tag{4}$$

The S/TRF model describes the probable structure of an environmental health process but not its actual structure. S/TRF models are used in many applications. The concept of an ensemble of possible states is more appropriate in certain systems than in others.

EXAMPLE 2: At the microscopic level quantum effects and the thermal motion of molecules allow particles to explore various states from a large ensemble. The state of individual particles can not be determined, but the behavior of the entire system is well described by ensemble averages. In the case of porous media and other environmental systems, we are interested in processes that do not have the dynamic freedom to explore states in the ensemble. In addition, it is possible to measure the properties of the medium at individual points in space-time. However, under ergodic conditions ensemble averages still provide useful estimates of spatially averaged (coarse grained) properties of the medium. In the case of strong fluctuations or long range correlations ensemble averages are not accurate estimators of coarse grained behavior. In this case, the pdf of the S/TRF provides a useful measure of the space-time variability that can be expected based on the available knowledge. □

Generally, the pdf must honor constraints imposed by the data. Constraining of the pdf is often referred to as *conditioning* to the data. Furthermore, the S/TRF model allows for the existence of an

Observation effect: *An S/TRF consists of various realizations that are assigned specific probabilities based on prior knowledge. After a specific observation takes place, the pdf must be updated to account for the new constraint.*

In particular, if an experimental value $X(\boldsymbol{p}_1)=\chi_1$ is observed at the point $\boldsymbol{p}_1$, the simplest conditioning approach is by collapsing that marginal pdf, i.e., by requiring that $f_X(\chi|\chi_1)=\delta(\chi-\chi_1)$. In other words, at the observation points the probability of the corresponding realization changes from less than certain to certain.

An S/TRF representing a natural process would be more appropriately denoted as

$$X(\boldsymbol{p},\boldsymbol{f}), \tag{5}$$

where $\boldsymbol{p}=(\boldsymbol{s},t)\in S\times T$ are the space-time coordinates and $\boldsymbol{f}$ is a vector of factors expressing the specific experimental conditions, scale and physical laws of the natural processes. In the case of a health process, the S/TRF may be written as

$$X(\boldsymbol{p},\, e,\boldsymbol{f}), \tag{6}$$

where e is the health condition-related event (disease, death, etc.), and $\boldsymbol{f}$ includes factors such as age, sex, race, marital status and duration of exposure (see, also, §I.4)

REMARK 1: Notation (6) makes explicit one of the main attractions of the S/TRF model in human exposure analysis, namely, that it can combine in a single formalism all three kinds of epidemiological studies, viz., cohort, case-control and ecological studies. This aspect is further discussed in Chapter IX. For simplicity, in the following e and f will usually be dropped.

2.2 Characterization in Terms of Ensemble Functions

The complete information about an S/TRF is included in the pdf. Indeed, if a pdf has a complicated shape, the only way to convey the information it contains regarding the S/TRF is to look at the complete picture as provided by Eq. (2). In many applications, however, the pdf structure is relatively simple, and a satisfactory characterization of $X(\boldsymbol{p})$ is provided in terms of *ensemble functions* (Thesis 2 of Table 1) of the general form

$$\overline{\Phi[X(\boldsymbol{p}_1),\ldots X(\boldsymbol{p}_m)]} = \int d\chi_1 \ldots \int d\chi_m \, \Phi[\chi_1,\ldots,\chi_m] f_x(\chi_1,\ldots,\chi_m) \,, \tag{7}$$

where $\overline{[\ldots]}$ denotes the operation of stochastic expectation and $\Phi[\cdot]$ is a functional of the $\chi_1,\ldots,\chi_m$. Usually, there is no need to specify limits of integration in Eq. (3), since if certain values of $\chi_1,\ldots,\chi_m$ are impossible, the pdf will be zero. Stochastic *moments* are obtained from Eq. (7) if the functional $\Phi[\cdot]$ is a polynomial in $\chi_1,\ldots,\chi_m$. Certain moments can often be estimated from the data or physical models, and they provide information regarding the properties of the pdf. Note, however, that it is not possible to represent all pdf models in terms of moments. Lévy processes, e.g., have infinite moments of order higher than two (Shlesinger *et al.*, 1994). Below, we present definitions of stochastic moments commonly used in applications.

Definition 2: *The space-time mean function is defined as*

$$m_X(\boldsymbol{p}) = \overline{X(\boldsymbol{p})} = \int d\chi \, \chi \, f_x(\chi) \tag{8}$$

for all $\boldsymbol{p} \in S \times T \subset R^n \times T$, *where* $f_x(\chi)$ *is a univariate pdf.*

The mean function expresses trends and systematic structures in space-time. In contrast, the covariance and semi-variogram functions defined below express correlations and dependencies between the points $\boldsymbol{p} = (\boldsymbol{s},t)$ and $\boldsymbol{p}' = (\boldsymbol{s}',t')$.

Definition 3: *The local S/TRF fluctuation is given by* $\delta X(\boldsymbol{p}) = X(\boldsymbol{p}) - m_X(\boldsymbol{p})$. *The space-time fluctuation covariance function is given by*

$$c_x(p_1,p_2) = \overline{\delta X(p_1)\,\delta X(p_2)} = \int\int d\chi_1\, d\chi_2 (\chi_1 - m_x)(\chi_2 - m_x') f_x(\chi_1,\chi_2;p_1,p_2) \tag{9}$$

for all $p_1,p_2 \in S\times T$, *where* $f_x(\chi_1,\chi_2;p_1,p_2)$ *denotes the bivariate pdf. The* $c_x(p_1,p_2)$ *is also called centered covariance.*

The *non-centered covariance*, which also includes the trends is often used. It is given by the following expectation

$$C_x(p_1,p_2) = \overline{X(p_1)X(p_2)} = \int\int d\chi_1\, d\chi_2 (\chi_1\chi_2) f_x(\chi_1,\chi_2;p_1,p_2). \tag{10}$$

The value of the centered covariance at zero lag is equal to the local variance of the fluctuations

$$\sigma_x^2(p) = c_x(p,p). \tag{11}$$

A useful measure of the relative strength of fluctuations is the local *coefficient of variation* (COV) which is equal to the ratio of the standard deviation over the mean value

$$\mu_x(p) = \sigma_x(p)/m_x(p). \tag{12}$$

Mathematical studies of environmental health processes in heterogeneous media are usually based on perturbation expansions of the fluctuations around the mean. The coefficient of variation emerges in such expansions as a dimensionless perturbation parameter. Then, first order perturbation includes only terms linear in the COV. Convergence of the perturbation series can be guaranteed only if the COV is everywhere less than one.

The *correlation function* is obtained by normalizing the centered covariance as follows

$$\rho_x(p_1,p_2) = c_x(p_1,p_2)/\sigma_x(p_1)\sigma_x(p_2). \tag{13}$$

Thus, the correlation function provides information about the spatial distribution of the fluctuation correlations but not about their strength.

Definition 4: *The space-time semi-variogram function is defined by*

$$\gamma_x(p_1,p_2) = \tfrac{1}{2}\overline{[X(p_1) - X(p_2)]^2} = \tfrac{1}{2}\int\int d\chi_1\, d\chi_2 (\chi_1 - \chi_2)^2 f_x(\chi_1,\chi_2;p_1,p_2) \tag{14}$$

for all $p_1, p_2 \in S \times T$.

Another two-point function that is used in practical applications is the *structure function* (also called *delta variance* or *variogram*), which is defined as follows

$$\Delta_x(p_1, p_2) = \overline{[X(p_1) - X(p_2)]^2}. \tag{15}$$

REMARK 2: Note that the covariance, the structure function, and the semi-variogram are all two-point functions, and thus they are interrelated. Depending on the application and the type of heterogeneity one of these functions may be preferred. In theoretical studies of homogeneous-stationary S/TRF the covariance is commonly used. On the other hand, the semi-variogram is often easier to estimate from experimental data. Finally, in studies involving fields with homogeneous-stationary increments, for example in turbulence, the structure function is preferred.

Definition 5: *The sysketogram function is defined as*

$$\begin{aligned}\beta_x(p_1, p_2) &= \varepsilon_x(p_1) - \varepsilon_x(p_1/p_2)\\ &= \int d\chi_1 [\int d\chi_2\, f_x(\chi_1, \chi_2) \ell og f_x(\chi_1/\chi_2) - f_x(\chi_1) \ell og f_x(\chi_1)]\end{aligned} \tag{16}$$

for all $p_1, p_2 \in S \times T$, *where* $f_x(\cdot/\cdot)$ *is the conditional pdf;* $\varepsilon_x(\cdot)$ *and* $\varepsilon_x(\cdot/\cdot)$ *denote the entropy and the conditional entropy, respectively.*

The sysketogram provides a measure of spatiotemporal correlation information, namely of the amount of information on $X(p_1)$ that is contained in $X(p_2)$. Note that, unlike the previous ensemble functions it is defined only in terms of the pdf. As we shall see below, ensemble properties can be instructive in determining a space-time metric.

Eqs. (7)-(16) are expressions for stochastic expectations that involve integrals of the pdf. If the pdf is a discrete function, the stochastic expectations involve summations over the discrete possibilities. The integral expressions above are still valid, if the pdf is replaced by

$$f_x(\chi) = \sum_j f_{x,j} \delta(\chi - \chi_j), \tag{17}$$

where $f_{x,j} = P[\chi = \chi_j]$ are the probabilities of the discrete events.

EXAMPLE 3: Consider Eq. (8) which is valid in continuous space; by using the above substitution we find

$$m_x(\boldsymbol{p}) = \sum_j f_{x,j} \int d\chi\, \chi\, \delta(\chi - \chi_j) = \sum_j \chi_j f_{x,j}\,, \tag{18}$$

which is the correct result in the discrete domain. □

To study probabilities and expectations of environmental health processes rather than specific realizations is in many cases a quite legitimate way of confronting theory with observations and making predictions (Theses 2 and 3 of Table 1). However, a word of caution is necessary here as expressed by the following

Scholium 1: *Probabilities and expectations make sense only when there is some understanding or working hypothesis about the phenomenon.*

In one sense, this rule is another formulation of the total knowledge principle (§I.2), which emphasizes the importance of incorporating physical laws and other kinds of data and information in the stochastic analysis of an uncertain situation. Physical laws should be incorporated in the stochastic analysis by means of models, and the models should respect the available information.

REMARK 3: In the following, the one vector $\boldsymbol{p}$-notation and the couple $(\boldsymbol{s},t)$-notation will be used interchangeably. Thus, models will be expressed in terms of $X(\boldsymbol{p})$, $c_x(\boldsymbol{p};\boldsymbol{p}')$, as well as $X(\boldsymbol{s},t)$, $c_x(\boldsymbol{s},\boldsymbol{s}';t,t')$.

3. S/TRF CLASSIFICATIONS

We now consider certain symmetry properties and their mathematical formulation that will be useful in the characterization of environmental health systems. A system is *translationally invariant* (in space or time) with respect to a certain attribute if translation by certain characteristic vectors (in space) or time lags (in time) does not change this attribute. Periodic systems such as crystals, planetary orbits and clocks belong in this category. A system is called *homogeneous* with respect to a specific attribute, if this attribute has a uniform value in space. We call this property *simple homogeneity* to distinguish it from *stochastic homogeneity* to be defined below. Even systems with high variability are modelled by means of simply homogeneous idealizations; for example, in groundwater hydrology deterministic models widely used by consulting firms and regulatory agencies assume a uniform average hydraulic conductivity in the whole domain, or in individual layers. Such assumptions are often sufficient for gaining physical insight, but they are usually inadequate for accurate estimation. A system is simply homogeneous if and only if it is translationally invariant for all possible translations.

In the case of stochastic systems the S/TRF realizations do not exhibit such convenient symmetry properties. However, it is often true that ensemble functions, e.g. the mean, the covariance and higher order moments of S/TRF possess certain symmetry properties. Hence, a process is said to exhibit *stochastic translation invariance* if its ensemble functions are translationally invariant. We continue with a discussion of the most important symmetry properties of stochastic moments.

Stochastic homogeneity-stationarity: *If the first two moments are invariant for all translations in space and time, the S/TRF is called stochastically homogeneous-stationary in the wide sense (also, second order homogeneous-stationary). If all multivariate pdf's are invariant for all translations in space and time the S/TRF is called stochastically homogeneous-stationary in the strict sense.*

Hence, strict homogeneity-stationarity requires that

$$F_x[\chi_1(s_1,t_1),\ldots,\chi_N(s_N,t_N)] = F_x[\chi_1(s_1+\boldsymbol{r},t_1+\tau),\ldots,\chi_N(s_N+\boldsymbol{r},t_N+\tau)] \tag{1}$$

for all $N = 1,2,\ldots,$ and for all spatial lags $\boldsymbol{r}$ and temporal lags τ. On the other hand, wide sense homogeneity-stationarity requires that the S/TRF $X(s,t)$ have a constant mean as follows

$$m_X(\boldsymbol{p}) = m_X; \tag{2}$$

and that the covariance be a function only of the spatial lag $\boldsymbol{r} = s_1 - s_2$ and temporal lag $\tau = t_1 - t_2$ between the two points, i.e.,

$$C_x(\boldsymbol{p}_1,\boldsymbol{p}_2) = C_x(s_1 - s_2; t_1 - t_2) = C_x(\boldsymbol{r},\tau), \tag{3}$$

EXAMPLE 1: A homogeneous-stationary field $X(s,t)$ in $R^3 \times T$ is characterized by the covariance

$$c_x(\boldsymbol{r},\tau)=\frac{\sin(\beta\tau)}{\pi^4\tau}\prod_{i=1}^{3}\frac{\sin(\alpha_i r_i)}{r_i}. \tag{4}$$

As $\alpha_i,\beta\to\infty$, the covariance is proportional to the delta function $\delta(\boldsymbol{r},\tau)$ and the S/TRF $X(\boldsymbol{s},t)$ tends to white noise. □

The same property also holds for the remaining two-point functions. It is also clear that the properties (2) and (3) also hold in the case of strict homogeneity-stationarity. The semi-variogram of a homogeneous-stationary S/TRF is related to the covariance by

$$\gamma_X(\boldsymbol{p})=\sigma_X^2-c_X(\boldsymbol{p}) \tag{5}$$

Eq. (5) is used for obtaining the covariance function from empirically determined semi-variograms. More results are presented in Chapter IV.

Note that we have focused on homogeneous-stationary S/TRF. However, S/TRF that are only homogeneous in space or only stationary in time are also possible. The properties of homogeneous (or stationary) random fields can be obtained from expressions that are similar to the above, but involve only the spatial (or temporal) lag. For example, the mean of homogeneous S/TRF, $m_X(\boldsymbol{s},t)=\overline{X}(t)$, is, in general a function of time, while the mean of a stationary S/TRF, $m_X(\boldsymbol{s},t)=\overline{X}(\boldsymbol{s})$, is in general a function of space.

Stochastic isotropy: *If the second order moments depend purely on the magnitude of the spatial lag vector, then the S/TRF is called stochastically isotropic in the wide sense. If all the multivariate pdf's depend purely on the lag magnitudes, the S/TRF is stochastically isotropic in the strict sense.*

Note that isotropy is a property related to purely spatial transformations of the S/TRF. If the S/TRF $X(\boldsymbol{p})$ is stochastically isotropic, it is necessarily stochastically homogeneous. The converse, however, is not true. Strict isotropy implies the following relation

$$F_x[\chi_1(\boldsymbol{p}_1),\ldots,\chi_N(\boldsymbol{p}_N)]=F_x[\chi_1(\boldsymbol{p}_1+r\boldsymbol{e}_1),\ldots,\chi_N(\boldsymbol{p}_N+r\boldsymbol{e}_N)], \tag{5}$$

for all unit vectors $\boldsymbol{e}_1,\ldots,\boldsymbol{e}_N$, where $r=|\boldsymbol{r}|$ denotes the Euclidean norm in R^n.

In the case of weak isotropy the mean is uniform in space, while the two-point functions depend only on the spatial lag (but time stationarity is not necessary); e.g.,

$$C_X(\boldsymbol{p}_1, \boldsymbol{p}_2) = C_X(r; t_1, t_2), \quad r = |\boldsymbol{r}|, \tag{6}$$

Stochastic isotropy does not constrain the S/TRF realizations of the random field, which may show anisotropic spatial dependence.

REMARK 1: To avoid possible confusion regarding the meaning of isotropy we emphasize that what we consider here is stochastic isotropy. The term "isotropic" is also used to denote tensors with principal components of the same magnitude in all directions. Hydraulic conductivity of porous media, e.g., is often assumed for modelling purposes to be locally isotropic and is represented by means of a scalar field. Stochastic isotropy, on the other hand, is a property that characterizes the pdf or the correlation function under rotations of the coordinate system. Large scale anisotropic structures in porous media are modelled using scalar random fields with anisotropic correlation functions.

Isotropic covariance functions are commonly used in turbulence (McComb, 1990), in statistical topography (Isichenko, 1992), in groundwater flow and transport (Dagan, 1989), in geostatistical applications (Deutsch and Journel, 1992), and in atmospheric pollution studies (Christakos and Hristopulos, 1996a). The rationale for the use of isotropic covariances is that in absence of macroscopic directional trends, averaging over the stochastic ensemble eliminates directional preferences. Isotropy is not always a reliable modelling assumption, since many environmental processes are definitely anisotropic at large scales. This is illustrated by groundwater flow in stratified media: the hydraulic conductivity covariance in such media depends significantly on the direction of flow. Nonetheless, it is possible to transform a stochastically anisotropic random field into an isotropic one by a simple rescaling of the axes as we show in §6 below.

Stochastic ergodicity: *If the second order ensemble moments are equal to the second order sample moments then the S/TRF is called stochastically ergodic in the wide sense. If all the ensemble and sample multivariate pdf's are equal to each other, the S/TRF is stochastically ergodic in the strict sense.*

While the ensemble moments are calculated in terms of the corresponding pdf's --Eqs. (7) through (15) of §2 above-- the *sample* moments are calculated on the basis of a single realization which is usually available in practice (explicit formulas for the calculation of the sample moments can be found, e.g., in Christakos, 1992).

Spatial random fields and temporal random functions (also called time series, random processes, or random sequences in the case of discrete time), which are considered as separate topics in most textbooks, are essentially special cases of the S/TRF concept. Hence, environmental health processes that are independent of time may be modeled using spatial random fields $X(\boldsymbol{s})$ in R^n. On the other hand, phenomena that have only a temporal component are modeled as temporal random functions $X(t)$ in T. In the first category we find random fields that represent the steady state of natural processes. In the second category we find global variables that represent spatial averages or processes localized in space; for example, the level in a water reservoir or the concentration of a toxic substance in a specific organ. However, for most natural phenomena of environmental health interest, both the spatial and temporal variabilities are important.

REMARK 2: The difference between a temporal random function and a random field (spatial or spatiotemporal) appears more significant from the viewpoint of fundamental physical theories. In particular, the evolution of a temporal random function $X(t)$ is described by ordinary differential equations (ODEs). The $X(t)$ may obey ODEs which represent mathematical expressions of fundamental physical laws governing the movement of particles as a function of time. The random fields $X(\boldsymbol{s})$ and $X(\boldsymbol{s},t)$, on the other hand, represent natural processes the evolution of which is expressed by means of PDEs. The fundamental physical laws that these fields obey are expressed by PDEs with three (spatial) or four (spatiotemporal) independent variables. Note that in many applications ODEs and PDEs constitute approximations rather than direct expressions of fundamental physical laws. In general, solution of ODEs is a much simpler operation, both mathematically and computationally, than the solution of PDEs.

Finally, it is sometimes useful to classify environmental health processes represented by S/TRFs in two categories based on the available information. For processes in the first category, only one sample of the S/TRF is available at different space-time points. Many geophysical, hydrological and atmospheric processes (radiative processes, precipitation, etc.) as well as health processes (epidemics, etc.) belong to this category. In the second category, multiple samples of the same process are available. Examples include toxicological processes (measurements of carcinogenic adduct concentrations in the DNA, etc.), in which samples from different individuals are available. The difference with epidemic processes is that in the latter exposure occurs at the population scale, where there is usually a single sample representing the space-time evolution of the disease. For toxicological processes the exposure characterizes a single individual and, thus, multiple samples can be obtained from different individuals (although, in many studies this may not be a valid approach due to intersubject variability; see Chapters II and IX). Processes of the first type can be sometimes mapped onto second type processes by considering generalized space-time increments as single realizations. This approach has been used to analyze one-dimensional S/TRF such as time records of natural processes (Hurst, 1965), and genomic DNA sequences (Peng *et al.*, 1994; Arneodo *et al.*, 1995). However, in spatiotemporal phenomena it is not always possible to determine unequivocally the order of the increment. Space-time analysis for environmental health processes of the first category entails determining the S/TRF from a single realization (see, also, Chapter IV).

Scholium 1: *The reader is reminded that both space-time notations [* $\boldsymbol{p}$, $X(\boldsymbol{p})$, $C_X(\boldsymbol{p};\boldsymbol{p}')$, $\gamma_X(\boldsymbol{p};\boldsymbol{p}')$, *etc.] and [* $(\boldsymbol{s},t)$, $X(\boldsymbol{s},t)$, $C_X(\boldsymbol{h};\tau)$, $\gamma_X(\boldsymbol{h};\tau)$, *etc.] will be used interchangeably throughout the book.*

4. SPACE-TIME METRICS

Metrics are mathematical expressions that define the concept of distance in the space-time continuum. Distance can not always be defined unambiguously in space-time. For example, it is not possible to decide, without additional information, how the distance d_{12} between two points $\boldsymbol{p}_1$ and $\boldsymbol{p}_2$ with a Euclidean spatial lag of three kilometers and a time lag of two days compares with the distance d'_{12} between points $\boldsymbol{p}'_1$ and $\boldsymbol{p}'_2$ with a two-kilometer spatial lag and a three-day time lag. It is possible to decide which of the two pairs of points has the larger separation distance by considering the outcome of a natural process. Consider, e.g., an experiment during which a tracer is released inside a porous medium at points $\boldsymbol{p}_1$ and $\boldsymbol{p}'_1$. If the tracer is detected at point $\boldsymbol{p}_2$ but not at point $\boldsymbol{p}'_2$, then we can claim that $d_{12} < d'_{12}$ with respect to the particular experiment. This approach provides a way of ordering distances between space-time points, but without further refinements a specific quantitative notion of space-time distances can not be obtained. Also note that distance defined above is not purely a geometric property of space-time, but it also depends on the medium's properties, and the particular natural phenomenon that we decide use in the measurement. For example, measuring distance by means of fluid tracer dispersion can lead to very different results than measuring distance by means of electromagnetic propagation. We might summarize the above considerations by means of the following

Basic idea: *The quantitative notions of space and time distances in S/TRFs come from physical experience (data, laws, etc.), can be made definite only by reference to physical experience, and are subject to change if a reconsideration of experience seems to warrant change.*

Following the above discussion, it is sometimes possible to define space-time metrics for homogeneous-stationary covariance functions of the form $\eta = g(r,\tau)$, where η denotes the space-time distance and $g(r,\tau)$ is a functional that depends on the space and time lags, and constant coefficients. It is necessary to emphasize that such forms must be adjudicated by the data. An example is provided by the metric of the form

$$\boldsymbol{\eta}^2 = (\boldsymbol{p}_1 - \boldsymbol{p}_2)^2 = \boldsymbol{r}^2 + \boldsymbol{v}^2 \tau^2, \tag{1}$$

where $\boldsymbol{v}$ is a suitable vector (e.g., $\boldsymbol{v}$ may denote the propagation velocity of the fluctuations in time). Fitting a covariance $c_X(\boldsymbol{r},\tau)$ to the available data can be helpful in determining the space-time metric. We illustrate a property of a covariance with metric (1) in

EXAMPLE 1: For illustration consider the covariance function (in $R^1 \times T$)

$$c_x(r;\tau) = c_0 \exp(-r^2/a^2 - \tau^2/b^2); \tag{2}$$

by evaluating partial derivatives it is shown that $(\partial c_x/\partial \tau)(\partial c_x/\partial r)^{-1} = v^2\, \tau/r$, where $v = a/b$. A valid metric is $\eta^2 = r^2 + v^2\tau^2 = r^2 + a^2\tau^2/b^2$ and the covariance can indeed be written as $c_x(r;\tau) = c_0 \exp(-\eta^2/a^2)$. □

If a physical law is known for the natural process $X(\boldsymbol{p})$, it can also play a fundamental role in establishing a space-time metric. For illustration, we present

EXAMPLE 2: Consider a process that satisfies the PDE

$$\partial X/\partial t + \boldsymbol{v} \cdot \nabla X = 0. \tag{3}$$

Solutions of this equation have the form $X(\boldsymbol{p}) = X(\boldsymbol{s},t) = X(\boldsymbol{s} - \boldsymbol{v}t)$. In this case the solution depends on the space-time vector $\eta = \boldsymbol{s} - \boldsymbol{v}t$. Hence, a plausible covariance model is a function that depends on the space-time lag via a space-time metric of the form

$$(\eta - \eta')^2 = (\boldsymbol{r} - \boldsymbol{v}\,\tau)^2, \tag{4}$$

where $\boldsymbol{v}$ is a velocity vector. □

On the other hand, in many applications of practical interest it is adequate to consider natural variations within a domain defined by the Cartesian product space$\times$time, and composite space-time metrics are not necessary.

EXAMPLE 3: A spatially isotropic and temporally stationary random field can be defined as having the covariance $c_x(\boldsymbol{p}_1, \boldsymbol{p}_2) = c_x(r, \tau)$, which depends separately on the spatial metric $r = |\boldsymbol{r}|$ and the temporal lag $\tau = t_1 - t_2$. □

5. SPATIOTEMPORAL CORRELATION MODELS

5.1 General Properties

First, we discuss some fundamental properties of covariance functions. Properties (a) through (d) below are general, and they are not restricted to homogeneous-stationary covariances. Property (e) is specific to homogeneous-stationary random fields. Before we proceed, we extend the definition of the covariance function given in §2 above to complex S/TRF.

Definition 1: *The centered covariance of a S/TRF $X(\boldsymbol{p}) \equiv X(\boldsymbol{s}, t)$ is defined by the following*

$$c_x(\boldsymbol{p}_1, \boldsymbol{p}_2) = \overline{[X(\boldsymbol{p}_1) - m_x(\boldsymbol{p}_1)][X^\dagger(\boldsymbol{p}_2) - m_x^\dagger(\boldsymbol{p}_2)]}, \tag{1}$$

where the dagger ($\dagger$) denotes the complex conjugate (complex conjugation is redundant for real-valued S/TRF).

A S/TRF is continuous in the mean square sense at the point $\boldsymbol{p}_0$, if and only if its covariance $c_x(\boldsymbol{p}, \boldsymbol{p}')$ is continuous at the point $\boldsymbol{p} = \boldsymbol{p}' = \boldsymbol{p}_0$ (Christakos, 1992). Most of the random field models used in environmental health science are mean square continuous. However, discontinuities are also encountered in environmental health processes. Permeabilities in porous media, e.g., jump discontinuously at lithotype boundaries. In such cases it is not advisable to use a continuous random field across the boundary, and simulations must account for the discontinuity.

Based on Definition 1 it can be shown that covariance functions satisfy the following general properties:

$$\textit{Property a}: \quad c_x(\boldsymbol{p}_1, \boldsymbol{p}_2) = c_x^{\dagger}(\boldsymbol{p}_2, \boldsymbol{p}_1). \tag{2}$$

In the case of real-valued and homogeneous-stationary random fields, while $c_x(\boldsymbol{r}, \tau)$ is generally nonsymmetric with respect to the $\boldsymbol{r}$ and τ axes, Eq. (2) implies a covariance that is centrally symmetric with respect to the space-time origin, i.e.,

$$\textit{Property b}: \quad \left.\begin{array}{l} c_x(\boldsymbol{r}, \tau) = c_x(-\boldsymbol{r}, -\tau) \\ c_x(\boldsymbol{r}, \tau) \neq c_x(\boldsymbol{r}, -\tau) \ \text{ for } \ \boldsymbol{r} \neq 0 \\ c_x(\boldsymbol{r}, \tau) \neq c_x(-\boldsymbol{r}, \tau) \ \text{ for } \ \tau \neq 0 \end{array}\right\}. \tag{3}$$

$$\textit{Property c}: \quad |c_x(\boldsymbol{p}_1, \boldsymbol{p}_2)| \leq \overline{|X(\boldsymbol{p}_1)|^2}\,\overline{|X(\boldsymbol{p}_2)|^2}. \tag{4}$$

Property (c) follows from the Schwartz inequality. It provides an upper bound for the covariance between any two points in space-time in terms of the local variances. In the case of homogeneous-stationary S/TRF, Eq. (4) leads to $c_x(\boldsymbol{p}_1 - \boldsymbol{p}_2) \leq \sigma_x^2$. Hence, in this case the covariance decreases monotonically with the space lag (time lag) when the time lag (space lag) is kept fixed.

Property d : The following bilinear form is non-negative for any number N of space-time points, and arbitrary complex numbers z_i, $i = 1, \ldots, N$,

$$\sum_{i=1}^{N} \sum_{j=1}^{N} c_x(\boldsymbol{p}_i, \boldsymbol{p}_j) z_i z_j^{\dagger} \geq 0. \tag{5}$$

Eq. (5) follows from the fact that the quantity $\overline{\left|\sum_{i=1}^{N} [X(\boldsymbol{p}_i) - \overline{X(\boldsymbol{p}_i)}] z_i\right|^2}$ is non-negative.

A continuous function that satisfies properties (a)-(d) above is called a *non-negative definite function*. Hence, the covariance functions of mean square continuous S/TRFs are non-negative definite functions. The converse can also be shown, namely that every non-negative definite function is a covariance function of a mean square continuous random field. As we further discuss in §5.2 on permissibility criteria below, useful conditions for a function to be non-negative definite are given by Bochner's theorem.

Property e : $\lim_{|r|\to\infty} c_x(\boldsymbol{r},\tau) = \lim_{|\tau|\to\infty} c_x(\boldsymbol{r},\tau) = 0$. (6)

Property (e) is valid only for homogeneous-stationary random fields. Its physical meaning is that the S/TRF values at two points that are widely separated in either space or time are uncorrelated.

Are all permissible covariances good candidates for an environmental health process? The answer is negative. Covariance models describe how the correlations behave in space and time and, therefore, they are inherently connected to the physical and biological laws governing the process. There is, for instance, a significant difference between short range (e.g., exponential or Gaussian models) and power law correlations. The latter indicate the existence of scaling in the system, which may arise due to a number of different physical or biological causes. It could possibly denote that the system is near so-called *critical points*, where correlation functions become invariant under scale transformations (see below for a more precise definition). Percolation type models, which have successfully been applied to various environmental health processes --including flow and transport in porous media and the spread of epidemics-- are among the systems that exhibit critical behavior.

EXAMPLE 1: In lattice percolation, two phases (say, red and blue) occupy the lattice sites with probabilities p (red) and $1-p$ (blue) respectively. At the critical threshold p_c a connected cluster of red sites spans the whole lattice. If the red phase represents the void space, the lattice becomes permeable at p_c. Near the percolation threshold the correlation functions behave like power laws with fractal exponents (Stauffer and Aharony, 1992), and the geometric structures on the lattice are fractal objects. □

5.2 Criteria of Permissibility

Criteria of permissibility are necessary and sufficient conditions satisfied by covariance functions. In the case of homogeneous-stationary fields, the permissibility conditions are obtained by means of the Bochner theorem (Bochner, 1959); and in the case of nonhomogeneous-nonstationary fields (ordinary or generalized) by the Bochner-Schwartz theorem (e.g., Gel'fand and Vilenkin, 1964; p.155-160). It is worthwhile to discuss the important role permissibility criteria play. Permissibility conditions would not be necessary, if the covariance functions could be determined from first principles. This is usually impossible for many practical problems investigated in environmental health science. In view of this situation two approaches are available:

The first approach involves obtaining the covariance function from the data. This is an ill-posed problem, since the determination of a covariance function involves infinitely many points, while the data provide only a finite set. In practice, this issue is simply resolved by fitting various empirical models to the data, and choosing the one that provides the best fit. For the purpose of determining various models for the fit, some knowledge of the mathematical properties of correlations based on a physical analysis of the system can be very helpful. Permissibility conditions are then required in order to constrain the parameters of the empirical models.

The second approach focuses on obtaining explicit expressions for the S/TRF $X(s,t)$ in terms of a known S/TRF $Y(s,t)$ via physical or biological modelling. Solutions of such models provide useful relations between different attributes of a process, which determine the covariance of $X(s,t)$ in terms of the known covariance of $Y(s,t)$. The $Y(s,t)$ may represent a natural gradient, e.g., temperature, and $X(s,t)$ the resulting flux. In several situations in practice, however, the S/TRF $Y(s,t)$ has multiple physical components, or the random field $X(s,t)$ satisfies a complicated dynamic equation, which can make explicit representation of $X(s,t)$ in terms of a well-characterized $Y(s,t)$ a rather difficult task. In any case, physical or biological modelling can be extremely valuable in providing information about the form of the covariance function that can be used by the first approach above. Indeed, fitting a physically or biologically based function to the data is a much better approach than using any function that merely provides a statistically good fit to the data. Hence, when the available knowledge makes it possible, a combination of both approaches is strongly suggested.

Space-time covariances are non-negative definite functions that satisfy the property (d) of §5.1 above. Property (d) is completely general, i.e., it is satisfied by space-time covariances regardless of nonhomogeneity or nonstationarity. In practice, confirming the validity of inequality (5) is awkward. Fortunately, Bochner's theorem (Bochner, 1933) provides useful conditions for functions to be non-negative definite. This theorem was initially formulated for a function $c_x(\boldsymbol{r})$ defined in a subspace of R^n with Euclidean distance measure $|\boldsymbol{r}| = (\sum_{i=1}^{n} r_i^2)^{1/2}$, but it can be extended to space-time functions because its validity does not depend on the definition of a space-time distance. Bochner's theorem can be used in the case of homogeneous-stationary covariances that depend purely on the space and time lags. For nonstationary-nonhomogeneous covariances the Bochner-Schwartz theorem (Chapter IV) is used.

Bochner's Theorem for S/TRF: *A continuous function $c_X(\boldsymbol{r},\tau)$ from $R^n \times T$ to the complex plane is non-negative definite if and only if it can be expressed as*

$$c_x(\boldsymbol{r},\tau) = \int_{\boldsymbol{\kappa}} \int_{\omega} \exp[i\,\boldsymbol{\kappa}\cdot\boldsymbol{r} - i\,\omega\,\tau]\tilde{c}_x(\boldsymbol{\kappa},\omega), \tag{7}$$

where the spectral density $\tilde{c}_x(\boldsymbol{\kappa},\omega)$ is a real-valued, integrable and non-negative function of the spatial frequency $\boldsymbol{\kappa}$ and the temporal frequency ω, $\int_{\boldsymbol{\kappa}} = (2\pi)^{-n}\int d\boldsymbol{\kappa}$ and $\int_{\omega} = (2\pi)^{-1}\int d\omega$.

REMARK 1: The spectral density must be integrable, but not necessarily bounded; hence, it may have singular points. For example, spectral densities that behave as $\tilde{c}_x(\boldsymbol{\kappa},\omega) \propto |\boldsymbol{\kappa}|^{-\alpha}$ for $\boldsymbol{\kappa}$ near zero are singular for $\alpha > 0$, but they are integrable if $n > \alpha$, where n denotes the space dimension. The theorem can also be expressed in terms of the spectral distribution function $\tilde{F}_x(\boldsymbol{\kappa},\omega)$ instead of the spectral density (e.g., Adler, 1981; p. 25). This formulation is useful if the spectral density does not exist, e.g., when the spectral distribution function has a jump; however, even in these cases the spectral density can be defined by means of generalized functions. The Bochner theorem conditions for the spectral distribution function are that it be real-valued, bounded, and that the measure $\mu(A) = \int_A d\tilde{F}_x(\boldsymbol{\kappa},\omega)$ be a non decreasing function.

5.3. Useful Tools for Constructing Spatiotemporal Correlation Functions

Empirical formulas for calculating spatiotemporal correlation functions in practice are given in Christakos (1992, Chapter 7). In addition, certain of the theoretical results obtained above can be used to built spatiotemporal correlation functions. As it was shown in Christakos (1984a, pp. 263-264), Eq. (7) can form the basis for deriving valid covariances from properly chosen spectral densities. Some other techniques for constructing correlation functions are discussed next.

Christakos (1992; pp. 262-265) used a method for deriving valid spatiotemporal correlation models by maximizing the relevant entropy function. Let $\{c_x(\boldsymbol{r},\tau),\ \tilde{c}_x(\boldsymbol{\kappa},\omega)\}$ be the spatiotemporal covariance-spectral density pair defined by Eq. (7). Assume that the knowledge available regarding spatiotemporal variability is expressed by the constraints

$$\int_{\boldsymbol{\kappa}\in I_{\boldsymbol{\kappa}}} \int_{\omega\in I_{\omega}} \exp[i\,\boldsymbol{\kappa}\cdot\boldsymbol{r}_{\ell} - i\,\omega\,\tau_{\ell}]\tilde{c}_x(\boldsymbol{\kappa},\omega) = c_x(\boldsymbol{r}_{\ell},\tau_{\ell}) \quad \text{for } \ell = 1,\ldots,L, \tag{8}$$

where $\int_{\kappa \in I_\kappa} = (2\pi)^{-n} \int d\kappa$ and $\int_{\omega \in I_\omega} = (2\pi)^{-1} \int d\omega$; the I_κ and I_ω are known frequency intervals. Constraints (8) express mathematically a common situation in practice, namely, covariance values are experimentally calculated for a set of space-time intervals (r_ℓ, τ_ℓ), $\ell = 1,\ldots,L$. Additional constraints may include knowledge regarding the behavior of $c_x(r,\tau)$ at larger space-time lags (e.g., spatial and temporal ranges of influence), etc.. Given these constraints, the problem of determining the shape of $c_x(r,\tau)$ is converted into the equivalent one of determining $\tilde{c}_x(\kappa,\omega)$. The latter is obtained by maximizing an entropy function, such as the *Burg-entropy* function

$$\varepsilon(\tilde{c}_x) = \int_{\kappa \in I_\kappa} \int_{\omega \in I_\omega} \ell og\, \tilde{c}_x(\kappa,\omega), \tag{9}$$

with respect to $\tilde{c}_x(\kappa,\omega)$ subject to constraints (8), etc.

Space transformation operators (ST; Chapter VII) offer another method that can be used to construct spatiotemporal correlation models from simpler models. If a correlation model is known in $R^1 \times T$, ST operators can be used to construct permissible correlation models in $R^n \times T$. The example below uses the Ψ_1^n-operator, but other choices are also available (Christakos, 1984b, 1992; Christakos and Panagopoulos, 1992).

EXAMPLE 2: If $c_{x,1}(r,\tau)$ is a permissible covariance in $R^1 \times T$, the

$$c_{x,n}(r,\tau) = \Psi_1^n[c_{x,1}(r,\tau)] = \frac{2\Gamma(n/2)}{\sqrt{\pi}\,\Gamma[(n-1)/2]} \int_0^1 du(1-u^2)^{(n-1)/2}\, c_{x,1}(ur,\tau), \tag{10}$$

is a permissible spatially isotropic covariance in $R^n \times T$. □

In the next section we discuss separable space-time covariance models and present some covariance models that are directly obtained from the solution of physical laws.

6. SEPARABLE SPACE-TIME COVARIANCE MODELS

6.1 Separability in the Space-Time Domain

Modelling of environmental health systems must account for the fact that the exact structure of covariance functions is usually unknown. Due to this indeterminacy, certain properties of covariance functions become the modeller's decision. A common choice is the use of *separable* space-time models. These are covariance models of the general form

$$C_x(s,t;s',t') = C_{x(1)}(s,s')C_{x(2)}(t,t'), \tag{1}$$

where $C_{x(1)}(s,s')$ is a purely spatial and $C_{x(2)}(t,t')$ a purely temporal covariance function. The assumption of separability is very convenient in both analytical and computational studies --because it simplifies the evaluation of stochastic moments that involve covariance integrals-- and the evaluation of generalized covariance for nonstationary-nonhomogeneous S/TRF. The separability implies that if we take snapshots of the system at different times the covariances will be similar, i.e., they will differ only by a constant factor of proportionality throughout the spatial domain. The same is true about the covariance of the time series obtained at two different locations in space. The decoupling of space and time coordinates also simplifies the mathematical analysis of separable models. For example, Bochner's theorem in the case of separable functions can be expressed as follows

Bochner's Theorem for Separability: *A separable continuous function* $c_x(\boldsymbol{r},\tau) = c_{x(1)}(\boldsymbol{r})c_{x(2)}(\tau)$ *from* $R^n \times T$ *to the complex plane is non-negative definite if and only if it can be expressed as*

$$c_x(\boldsymbol{r},\tau) = \int_{\boldsymbol{\kappa}} \exp(i\,\boldsymbol{\kappa}\cdot\boldsymbol{r})\tilde{c}_{x(1)}(\boldsymbol{\kappa})\int_{\omega} \exp(-i\,\omega\,\tau)\tilde{c}_{x(2)}(\omega), \tag{2}$$

where the spectral densities $\tilde{c}_{x(1)}(\boldsymbol{\kappa})$ *and* $\tilde{c}_{x(2)}(\omega)$ *are real-valued, integrable and non-negative function of the wavevector* $\boldsymbol{\kappa}$ *and the frequency* ω, *respectively.*

In light of the separability assumption, we present next certain spatial covariance models that are useful in applications. Short range models are routinely used in environmental sciences; long range models have also become important in light of recent investigations.

6.2 Classes of Spatial Covariance Models

Below, we concentrate primarily on homogeneous covariance models. Spatial covariances of higher continuity orders can be constructed based on homogeneous residuals.

Short Range Models: These models have correlations that decay to zero fast enough for the $n-$dimensional integral of the covariance function to exist. Examples include the spherical, the exponential, the cubic and the Gaussian models. The correlations of short range models become negligible when the lag increases beyond a certain value. For many short range models the decay of the correlation function is determined by a single scale, the correlation length. In this case a dimensionless lag can be defined by dividing the actual lag with the correlation length. Consequently, the Fourier transform (FT) of single-scale models admits the following general expression

$$\tilde{c}_x(\boldsymbol{\kappa}) = \sigma_x^2\, \xi^n\, \tilde{f}(\kappa\xi), \tag{3}$$

where ξ denotes the correlation length and $\kappa\xi$ is a dimensionless variable.

EXAMPLE 1: The *exponential* covariance model is a short range model given by

$$c_x(\boldsymbol{r}) = \sigma_x^2 \exp(-r/\xi); \tag{4}$$

its FT is given by

$$\tilde{c}_x(\kappa) = \frac{2^n\, \xi^n\, \pi^{n/2}}{\Gamma(2-n/2)} \frac{\sigma_x^2}{(1+\xi^2\kappa^2)^{(n+1)/2}}, \tag{5}$$

where $\Gamma(\cdot)$ denotes the Gamma function; the latter is defined by means of the recursive relationship $\Gamma(\alpha+1) = \alpha\,\Gamma(\alpha)$ and the initial conditions $\Gamma(1/2) = \sqrt{\pi}$; $\Gamma(1) = 1$. The exponential model is also called *Markovian*, because it can be shown that Gaussian stationary processes are Markovian only if the covariance function is exponential. □

Space Separable Models: The correlation functions of separable models can be expressed in terms of a product of any permissible unidimensional correlation functions $\rho_{x,i}(r_i)$ as follows

$$\rho_x(\boldsymbol{r}) = \prod_{i=1}^{n} \rho_{x,i}(r_i); \tag{6}$$

its FT is

$$\tilde{c}_x(\boldsymbol{\kappa}) = \sigma_x^2 \prod_{i=1}^{n} \tilde{\rho}_{x,i}(\kappa_i). \tag{7}$$

EXAMPLE 2: The *Gaussian* model is a separable, single-scale model given by

$$c_x(\boldsymbol{r}) = \sigma_x^2 \exp(-r^2/\xi^2), \tag{8}$$

where

$$\tilde{c}_x(\boldsymbol{\kappa}) = \sigma_x^2 (\xi\sqrt{\pi})^n \exp(-\kappa^2 \xi^2/4). \tag{9}$$

is its FT. □

Fractal Models: Certain processes, including fractional noises and fractional Brownian motions (Mandelbrot, 1982) are characterized by power law correlations, i.e., $c_x(\lambda r) = \lambda^{2H} c_x(r)$, where H is the scaling (fractal) exponent. Following Mandelbrot's work on fractals, many data sets that were previously thought to represent incoherent and structureless noise have been well-characterized by means of fractal correlation models (e.g., Bak and Chen, 1989). The apparent irregularity of such processes was shown to derive from the long range nature of the power law correlations among individual events. Power law correlations have been observed in environmental and economic processes. A recent analysis of the epidemic size and duration of measles in small, isolated communities showed that they also exist in biological systems and have important implications for human health (Rhodes and Anderson, 1996).

Power law correlations can be classified as short range, if they are integrable and do not change the asymptotic scaling properties of the system, and long range when they lead to new types of asymptotic behavior at large distances or times. For example, in the case of diffusion in random media, long range correlations in the velocity field lead to anomalous, non-Fickian diffusion (e.g., Bouchaud and Georges, 1990). Anomalous diffusion has been observed in porous rocks, and it has been studied by means of a two-dimensional layered media model (Matheron and de Marsily, 1980).

EXAMPLE 3: *Isotropic* fractal processes are self-similar; thus they are characterized by a spherically symmetric spectral density

$$\tilde{\varpi}_x(\kappa) = (2\pi)^n A/\kappa^{\gamma}, \tag{10}$$

where κ denotes the spatial frequency. The expression above is valid within a finite frequency range $\kappa_{\mathrm{m}} \leq \kappa \leq \kappa_0$. This leads to a finite range of distances within which the system is characterized by fractal behavior (Feder, 1988). We have not specified yet what the spectral density exactly represents. If $\tilde{\varpi}_x(\kappa)$ is integrable, it represents the spectral density of a homogeneous covariance. The real-space covariance function can be obtained from the inverse FT of Eq. (10), and it behaves as a power law within the fractal range. The integrability of $\tilde{\varpi}_x(\kappa)$ depends on the value of the exponent γ, as we discuss in more detail below. If $\tilde{\varpi}_x(\kappa)$ is not integrable, it represents the semi-variogram spectral density of a random field with homogeneous increments. □

EXAMPLE 4: *Homogeneous* fractal processes are obtained if the spectral density exponent is within the range

$$(n-1)/2 < \gamma < n, \tag{11}$$

where n denotes the spatial dimension. Within the fractal range the covariance function has power law behavior, i.e.,

$$c_x(r) = \zeta_n A r^{2H}, \quad \kappa_0^{-1} << r_0 \leq r \leq r_m << \kappa_m^{-1}, \tag{12}$$

where the ζ_n are constant numbers (Isichenko, 1992). The parameter H represents the *Hurst exponent*, which is related to the spectral density exponent via the equation

$$H = (\gamma - n)/2 \tag{13}$$

The FT of Eq. (12) exists and is given by the spectral density $\tilde{Z}_x(\kappa)$, only if the Hurst exponent lies in the range

$$-(n+1)/4 < H < 0. \tag{14}$$

Power law processes with Hurst exponents that satisfy Eq. (14) are called *fractional noises* (Mandelbrot and Van Ness, 1968). Note that Eq. (12) holds within the fractal window $r_0 \leq r \leq r_m$. If it is extrapolated outside the scaling range, it leads to a divergence near zero due to the negative value of the Hurst exponent. This behavior is not meaningful, unless the system is in a critical state that has correlations extending throughout the entire system. Covariance models with non-singular behavior at the origin and asymptotic power law behavior can be constructed; such a model is given by means of the expression

$$c_x(r) = \sigma_x^2 (1 + r^2/\xi^2)^H. \tag{15}$$

The model is asymptotically scaling, i.e., $c_x(\lambda r) = \lambda^{2H} c_x(r)$ for $r >> \xi$. □

EXAMPLE 5: An interesting situation are fractal models with *homogeneous increments*. The covariance offers a useful two-point correlation function for a negative Hurst exponent, i.e., as in Eq. (14). If the Hurst exponent is positive ($\gamma > n$), the spectral density is not integrable due to the infrared divergence. When the spectral density exponent satisfies

$$n < \gamma < n+2, \tag{16}$$

$\tilde{\varpi}_x(\kappa)$ is a permissible spectral density for the semi-variogram of a S/TRF with homogeneous increments. The semi-variogram is obtained from the spectral density by means of the following expression

$$\gamma_x(\boldsymbol{r}) = \int_{\boldsymbol{\kappa}} \tilde{\varpi}_x(\kappa)[1 - \exp(i\,\boldsymbol{\kappa} \cdot \boldsymbol{r})]. \tag{17}$$

The integral in Eq. (17) converges at both limits if the exponent of the spectral density satisfies the inequality (16). Hence, the corresponding range of the Hurst exponent is

$$0 < H < 1. \tag{18}$$

A Hurst exponent in the above range is characteristic of fractional Brownian motions. The real space semi-variogram obtained from the spectral density is a power law

$$\gamma_x(\boldsymbol{r}) = \eta_n\, A\, r^{2H}, \quad r_0 \leq r \leq r_m,, \tag{19}$$

where the η_n are constant numbers (Isichenko, 1992). □

Anisotropic Models: Anisotropic covariance functions are used to model directional heterogeneity. The FT of such models is expressed as follows

$$\tilde{c}_x(\boldsymbol{\kappa}) = \sigma_x^2 (\textstyle\prod_{i=1}^{n} \xi_i) \tilde{f}(\kappa_1\, \xi_1, \ldots, \kappa_n\, \xi_n) \tag{20}$$

where the ξ_i, $i = 1,\ldots,n$ are the correlation lengths in the principal directions. By means of the rescaling transformations $\xi_i / \xi_1 = \lambda_i$ and $\kappa_1' = \kappa_1$, $\kappa_i' = \lambda_i\, \kappa_i$ for $i = 2,\ldots,n$, the anisotropic covariance can be expressed as the following isotropic covariance function

$$\tilde{c}_x(\boldsymbol{\kappa}') = \sigma_x^2 \, \xi_1^n \left(\prod_{i=2}^n \lambda_i\right) \tilde{f}'(\kappa' \xi_i), \tag{21}$$

where $\tilde{f}'(\kappa' \xi_i) = \tilde{f}(\kappa \xi_i)$. Isotropic covariance functions are more convenient for analytical calculations.

EXAMPLE 6: Anisotropic covariances are direct extensions of the isotropic models. As an example we apply the rescaling transformation to the three-dimensional anisotropic exponential covariance

$$\tilde{c}_x(\kappa) = \frac{8\pi \sigma_x^2 \prod_{i=1}^3 \xi_i}{(1+\sum_{i=1}^3 \kappa_i^2 \, \xi_i^2)^2}. \tag{22}$$

We set $\kappa_1' = \kappa_1$, $\xi_2/\xi_1 = \lambda_2$, $\xi_3/\xi_1 = \lambda_3$, $\kappa_2' = \lambda_2 \, \kappa_2$ and $\kappa_3' = \lambda_3 \, \kappa_3$, in view of which the covariance becomes isotropic in the primed coordinate system

$$\tilde{c}_x(\kappa') = \lambda_2 \, \lambda_3 \frac{8\pi \xi_1^3 \, \sigma_x^2}{(1+\kappa'^2 \, \xi^2)^2} \tag{23}$$

An application of the rescaling technique in calculations of the effective hydraulic conductivity in porous media is found in Hristopulos and Christakos (1997). □

Anisotropic Fractal Models: An extension of Eq. (15) leads to the following anisotropic covariance model

$$c_x(r) = \sigma_x^2 (1+\sum_{i=1}^n r_i^2/\xi_i^2)^H, \tag{24}$$

where $-(n+1)/4 < H < 0$. In Eq. (24) a single exponent characterizes the asymptotic scaling. It is also possible for the power-law exponents to vary in different directions. A covariance model of this type has been used in Makse *et al.* (1996a), namely,

$$c_x(r) = \sigma_x^2 \sum_{i=1}^n (1/n + r_i^2)^{-\beta_i/2}, \tag{25}$$

where the β_i are the (positive) directional exponents. The model (24) is asymptotically scaling, i.e., it satisfies the relation $c_x(\lambda \, r) = \lambda^{2H} \, c_x(r)$ --if distances are measured in the rescaled coordinate system. This is not true of the second model (25), because the power-law exponents vary with the direction. Models that have different scaling exponents in different directions are called *self-affine* (Mandelbrot, 1982).

Above we have presented a number of covariance functions that are useful in modelling environmental health processes. In the next section we present two examples of covariance functions that are obtained explicitly by solving physical models.

7. NATURAL MODELS AND COVARIANCE FUNCTIONS

The separable covariance model, in spite of its simplicity, is not always physically motivated. However, even in cases where separability is supported by explicit covariance calculations, it may still provide a practical approximation. In the following, we discuss separability within the framework of the diffusion equation and the invasion percolation model.

Let us consider an S/TRF $X(s,t)$ which satisfies the following stochastic PDE (SPDE)

$$\partial X(s,t)/\partial t = \mathcal{L}_s[X(s,t)], \tag{1}$$

where $\mathcal{L}_s$ is a linear spatial differential operator. We seek solutions $X(s,t)$ of this equation in the form $X(s,t) = X_1(s)X_2(t)$, that are given in general by

$$X(s,t) = \sum_n A_n \chi_{1n}(s)\chi_{2n}(t), \tag{2}$$

where $\chi_{1n}(s)$ and $\chi_{2n}(t)$ represent eigenfunctions (modes) of the SPDE, and A_n are random variables to be determined from the initial and boundary conditions. Randomness in the above example can be introduced by: (i) the initial or boundary conditions leading to random coefficients A_n; (ii) the differential operator $\mathcal{L}_s$ leading to random eigenfunctions $\chi_{1n}(s)$; and (iii) by both of the above. The covariances in each case are given by:

$$\text{Case I:} \quad C_x(s,t;s',t') = \sum_{n,m} c_{nm}\chi_{1n}(s)\chi_{1m}(s')\chi_{2n}(t)\chi_{2m}(t'); \tag{3}$$

the coefficients c_{nm} represent correlations of the mode coefficients, i.e., $c_{nm} = \overline{A_n A_m}$.

$$\text{Case II:} \quad C_x(s,t;s',t') = \sum_{n,m} A_n A_m c_{\chi(n,m)}(s,s')\chi_{2n}(t)\chi_{2m}(t'); \tag{4}$$

the two-point function $c_{\chi(n,m)}(s,s')$ denotes the correlation $\overline{\chi_{1n}(s)\chi_{1m}(s')}$.

Case III: $C_x(s,t;s',t') = \sum_{n,m} \overline{A_n A_m \chi_{1n}(s)\chi_{1m}(s')}\chi_{2n}(t)\chi_{2m}(t'),$ (5)

The solutions may not be homogeneous-stationary due to a number of reasons including the boundary and initial conditions. Naturally, the departure from homogeneity-stationarity is more pronounced near the boundaries. In all three cases above the covariance functions are expressed as a sum over modes, each mode including different functions with separable spatial and temporal components. If one mode is dominant, it is justified to use a separable covariance model. Even if a single mode representation is not accurate, the covariance function may in certain cases still be approximated in terms of separable components.

EXAMPLE 1: We examine the one-dimensional *diffusion* equation in a domain of length L

$$\partial X(s,t)/\partial t = D\, \partial^2 X(s,t)/\partial s^2, \tag{6}$$

where $X(s,t)$ denotes the concentration, and D denotes the diffusion coefficient. We assume the boundaries do not permit the concentration to escape (impervious BC), i.e,

$$\partial X(L,t)/\partial s = \partial X(0,t)/\partial s = 0, \tag{7}$$

and an initial concentration profile determined by the following function

$$X(s,0) = f(s). \tag{8}$$

The concentration $X(s,t)$ should remain bounded for large times. The solution of this diffusion equation is given by (Pipes and Harvill, 1970, pp. 412-416)

$$X(s,t) = \sum_{n=0}^{\infty} A_n e^{-(Dn^2\pi^2 t/L^2)} \cos(n\pi s/L), \tag{9}$$

where the mode coefficients A_n are determined from the initial profile by means of

$$A_n = 2[\int_0^L f(s)\cos(n\pi s/L)ds]/L, \tag{10}$$

for all mode numbers $n \geq 1$, and

$$A_0 = \int_0^L f(s)ds/L. \tag{11}$$

For all practical purposes, the infinite series for the concentration field can be truncated at a finite number of modes. At $t = 0$ the number of modes is determined by the accuracy that we require for the approximation of $f(s)$. As the time increases the number of significant modes is reduced, because modes decay exponentially and mode numbers $n >> L(t\,D)^{-1/2}/\pi$ become insignificant. As a specific example, we investigate a parabolic initial concentration profile

$$f(s) = c_0\,[1/12 + (s^2/L^2)(2 - s/3L)]. \qquad (12)$$

The mode coefficients for the above profile are given by $A_n = \alpha_n\, c_0$, where

$$\alpha_0 = 2/3, \text{ and } \alpha_n = 6(-1)^n/(n^2\,\pi^2) - 8\,\delta_{n,2\ell+1}/(n^4\,\pi^4), \quad \text{for } n \neq 1. \qquad (13)$$

In Fig. 2 we plot the initial concentration profile and approximations by means of truncated

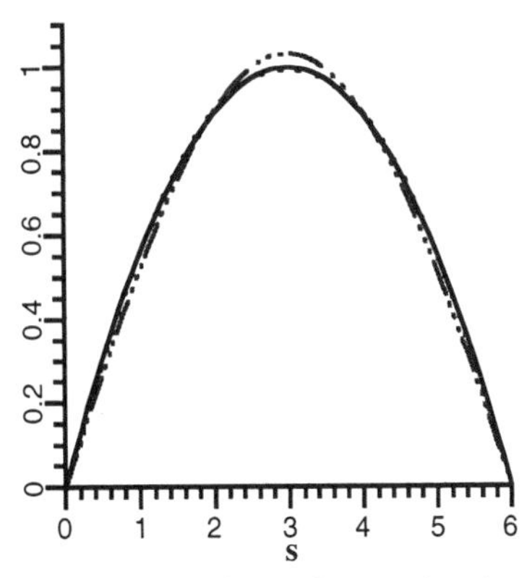

FIGURE 2: Plot of initial concentration profile, and approximations using a truncated cosine series with three (-···-) and ten modes (···); $c_0 = 1$.

cosine series: the concentration profile is represented quite well by using three modes, while the ten-mode approximation is very accurate. In Fig. 3 we show the effect of diffusion on the initial profile by plotting the concentration at three different time instants. Fig. 3 also shows that diffusion tends to homogenize the initial concentration profile. Consider a random initial profile, where c_0 is a random variable with second moment given by $c_2 > 0$. The concentration covariance is then given by

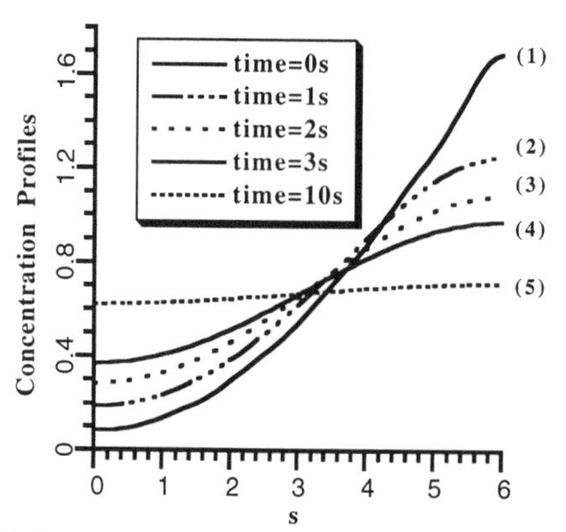

FIGURE 3: Plot of the diffusing concentration profile for five different time instants using a ten mode approximation; $D = 1cm^2 s^{-1}$.

$$C_x(s,s';t,t') = c_2 \sum_{n,m} \alpha_n \alpha_m \cos(n\pi s/L)\cos(m\pi s'/L) e^{-(D\pi^2 t_{nm}/L^2)}, \tag{14}$$

where $t_{nm} = tn^2 + t'm^2$. Next, we consider a uniformly distributed random diffusion coefficient $D \in [D_1, D_2 > D_1]$. The covariance in this case is

$$C_x(s,t;s',t') = \frac{c_0^2 L^2}{2\pi^2 \delta D} \sum_{n,m} \alpha_n \alpha_m \cos(n\pi s/L)\cos(m\pi s'/L)\; e^{-(\overline{D}\pi^2 t_{nm}/L^2)}$$
$$\sinh(\pi^2 \overline{D}\, t_{nm}/L^2)/t_{nm} \tag{15}$$

for all $t,t' > 0$, where $\overline{D} = (D_1 + D_2)/2$ and $\delta D = (D_2 - D_1)/2$. For the purpose of comparing with the previous case we assume that $\overline{D} = D$ and $c_2 = c_0^2(\overline{D}/\delta D)$. Since $c_2 \geq c_0^2$, the last equation constrains the diffusion coefficient to have a COV less than one. The difference between the two expansions is that Eq. (23) involves the mode-dependent factor $g_{nm}(t,t')$

$$g_{nm}(t,t') = \sinh(\lambda t_{nm})/(2\lambda t_{nm}), \tag{16}$$

where $\lambda = \pi^2 \overline{D}/L^2$. Since $g_{nm}(t,t') \geq 1$, a random diffusion coefficient slows down the homogenization of the initial concentration profile. □

EXAMPLE 2: A different type of S/TRF includes models of growth and pattern formation in which the spatiotemporal evolution is governed by a set of dynamic rules instead of a differential equation; (for a collection of essays on models of growth and pattern formation see Stanley and Ostrowski, 1986). Invasion percolation (Wilkinson and Willemsen, 1983) which describes the displacement of a defending viscous fluid in a porous medium (e.g., oil) by an invading fluid of lower viscosity (e.g., water) is an example of such a dynamic model with applications in oil recovery. When the flow rate is slow the invasion process is dominated by capillary forces. In a lattice simulation of invasion percolation each site is assigned a random number which represents the pore size and is inversely proportional to the capillary pressure. At each time step the sites occupied by the defending fluid (e.g., oil) that have neighbor sites occupied by the invader (e.g., water) are identified as growth sites. The invading fluid advances to the growth site with the smallest pore size. The invading water phase may completely surround sites of the oil phase. Since oil is essentially incompressible, trapped oil can not be removed by the invading fluid, and thus a residual oil saturation forms. For a stochastic analysis of the invasion percolation process we use the indicator function $I(\boldsymbol{s},t)$. If the site $\boldsymbol{s}$ is invaded at time t, then $I(\boldsymbol{s},t)=1$ and $I(\boldsymbol{s},t')=0$ for all times $t \neq t'$. Furuberg *et al.* (1988) have estimated, based on computer simulations, the conditional probability $C_{I-P}(\boldsymbol{s},\boldsymbol{s}_0;t,t_0)=P[I(\boldsymbol{s},t)=1 | I(\boldsymbol{s}_0,t_0)=1]$ that the site $\boldsymbol{s}$ is invaded at time t given that a reference site $\boldsymbol{s}_0$ was invaded at time t_0. They found that $C_{I-P}(\boldsymbol{s},\boldsymbol{s}_0;t,t_0)$ satisfies the dynamic scaling form

$$C_{I-P}(\boldsymbol{r},\tau)=r^{-1}\,g(r^z/\tau), \tag{17}$$

where $r=|\boldsymbol{s}-\boldsymbol{s}_0|$ and $z=1.82$. The function $g(x)$ peaks at $x \cong 1$ and scales asymptotically as a power-law

$$g(x) \sim \begin{cases} x^{\alpha}, & x<<1 \\ x^{-b}, & x>>1 \end{cases}, \tag{18}$$

with exponents $\alpha \cong 1.4$ and $b \cong 0.6$. We approximate $g(x)$ with the following function

$$g'(x)=c_1\,x^{\alpha}\,\theta(x_0-x)+c_1\,x^{-b}\,\theta(x-x_0), \tag{19}$$

such that $g'(x)$ peaks at $x=x_0$ and $g'(x)=g(x)$ for large (small) x. The function $g'(x)$ has a cusp at $x=x_0$ and two branches with different exponents around the cusp (Fig. 4). At first glance one would think that the above covariance is separable, since any power-law

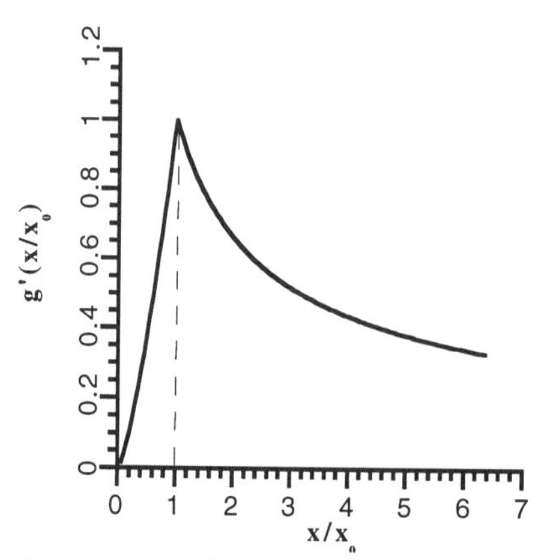

FIGURE 4: Plot of the function $g'(x)$ vs. the normalized scaling variable x/x_0.

in x where $x = r^z/\tau$, can be expressed as the product of power laws in r and τ. At first glance one would think that the above covariance is separable, since any power-law in x where $x = r^z/\tau$, can be expressed as the product of power laws in r and τ. However, the separability breaks down due to the presence of the cusp: If τ is held fixed and r is increased so that the x crosses over the cusp from the left to the right branch, the characteristic exponent of the spatial power law changes. Hence, the covariance function can not be expressed as the product of two separable components. □

8. REGRESSION S/TRF MODEL

8.1. Mathematical Formulation

The regression S/TRF model is defined as (Bogaert and Christakos, 1997a and 1997b)

$$X(s,t) = Y(s,t) + M_1(s) + M_2(t) + \mu(s,t), \tag{1}$$

where $Y(s,t)$ is a space homogeneous and time stationary random component, $M_1(s)$ is a purely spatial random component, $M_2(t)$ is a purely temporal random component, and $\mu(s,t)$ denotes a deterministic function of space and time. The three random components have zero mean and are assumed to be mutually independent. The deterministic component $\mu(s,t)$, which represents the mean value of the process, will be expressed in terms of a linear regression model, the parameters of which are functions of the space-time coordinates. Moreover, for estimation purposes the $M_1(s)$ and $M_2(t)$ components are assumed to have space homogeneous and time stationary increments respectively, whereas the $Y(s,t)$ component has a space-time separable covariance function of the form

$$c_y(\boldsymbol{r},\tau)=\sigma_y^2\,\rho_{y,s}(\boldsymbol{r}_{ij})\,\rho_{y,t}(\tau_{k\ell}), \tag{2}$$

where $\rho_{y,s}(\boldsymbol{r}_{ij})$ and $\rho_{y,t}(\tau_{k\ell})$ are spatial and temporal correlation functions, respectively, and $\boldsymbol{r}_{ij}=\boldsymbol{s}_i-\boldsymbol{s}_j$, $\tau_{k\ell}=t_k-t_\ell$ are spatial and temporal lags, respectively.

If the $M_1(s)$ is a homogeneous and $M_2(t)$ is a stationary random field, the space-time covariance function $c_x(\boldsymbol{r},\tau)$ is considered; otherwise, the semi-variogram (structure) function $\gamma_x(\boldsymbol{r},\tau)$ will be used. In light of Eq. (1), these two functions are given by

$$c_x(\boldsymbol{r},\tau)=c_y(\boldsymbol{r},\tau)+\mathrm{c}_{\mathrm{M}_1}(\boldsymbol{r})+\mathrm{c}_{\mathrm{M}_2}(\tau), \tag{3}$$

and

$$\gamma_x(\boldsymbol{r},\tau)=\gamma_y(\boldsymbol{r},\tau)+\gamma_{\mathrm{M}_1}(\boldsymbol{r})+\gamma_{\mathrm{M}_2}(\tau), \tag{4}$$

respectively. Model (1) includes, as special cases, certain existing space-time models (such as those proposed by Bilonick, 1985). Furthermore, model (1) takes into consideration interesting features of physical data, which cannot be accounted for by some other of the existing models.

EXAMPLE 1: In many applications the variances computed along the space and time axes differ by one order of magnitude. To deal with the situation, attempts have been made to establish a joint representation of the spatial and temporal dependencies using models of anisotropy between the space and time axes. This approach, however, is very restrictive and often difficult to apply. The fact that variances appear to be different along the space and time axes has led some authors to consider superpositions of purely spatial and purely temporal processes. Unfortunately, the usefulness of this model is limited in practice, for it assumes a very specific space-time decomposition of variability that is not representative of most physical data sets. Moreover, it can lead to serious problems when used for space-time estimation purposes. The regression S/TRF model (1), on the other hand, incorporates three different random components and a spatiotemporal mean, which allow it to handle physical data sets with significantly different spatial and temporal variances. Each of the components in model (1) can be associated with a particular aspect of the space-time variation. The purely temporal $M_2(t)$ component can be viewed as the result of a physical mechanism acting at such a large spatial scale that no significant spatial variability is observed. The purely spatial $M_1(s)$ component is associated with effects occurring at large time scales, so that the temporal variation is negligible over the time period of interest. The space-time $Y(s,t)$ component accounts for interactions at different spatial locations and time instants. The mean $\mu(s,t)$ is generally a function of space and time coordinates. It can, thus, account for systematic spatial effects, seasonal fluctuations, etc. of the observed natural processes. In practice, $\mu(s,t)$ is usually expressed in terms of space-time polynomials, but other functions can be used, as well. □

For certain data sets, the separation into spatial and temporal random variations may be inappropriate. But model (1) should still be useful, as the comparison of the relative importance of these components could provide information about the natural mechanism involved at the spatial and the temporal scales.

EXAMPLE 2: Consider the case where very similar time series are observed at different spatial locations, regardless of the distances between them. Fitting the model (1) to the data will reveal a dominant temporal $M_2(t)$ component, implying that the natural process exhibits significant variation in time but changes little over space. □

Finally, while the assumptions underlying model (1) may restrict its general applicability, they reduce considerably the computational demands of space-time analysis, by allowing simplifications in the calculation of the correlation functions and the formulation of the estimation systems.

8.2. Statistical Inference Issues

When dealing with real data, the components of the S/TRF regression model are not directly distinguishable. One observes only the resulting S/TRF $X(s,t)$ and, hence, it is impossible to infer the semi-variogram of the various components directly from the data. An additional problem stems from the fact that the mean $\mu(s,t)$ is unknown, even if we assume a known parametric form. Nevertheless, under certain conditions in practice it is possible to construct unbiased estimators of the component semi-variograms. Below, we assume that the n_x distinct spatial locations remain the same for all n_t distinct time instants.

We will start by studying the semi-variogram $\gamma_y(\boldsymbol{r},\tau)$ of the space-time component $Y(\boldsymbol{s},t)$. Let us focus on the spatial dependence obtained by setting $\tau=0$, in order to get $\gamma_y(\boldsymbol{r},0)$. An unbiased estimator $\hat{\gamma}_y(\boldsymbol{r},0)$ of $\gamma_y(\boldsymbol{r},0)$ can be obtained by

$$\hat{\gamma}_y(\boldsymbol{r},0)=\frac{1}{2\kappa_t n_t|N(\boldsymbol{r})|}\sum_{j=1}^{n_t}\sum_{pairs=1}^{|N(\boldsymbol{r})|}\left[\hat{Y}(\boldsymbol{s}+\boldsymbol{r},t_j)-\hat{Y}(\boldsymbol{s},t_j)\right]^2, \tag{5}$$

where κ_t is a correction factor, $N(\boldsymbol{r})$ is the number of point pairs separated in space by the vector $\boldsymbol{r}$, and $\hat{Y}(\boldsymbol{s}+\boldsymbol{r},t_j)$, $\hat{Y}(\boldsymbol{s},t_j)$ are residuals. The difference $\hat{Y}(\boldsymbol{s}+\boldsymbol{r},t_j)-\hat{Y}(\boldsymbol{s},t_j)$ is defined as

$$\hat{Y}(\boldsymbol{s}+\boldsymbol{r},t_j)-\hat{Y}(\boldsymbol{s},t_j)=\left(X(\boldsymbol{s}+\boldsymbol{r},t_j)-X(\boldsymbol{s},t_j)\right)-\left(\hat{v}(t_j|\boldsymbol{s}+\boldsymbol{r})-\hat{v}(t_j|\boldsymbol{s})\right), \tag{6}$$

where the $\hat{v}(t_j|\boldsymbol{s})$ is a linear combination of the $X(\boldsymbol{s},t)$-values along the time axis, i.e.,

$$\hat{v}(t_j|\boldsymbol{s})=\sum_{j'=1}^{n_t}w_{jj'}X(\boldsymbol{s},t_{j'}), \tag{7}$$

with $\hat{v}(t_j|\boldsymbol{s})=\{M_1(\boldsymbol{s})+\mu(\boldsymbol{s},t_j)\}$, where the brackets "{.}" mean that the two components cannot be estimated separately along the time axis. Unbiased estimates require that $E_t[\hat{v}(t_j|\boldsymbol{s})]=v(t_j|\boldsymbol{s})$, where $E_t[.]$ designates expectation of the time distribution; this leads to the following conditions on the weights,

$$\left.\begin{aligned}&\sum_{j'=1}^{n_t}w_{jj'}=1\\&\sum_{j'=1}^{n_t}w_{jj'}\mu(\boldsymbol{s},t_{j'})=\mu(\boldsymbol{s},t_j)\end{aligned}\right\}. \tag{8}$$

The first condition is linked to the $M_1(\boldsymbol{s})$ component, which is constant along the time axis. The second condition is due to the nonhomogeneity-nonstationarity of the mean function $\mu(\boldsymbol{s},t)$. The term $M_2(t_j)$ has been filtered out in Eq. (6). The value of κ_t is determined in terms of the bias introduced in the estimation of $\gamma_y(\boldsymbol{r},0)$ due to use of the estimator $\hat{Y}(\boldsymbol{s},t_j)$ instead of the actual $Y(\boldsymbol{s},t_j)$. Indeed, since the $Y(\boldsymbol{s},t)$ component has a space-time separable covariance, it can be shown that (Bogaert and Christakos, 1997a and 1997b)

$$\kappa_t=1-trace\left(\boldsymbol{\Sigma}_{y,\tau}\boldsymbol{W}_t\right)/n_t, \tag{9}$$

where W_t is a matrix of suitable weights and $\boldsymbol{\Sigma}_{y,\tau}$ is a matrix of temporal correlations. Taking advantage of the symmetries, an unbiased estimate $\hat{\gamma}_y(\boldsymbol{0},\tau)$ of the temporal semi-variogram $\gamma_y(\boldsymbol{0},\tau)$ is obtained in a completely analogous way using a correction factor κ_s.

After the estimates $\hat{\gamma}_y(\boldsymbol{r},0)$ and $\hat{\gamma}_y(\boldsymbol{0},\tau)$ have been obtained, separate parametric models can be assumed for the spatial and temporal semi-variograms $\gamma_y(\boldsymbol{r},0;\boldsymbol{\theta}_r)$ and $\gamma_y(\boldsymbol{0},\tau;\boldsymbol{\theta}_\tau)$, respectively, where $\boldsymbol{\theta}_r$ and θ_τ are vectors of parameters. Similarly, one can define the spatial and temporal covariance functions $\rho_y(\boldsymbol{r},0;\boldsymbol{\theta}_r)$ and $\rho_y(\boldsymbol{0},\tau;\boldsymbol{\theta}_\tau)$. According to the separability hypothesis, a parametric S/T covariance function model $c_y(\boldsymbol{r},\tau;\boldsymbol{\theta})=\sigma_y^2\rho_y(\boldsymbol{r},0;\boldsymbol{\theta}_r)\rho_y(\boldsymbol{0},\tau;\boldsymbol{\theta}_\tau)$ can be used.

The fact that the mean is generally nonhomogeneous-nonstationary makes the estimation of the $M_1(s)$ and $M_2(t)$ components difficult. A parametric approach using the Maximum Likelihood (ML) method (Kitanidis, 1983) along the space and time axes can provide unbiased estimates. In this case, the fitted estimates $\gamma_y(\boldsymbol{r},0;\hat{\boldsymbol{\theta}}_r)$ and $\gamma_y(\boldsymbol{0},\tau;\hat{\boldsymbol{\theta}}_\tau)$ of the spatial and temporal semi-variograms $\gamma_y(\boldsymbol{r},0)$ and $\gamma_y(\boldsymbol{0},\tau)$ can be included as fixed terms in the sums $\gamma_{M_1}(\boldsymbol{r},\boldsymbol{\theta}_1)+\gamma_y(\boldsymbol{r},0;\hat{\boldsymbol{\theta}}_r)$ and $\gamma_{M_2}(\tau,\boldsymbol{\theta}_2)+\gamma_y(\boldsymbol{0},\tau,\hat{\boldsymbol{\theta}}_\tau)$, respectively. The estimates $\hat{\boldsymbol{\theta}}_1$ and $\hat{\boldsymbol{\theta}}_2$ for the sums can be obtained conditionally to each space location and time instant, respectively (a hint regarding the model shape is provided by the semi-variogram of residuals), and the estimated parameters are averaged over space and time, respectively. The models $\gamma_{M_1}(\boldsymbol{r},\hat{\boldsymbol{\theta}}_1)$ and $\gamma_{M_2}(\tau,\hat{\boldsymbol{\theta}}_2)$ then follow.

In conclusion, in its most general form the S/TRF regression model requires the combination of a parametric model for the mean along the space and time axes, a non-parametric estimator for the semi-variogram of the $Y(s,t)$ component, and a ML approach for the estimation of the semi-variograms of the $M_1(s)$ and $M_2(t)$ components.

9. WAVE REPRESENTATION OF S/TRF

A travelling wave S/TRF representation (also known as the frozen field model; Johnson and Dudgeon, 1993). is given by

$$X(s,t)=X(s-vt,0)=Y(s-vt), \tag{1}$$

where v is a constant vector model parameter to be determined from the data. A continuous operation $\mathcal{Q}=\partial/\partial t+v\cdot\nabla$ can be associated with Eq. (1) so that $\mathcal{Q}[X]=0$.

A physical S/TRF theory implies that field equations such as the above constrain the values of environmental health processes in space-time.

The space homogeneous-time stationary covariance function associated with the S/TRF representation (1) is given by

$$c_x(\boldsymbol{r},\tau) = c_x(\boldsymbol{r}-\boldsymbol{v}\tau,0) = c_y(\boldsymbol{r}-\boldsymbol{v}\tau). \tag{2}$$

This implies that the covariance $c_x(\boldsymbol{r},\tau)$ satisfies the equation

$$\boldsymbol{H}_c\boldsymbol{v} = -\nabla\,\partial c_x(\boldsymbol{r},\tau)/\partial\tau, \tag{3}$$

where $\boldsymbol{H}_c$ is the Hessian matrix of the spatial derivatives of $c_x(\boldsymbol{r},\tau)$. The Eq. (3) permits to determine the vector $\boldsymbol{v}$ from the covariance function.

EXAMPLE 1: In $R^2 \times T$, Eq. (3) yields

$$\boldsymbol{v} = \begin{bmatrix} v_1 \\ v_2 \end{bmatrix} = -\begin{bmatrix} \partial^2 c_x(\boldsymbol{r},\tau)/\partial r_1^2 & \partial^2 c_x(\boldsymbol{r},\tau)/\partial r_2\partial r_1 \\ \partial^2 c_x(\boldsymbol{r},\tau)/\partial r_2\partial r_1 & \partial^2 c_x(\boldsymbol{r},\tau)/\partial r_2^2 \end{bmatrix}^{-1} \begin{bmatrix} \partial^2 c_x(\boldsymbol{r},\tau)/\partial\tau\partial r_1 \\ \partial^2 c_x(\boldsymbol{r},\tau)/\partial\tau\partial r_2 \end{bmatrix}. \tag{4}$$

The vector $\boldsymbol{v}$ can be calculated from the Eq. (4) anywhere on the covariance surface $c_x(\boldsymbol{r},\tau)$, except where the Hessian determinant, $Det\,\boldsymbol{H}_c$, vanishes. For the covariance to be in the form of Eq. (2) the result of Eq. (4) must be independent of space and time. □

Chapter IV: Modelling Exposure Heterogeneities

"The mind is not a vessel to be filled but a fire to be kindled".
Plutarch

1. A CLASS OF SPATIOTEMPORAL RANDOM FIELDS

1.1 The Central Idea and Definitions

Most scientists would agree if we describe models, by paraphrasing Picasso's words, as "lies that help us to see the truth". Models are incomplete representations of reality, but they help to reveal important features of natural and biological phenomena and, in certain cases, they lead to useful predictions. Successful models are based on physically meaningful concepts and adequate mathematical formulations. Reliable prediction requires accurate estimation of model parameters from the available information. The S/TRF model provides a flexible representation of environmental health processes that incorporates the natural and biological variabilities of such processes with mathematical power.

Space-time analysis for environmental health processes with complicated space-time patterns entails determining the S/TRF from a single realization. This problem is not well defined mathematically, since there is no *a priori* criterion for distinguishing between space-time trends and fluctuations. If other information about the trends is not available, variability due to trends may be modeled as fluctuations, thus leading to an artificial increase in the fluctuation strength, as we saw in a previous example (Chapter III). In light of this statistical indeterminacy, trend-free (homogeneous and stationary) S/TRF have been assumed widely in environmental health modelling, because they are efficient for explicit and numerical calculations. However, the homogeneous-stationary model is not the best option for phenomena with large and complicated space-time variabilities. In light of these issues, a class of S/TRFs is discussed in this Chapter, that is considerably more general than the restricted class of homogeneous-stationary S/TRF. This new class is capable of handling complicated space-time variabilities of any size in a mathematically rigorous and physically meaningful manner. The mathematical theory of the nonhomogeneous-nonstationary S/TRF to be presented below is based on the following central idea.

Central idea: *The variability of an S/TRF can be characterized by means of its degree of departure from homogeneity and stationarity.*

Essentially, what we are saying is that the hypothesis of homogeneity-stationarity is very useful for explicit calculations, and we would like to be able to exploit it, to a certain extent, even in the study of S/TRF that lack such symmetries. This can be accomplished in practice by means of a mathematical operation on the S/TRF that eliminates the nonhomogeneous and nonstationary parts. The departure of a S/TRF from homogeneity-stationarity determines the order of this mathematical operation, in a sense that will be defined below. At this point, we introduce a space-time *detrending operator* Q that transforms the S/TRF $X(s,t)$ into a homogeneous-stationary field $Y(s,t)$, i.e.,

$$Q[X(s,t)] = Y(s,t). \tag{1}$$

Hence, the Q-operator removes spatial and temporal trends from the random field. If the multivariate probability distribution of $X(s,t)$ is known *a priori*, it is possible, at least in principle, to construct the Q-operator that yields a strictly homogeneous-stationary S/TRF $Y(s,t)$. In reality, of course, the multivariate distribution is never available, and in most applications a second-order representation of the random field $X(s,t)$ is employed. Then, the Q-operator generates a second-order (wide sense) homogeneous-stationary random field $Y(s,t)$. For environmental health processes that evolve within domains containing complicated boundaries, physical trends and noise sources, the departure of the random field from homogeneity-stationarity is expected to vary locally. It is, thus, more meaningful to construct local Q-operators that generate homogeneous-stationary random fields $Y(s,t)$ within local neighborhoods Λ, instead of seeking global representations. The size of the space-time neighborhood should be determined based on the existing data for the environmental health process (this is discussed in more detail in Chapters V and VI).

The spatiotemporal trend of $X(s,t)$ inside a local neighborhood Λ may be represented using polynomials (although other functions can be used, as well). The coefficients of the polynomials are constant within the local neighborhood. The partial detrending operator removes these polynomial trends, in order to generate a homogeneous-stationary field $Y(s,t)$. For continuum S/TRF the local detrending Q-operator is represented by a linear differential operator that eliminates the polynomial trends in $\overline{X(s,t)}$. In the case of lattice S/TRF --which are routinely used in simulations-- the differential operators are replaced with space-time finite differences.

In order to implement these ideas, some rather sophisticated mathematical analysis is necessary. But first, we discuss briefly some matters of notation.

- The variability of S/TRF is in general characterized by two integer indices, ν for space and μ for time, which are called *continuity orders*. Their values determine the degree of departure from homogeneity and stationarity.
- Polynomial functions appear naturally in the formalism of S/TRF. We use the symbol $p_{\nu/\mu}(s,t)$ to denote a polynomial of degrees ν in space and μ in time. By standard convention, the index values $\nu = \mu = -1$ correspond to spatially homogeneous-temporally stationary random fields (no trends). The case $\nu = \mu = 0$ denotes a S/TRF with homogeneous-stationary increments; random fields of this type may have linear trends in space and time that are due to the mean of the increment $Y(s,t)$. If the mean of the increment is zero, $\overline{X(s,t)}$ is constant.
- For S/TRF with continuity orders ν/μ the Q-operator involves spatial derivatives of order $\nu+1$ and time derivatives of order $\mu+1$. Usually, the Q-operator is local, i.e., it operates at a specific point in space and time. The symbol $Q_{(s,t)}^{(\nu+1/\mu+1)}[\cdot]$ will be used to denote a Q-operator that involves spatial derivatives of order $\nu+1$ and time derivatives of order $\mu+1$ operating at the point (s,t).

In the following, the symbols for the continuity orders and the point of operation are dropped if their values are obvious from context. A general definition of an S/TRF of order ν/μ follows (Christakos, 1991; 1992).

Definition 1: *If a Q-operator exists so that all the random fields*

$$Y(s,t) = Q_{(s,t)}^{(\nu+1/\mu+1)}[X(s,t)] \tag{2}$$

are homogeneous-stationary, the random field $X(s,t)$ is said to be an S/TRF of order ν/μ (S/TRF-ν/μ). Without loss of generality, the residual random field $Y(s,t)$ is usually assumed to have zero mean.

Certain interesting questions are raised by this definition. What is the nature of Q-operators? As we discussed above, we are interested in differential operators that eliminate the trends. Is the detrending operator unique? The answer to this is negative: if one detrending operator exists, then differential operators of higher order will also generate homogeneous-stationary residuals upon acting on $X(s,t)$. Linear operators are more convenient to work with, although other Q-operators are also possible.

Assume that Q is the linear and homogeneous space-time differential operator of order ν/μ given by

$$Q^{(\nu+1/\mu+1)} = \sum_{|\boldsymbol{\rho}|=\nu+1} \alpha_{\boldsymbol{\rho}/\mu}\, \partial^{|\boldsymbol{\rho}|+\mu+1} \big/ \partial s^{|\boldsymbol{\rho}|}\, \partial t^{\mu+1} . \tag{3}$$

In the above, $\boldsymbol{\rho}$ denotes the set of integers $\{\rho_1, \rho_2, \ldots, \rho_n\}$, ρ_i denotes the order of the partial derivative with respect to s_i, and $|\boldsymbol{\rho}|$ is the order of the differential operator, $|\boldsymbol{\rho}| = \sum_{i=1}^{n} \rho_i$; the $\alpha_{\boldsymbol{\rho}/\mu}$ denote constant coefficients, and $\partial^{|\boldsymbol{\rho}|}/\partial s^{|\boldsymbol{\rho}|}$ is a composite spatial differential operator given by

$$\partial^{|\boldsymbol{\rho}|}\big/\partial s^{|\boldsymbol{\rho}|} = \partial^{|\boldsymbol{\rho}|}\big/\partial s_1^{\rho_1}\, \partial s_2^{\rho_2} \ldots \partial s_n^{\rho_n} , \tag{4}$$

The differential operator (3) eliminates space-time polynomials of degree ν/μ. If the partial point derivatives of the $X(\boldsymbol{s},t)$ do not exist, the definition (3) can be extended using generalized random fields.

EXAMPLE 1: Detrending operators that are special cases of Eq. (3) include

$$Q^{(\nu+1/\mu+1)} = \partial^{\nu+\mu+2} \big/ \partial s_1^{\nu_1}\, \partial s_2^{\nu_2} \ldots \partial s_n^{\nu_n}\, \partial t^{\mu+1} , \tag{5}$$

where $\sum_{i=1}^{n} \nu_i = \nu+1$;

$$Q^{(\nu+1/\mu+1)} = \sum_{i=1}^{n} \partial^{\nu+\mu+2} \big/ \partial s_i^{\nu+1}\, \partial t^{\mu+1} ; \tag{6}$$

and

$$Q^{(\nu+1/\mu+1)} = \partial^{\mu+1}\big/\partial t^{\mu+1} + \sum_{i=1}^{n} \partial^{\nu+1}\big/\partial s_i^{\nu+1} . \tag{7}$$

They all eliminate space-time polynomials of degree ν/μ. □

EXAMPLE 2: Consider that $n=2$, $\nu=2$ and $\mu=1$. The operator $Q^{(3/2)}$ in Eq. (6) above is expressed as

$$Q^{(3/2)} = (\partial^3/\partial s_1^3 + \partial^3/\partial s_2^3)\, \partial^2/\partial t^2 .$$ □

REMARK 1: The Q-operators are not uniquely defined: If $Q^{(\nu+1/\mu+1)}$ is a differential operator of spatial order ν and temporal order μ, all operators $Q^{(\nu'+1/\mu'+1)}$ such that $\nu' > \nu$ and $\mu' > \mu$ are also admissible Q-operators.

1.2 Determination of the Q-Operator in Practice and its Physical Significance

As it was mentioned above, the physical significance of the Q-operator is the elimination of trends from the S/TRF $X(s,t)$. How is the form of the Q-operator determined in practice? In several situations the Q-operator can be derived on the basis of the physical law governing the phenomenon of interest. To gain some insight, let us study the following example.

EXAMPLE 3: Consider groundwater flow in a heterogeneous aquifer. The principle of mass conservation is expressed as (Bear, 1972)

$$\nabla \cdot [K(s)\nabla H(s,t)] = S_s \,\partial H(s,t)/\partial t + \varphi(s,t), \tag{8}$$

where $K(s)$ is the hydraulic conductivity, $H(s,t)$ the hydraulic head, S_s the specific storage coefficient, and $\varphi(s,t)$ is a source. In the steady state with no source term the mean hydraulic head is not renormalized by the hydraulic conductivity fluctuations (Hristopulos and Christakos, 1997); thus it satisfies the Laplace equation

$$\nabla^2 \overline{H(s)} = 0. \tag{9}$$

This implies that $H(s)$ is nonhomogeneous. Eq. (9) leads to

$$\nabla^2 H(s) = Y(s), \tag{10}$$

where $h(s) = H(s) - \overline{H(s)}$ and $Y(s) = \nabla^2 h(s)$ is a homogeneous residual. Hence, the Q-operator is given by the Laplacian ∇^2. The simplest solution that satisfies Eq. (9) is

$$\overline{H(s)} = H_0 + \boldsymbol{J} \cdot \boldsymbol{s}. \tag{11}$$

Nonpolynomial solutions of the Laplace equation are also possible for appropriate boundary conditions. □

Determining explicitly the detrending operator is not always possible as in the above example. In many situations the trends vary considerably over large domains or time intervals due to the interaction of various physical and chemical processes, complicated boundary conditions, and time-varying inputs. Then, the form of the detrending Q-operator changes in space-time and it can be considered invariant only within local neighborhoods. Due to this natural complexity, it is impossible to determine these neighborhoods and the form of the detrending operator without taking the data into account. Hence, in practice the neighborhoods and the local form of the Q-operator are provided by the model that leads to the optimal fit with the control data (the approach is discussed in more detail in a later section). Statisticians call this a "non-parametric approach", stressing the fact that no assumptions regarding the form of the Q-operator are made. This approach is powerful in exploring relationships between environmental exposure and health effects (e.g., ozone concentration-respiratory problems, particulate matter concentration-increased mortality rates), and it may also provide useful diagnostic tools based on these relationships. In the same context, the Q-operation can be viewed as a spatiotemporal *filter* at the observation scale Λ determined by the local neighborhood.

EXAMPLE 4: The operation

$$Y = Q[X] \tag{12}$$

is a high-pass filter that annihilates trends and enhances detail of the space-time pattern. Conversely, the

$$Q'[X] = X - Q[X] \tag{13}$$

is a low-pass filter (containing long-term trends and seasonal effects). □

In conclusion, the Q-operator is determined in practice either on the basis of a physical law (when available), or by fitting S/TRF models locally to the data. In the second case -- which is most common in practice-- Q is not directly related to an explicit physical mechanism (e.g., diffusion or convection). In fact, it may account for the effects of more than one mechanisms acting simultaneously. In such cases, when prior information regarding the laws that govern the process is insufficient, the only feasible approach is to let the data speak for themselves. At a later stage, it may be possible by examining the spatiotemporal patterns revealed by the data to formulate mechanistic hypotheses. This approach of starting with a mathematical structure which is later given physical meaning is common in science (e.g., Stratton, 1941; Longair, 1984).

1.3 An S/TRF Decomposition

To gain more understanding of the properties of an S/TRF- ν / μ consider the following

Decomposition: *Given the* S/TRF- ν / μ $X(s,t)$, *an S/TRF* $X_r(s,t)$ *with continuity orders less than or equal to* ν / μ *exists such that*

$$X(s,t) = X_r(s,t) + p_{\nu/\mu}(s,t). \tag{14}$$

The above decomposition puts the remarks in Example 3 --regarding the decomposition of the hydraulic head into mean and fluctuation components-- on a more formal basis. In light of Eq. (2), the Q-operator acting on $X_r(s,t)$ leads to the residual $Y(s,t)$ by eliminating the polynomial term, i.e.,

$$Q[X_r(s,t)] = Y(s,t), \tag{15}$$

$$Q[p_{\nu/\mu}(s,t)] = 0. \tag{16}$$

The polynomial $p_{\nu/\mu}(s,t)$ is given by

$$p_{\nu/\mu}(s,t) = \sum_{\zeta=0}^{\mu} \sum_{|\rho|=0}^{\nu} a_{\rho/\zeta}\, s_1^{\rho_1} s_2^{\rho_2} \ldots s_n^{\rho_n}\, t^{\zeta}, \tag{17}$$

where ν, μ, ζ and ρ_i are all integers. The coefficients of the space-time monomial $a_{\rho/\zeta}$ are, in general, random variables with mean value $\overline{a_{\rho/\zeta}} = m_{\rho/\zeta}$ and correlation $\overline{a_{\rho/\zeta}\, a_{\rho'/\zeta'}}$. In addition, the polynomial coefficients are not correlated with the S/TRF $X_r(s,t)$, i.e.,

$$\overline{a_{\rho/\zeta}\, X_r(s,t)} = 0 \text{ for all } \rho \text{ and } \zeta. \tag{18}$$

2. SPATIOTEMPORAL POLYNOMIAL NOTATION

We must digress for a moment to reflect upon the space-time polynomials introduced in the last section. The notation used in Eq. (1.17) above shows explicitly the degrees of all spatial and temporal monomials, but it is cumbersome, since it requires keeping track of all the power exponents ρ_i, $i = 1,...,n$ and ζ. The following alternative notation that is based on the monomial functions $g_\alpha(s,t)$ is more efficient. Let us define

$$g_\alpha(s,t) = (\prod_{i=1}^{n} s_i^{\rho_i})t^\zeta , \tag{1}$$

where $\sum_{i=1}^{n} \rho_i \leq \nu$, $\zeta \leq \mu$, and $\alpha = 1,...,N_n(\nu/\mu)$ -- $N_n(\nu/\mu)$ being the number of monomials. In view of the above, a polynomial of degree ν/μ can be expressed as

$$p_{\nu/\mu}(s,t) = \sum_{\alpha=1}^{N_n(\nu/\mu)} a_\alpha\, g_\alpha(s,t). \tag{2}$$

This notation is more compact than the one used in Eq. (1.17) above and involves only a single index α instead of the $n+1$ space-time indices $(\rho_1,...,\rho_n,\zeta)$. The number $N_n(\nu/\mu)$ depends on the spatial dimension n and the continuity orders ν/μ.

EXAMPLE 1: Consider the special case of $n = 2$ spatial dimensions, let (s_1, s_2) denote the Cartesian coordinates of the vector s in R^2, and let $\nu/\mu = 2/1$ be the space-time orders of continuity. The corresponding monomials are given in Table 1. □

TABLE 1: Monomials in R^2 with $\nu/\mu = 2/1$.

$\nu/\mu = 2/1$	ρ	ζ	Monomials
$g_1(s,t)$	0	0	$s_1^0\, s_2^0\, t^0$
$g_2(s,t)$	1	0	$s_1^1\, s_2^0\, t^0$
$g_3(s,t)$	1	0	$s_1^0\, s_2^1\, t^0$
$g_4(s,t)$	0	1	$s_1^0\, s_2^0\, t^1$
$g_5(s,t)$	1	1	$s_1^1\, s_2^0\, t^1$
$g_6(s,t)$	1	1	$s_1^0\, s_2^1\, t^1$
$g_7(s,t)$	2	0	$s_1^2\, s_2^0\, t^0$
$g_8(s,t)$	2	0	$s_1^0\, s_2^2\, t^0$
$g_9(s,t)$	2	0	$s_1^1\, s_2^1\, t^0$
$g_{10}(s,t)$	2	1	$s_1^1\, s_2^1\, t^1$
$g_{11}(s,t)$	2	1	$s_1^1\, s_2^0\, t^1$
$g_{12}(s,t)$	2	1	$s_1^0\, s_2^2\, t^1$

The $N_n(\nu/\mu)$ for any n and ν/μ is determined as follows: For $|\mathbf{p}| \leq \nu$ number of monomials is equal to the permutations of n integers from $\{0,1,\ldots,\nu\}$ that add up to $|\mathbf{p}|$, i.e., $(|\mathbf{p}|+n-1)!/|\mathbf{p}|!(n-1)!$ (Kubo *et al.*, 1981; p. 37). Hence, $N_n(\nu/\mu)$ is

$$N_n(\nu/\mu) = (\mu+1)\sum_{|\mathbf{p}|=0}^{\nu} \frac{(|\mathbf{p}|+n-1)!}{|\mathbf{p}|!(n-1)!}. \tag{3}$$

EXAMPLE 2: In Table 2 we show the number of monomials for the ν/μ combinations

TABLE 2: Number of monomials for $n = 2,3$ and different combinations of continuity orders.

ν	μ	$N_2(\nu/\mu)$	$N_3(\nu/\mu)$
0	0	1	1
1	0	3	4
0	1	2	2
1	1	6	8
2	0	6	10
2	1	12	20
1	2	9	12
2	2	18	30

most commonly used in applications. □

3. ORDERS OF SPATIOTEMPORAL CONTINUITY ν/μ

In practice, different space-time neighborhoods have different degrees of heterogeneity, which can be described only by means of S/TRF with varying orders of spatiotemporal continuity. As a consequence, the form of the spatiotemporal covariance function also varies in different neighborhoods. The neighborhoods should be chosen either objectively by means of the accuracy of the estimates they lead to, or subjectively by data inference.

REMARK 1: Strictly speaking, determining the orders of continuity ν and μ requires estimating the S/TRF properties based on a unique realization of $X(\mathbf{s},t)$. This is impossible for nonhomogeneous-nonstationary S/TRF which are nonergodic. Thus, a rigorous distinction between trends and fluctuations based on a single sample is not feasible. This indeterminacy is better understood in terms of an example.

EXAMPLE 1: Fig. 1 presents a least squares fit of a nonstationary random field realization in R^1 with a nonrandom polynomial model. The random field consists of white noise and

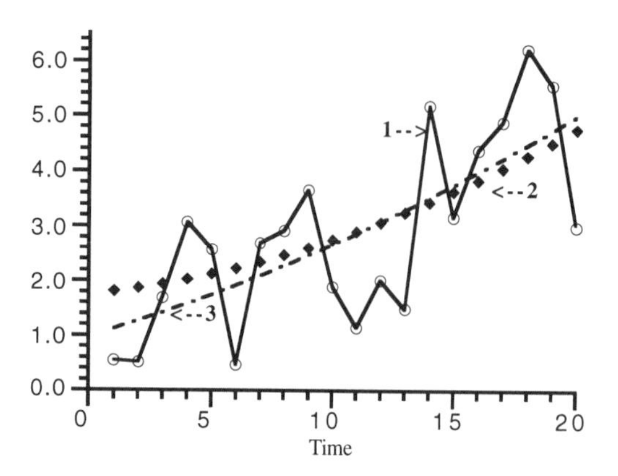

FIGURE 1: Plots of an unidimensional random field realization (curve 1), its trend (diamonds) and the best 2nd-degree polynomial fit to the trend (curve 3).

2nd-degree polynomial trend. The best fit obtained with a 2nd-degree polynomial does not reproduce the trend accurately, since the model attempts to fit both the trend and the fluctuations. In addition, the quality of the fit slightly improves if the data are fitted with higher-degree polynomials. Nonetheless, this is a plausible model when no other information is provided besides the data. □

In view of the lack of information regarding the S/TRF trends in real case studies, the continuity orders can only be justified *a posteriori* based on the accuracy of the estimates they provide (Chapter VI). High orders of continuity should be avoided, because they imply large degree polynomials that may be due to fluctuations instead of trends. Hence, in practical situations the numerical search is usually limited to third degree polynomials. Regional S/TRF parameters ν and μ offer meaningful insight into local space-time variations. The calculation of these parameters in practice is made in terms of a computational procedure discussed in detail in (SANLIB, 1995). A real-world situation is discussed in

EXAMPLE 2: Ozone (O_3) is an inorganic gaseous pollutant. O_3 forms in the upper or the lower atmosphere. From an environmental health perspective, the most important source of O_3 is that found in photochemical smog. A map showing the regionalized $\mu - \nu$

distribution of O_3 concentrations (in ppm) over Eastern U.S. (July 16, 1995) is shown in Fig. 2. This distribution provides information about the relative trends of ozone in space

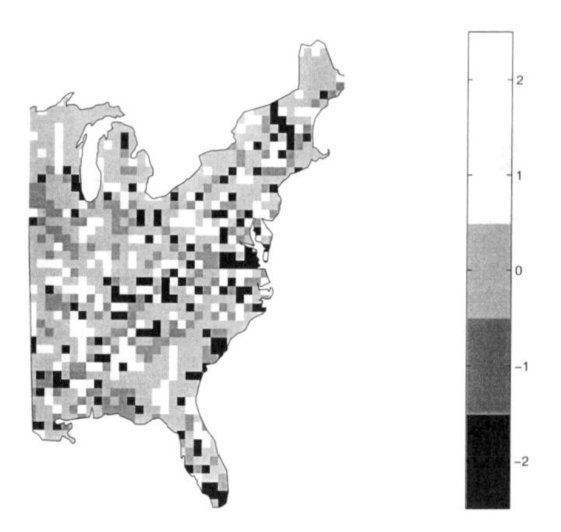

FIGURE 2: Map of $\mu - \nu$ distribution of ozone concentrations (ppm) over Eastern U.S. (July 16, 1995; from Christakos and Vyas, 1998a).

and time (a negative difference $\mu - \nu$ implies that the spatial trends are higher degree polynomials; a positive $\mu - \nu$ implies that temporal trends are of higher degree). The temporal trends are mainly due to weather patterns and the spatial trends to dispersion of pollutants. Clearly, the spatial continuity order ν changes in space, and the temporal continuity order μ varies in space, as well. Spatial and temporal trends are interrelated, because dispersion is governed by temporal processes --such as temporal dependence of pollution sources, amount of sunlight available for photolytic reactions and wind patterns. We further explore the role of the model parameters ν and μ in trend determination in the following section. □

EXAMPLE 3: Data on breast cancer incidence (i.e., the rate of diagnosis of new cancer cases at a particular time) offer a clear idea of the burden of breast cancer in a population. Data of breast cancer incidence per 100,000 population are provided by the Department of Environment, Health and Natural Resources (DEHNR) and include data at 100 counties throughout North Carolina. These data were studied in detail by Christakos and Lai (1997). A map of the order difference $\nu - \mu$ of breast cancer incidence over the state of North Carolina for the year 1992 is shown in Fig. 3. The scale considered was $\Lambda = D/T$=60 miles/3 years (i.e., an observation neighborhood including all data within a space-time cylinder with diameter D=60 km and height T=3 years was considered; see Fig. V.3 for a geometrical explanation). Similar maps --with some small changes mainly

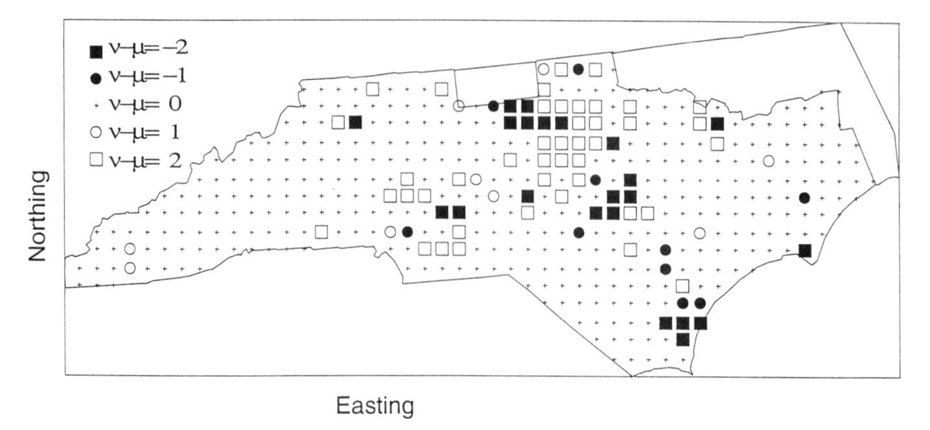

FIGURE 3: Map of $\nu-\mu$ of the breast cancer incidence (North Carolina, year 1992; Λ =60 miles/3 years).

in the Eastern and Western parts of the state-- were obtained for the years 1990 and 1991. The maps show that the variation of breast cancer incidence has a nonhomogeneous-nonstationary structure (corresponding to the varying orders ν and μ). Relatively high temporal orders μ were observed in some metropolitan areas --possibly associated with population changes during the years. The higher the value of ν (or μ) is, the higher the spatial (or temporal) order of the Q-operator. Spatial variation is more significant than temporal variation during the limited 3 year-period for which data are available. This is reflected in the preponderance of positive values in the $\nu-\mu$ map, which also provides a quantitative assessment of the relative rate of change of the space-time breast cancer incidence trends. □

4. CONTINUUM REPRESENTATIONS OF S/TRF- ν/μ

The S/TRF $X(s,t)$ represents the spatial distribution and temporal evolution of an environmental health process. If the dynamical laws that govern the specific process are known and can be solved the random field $X(s,t)$ is fully determined. However, for many environmental health problems the dynamical laws are either unknown, or their solution exceeds the capabilities of the computational resources currently available. In such cases, it is desirable to estimate the properties of the random field $X(s,t)$ based on available information from measurements or soft data as shown in Chapter V, and to generate efficiently random field realizations. Simulation requires a general procedure for generating S/TRF- ν/μ that sufficiently represent the spatiotemporal variability of the data.

General S/TRF models can be generated from homogeneous-stationary random fields $Y(s,t)$ by inverting Eq. (1.2). Since the Q-operator is not uniquely defined, more than one $X(s,t)$ can be generated from the same residual $Y(s,t)$. Conversely, more than one homogeneous-stationary fields $Y(s,t)$ can be obtained from the same $X(s,t)$ using different Q-operators. This one to many correspondence does not represent a theoretical deficiency of the model. It is simply evidence that (i) multiple homogeneous-stationary residuals $Y(s,t)$ can be obtained by applying different Q-operators to the same $X(s,t)$, and (ii) the residual field is not sufficient for determining the random field $X(s,t)$. In addition, in applications the residual is not known *a priori*. Hence, we aim to generate a class of permissible random field models, and to determine from this class the optimal model based on the existing data. We illustrate this point using the familiar example of homogeneous-stationary processes. In this case, the mean and the variance are determined from existing data by invoking the ergodicity assumption and using sample means to estimate expectations. However, the form of the covariance function is not uniquely determined by the homogeneity-stationarity assumption. Instead, the values of the covariance at discrete locations and times are estimated numerically from the data (Christakos, 1992). Then, the empirical covariance is fitted with various permissible models, and the model that leads to the best fit is selected. Similarly, the statistical features of nonhomogeneous-nonstationary S/TRF are determined based on fits of empirical covariances to specific covariance models.

In the following, we study representations of S/TRF that are obtained by inverting the detrending Q-operators. Based on the definition of S/TRF-ν/μ, the Q-operators that involve partial derivatives of the spatial coordinate s_i ($i=1,...,n$) generate homogeneous-stationary random fields $Y_i(s,t)$ when they operate on $X(s,t)$.

EXAMPLE 1: Consider the following partial residuals $Y_i(s,t)$, $i=1,...,n$, of the S/TRF $X(s,t)$ generated by means of the following Q-operators

$$Y_i(s,t) = Q_{(s,t)}^{(\nu+1/\mu+1)}[X(s,t)] = \partial^{\nu+\mu+2} X(s,t) \big/ \partial s_i^{\nu+1}\, \partial t^{\mu+1}. \tag{1}$$

By exploiting the linearity of the Q-operators, the following relation is obtained for the composite Q-operator

$$\frac{\partial^{\mu+1}}{\partial t^{\mu+1}} \sum_{i=1}^{n} \frac{\partial^{\nu+1} X(s,t)}{\partial s_i^{\nu+1}} = \sum_{i=1}^{n} Y_i(s,t) = Y(s,t). \tag{2}$$

Given the residual $Y(s,t)$, the random field $X(s,t)$ is obtained by solving the above PDE. Let us define the S/TRF $Z(s,t)$ as the solution of the following, purely temporal ODE

$$\partial^{\mu+1} Z(s,t)/\partial t^{\mu+1} = y(s,t) + \overline{Y}, \tag{3}$$

where $\overline{Y}$ denotes the constant mean of the residual $Y(s,t)$ and $y(s,t)$ the fluctuations. In view of Eqs. (2) and (3) it follows that

$$Z(s,t) = \sum_{i=1}^{n} \partial^{\nu+1} X(s,t)/\partial s_i^{\nu+1}. \tag{4}$$

The solution of Eq. (3) is

$$Z(s,t) = p_\mu(t) + \overline{Y}\, t^{\mu+1}/(\mu+1)! + z(s,t), \tag{5}$$

where the S/TRF $z(s,t)$ represents the contribution due to the zero-mean residual

$$z(s,t) = \int_{-\infty}^{t} dt'(t-t')^{\mu}\, y(s,t')/\mu!, \tag{6}$$

and $p_\mu(t)$ is a polynomial of degree μ in t. Let $G_0^{(\nu+1)}(s,s')$ denote the spatial Green's function of the Q-operator that satisfies

$$\sum_{i=1}^{n} \partial^{\nu+1} G_0^{(\nu+1)}(s,s')/\partial s_i^{\nu+1} = \delta(s-s'). \tag{7}$$

In light of Eq. (7), the solution of Eq. (4) is

$$X(s,t) = \beta\overline{Y}\,(\sum_{i=1}^{n} \theta_i^2\, s_i^{\nu+1})\,t^{\mu+1} + p_{\nu/\mu}(s,t) + \int \mathbf{ds}'\, G_0^{(\nu+1)}(s,s')\,z(s',t), \tag{8}$$

where the θ_i represent the direction cosines of an arbitrary unit vector $\boldsymbol{\theta}$, i.e., $\sum_{i=1}^{n} \theta_i^2 = 1$, and $\beta = [(\mu+1)!(\nu+1)!]^{-1}$. □

REMARK 1: Several comments are in order here: First of all, the polynomial $p_{\nu/\mu}(s,t)$ represents partially the trends. The first term on the right hand side is a monomial of degree $\nu+1/\mu+1$ generated by the mean of the residual $Y(s,t)$. This term includes an arbitrary spatial dependence through the direction cosines which is filtered out, and thus lost in the residual. Additional polynomial terms or combinations of polynomials with other functions may result from the fluctuation integral (the third term on the right hand-side). Finally, different S/TRF models $X(s,t)$ are obtained using different Q-operators.

EXAMPLE 2: For the Q-operator

$$[\partial^{\mu+1}/\partial t^{\mu+1} + \sum_{i=1}^{n} \partial^{\nu+1}/\partial s_i^{\nu+1}]X(s,t) = Y(s,t), \tag{9}$$

we obtain

$$X(s,t) = \int ds'\, G_0^{(\nu+1/\mu+1)}(s,s';t,t')\, y(s',t') + p_{\nu/\mu}(s,t)$$
$$+\bar{Y}[\sum_{i=1}^{n} \theta_i^2\, s_i^{\nu+1}/(\nu+1)! + \theta_{n+1}^2\, t^{\mu+1}/(\mu+1)!], \tag{10}$$

where the Green's function satisfies the equation

$$\left(\partial^{\mu+1}/\partial t^{\mu+1} + \sum_{i=1}^{n} \partial^{\nu+1}/\partial s_i^{\nu+1}\right) G_0^{(\nu+1/\mu+1)}(s,s';t,t') = \delta(s-s')\delta(t-t'), \tag{11}$$

and $\sum_{i=1}^{n} \theta_i^2 = 1$. □

5. DISCRETE REPRESENTATIONS: SPATIOTEMPORAL INCREMENTS OF ORDERS ν/μ

In numerical studies the random field is simulated on discrete lattices. Hence, the derivatives involved in the continuum representation of the detrending operator are replaced with difference operators. The continuity orders of the random field may vary within the numerical domain due to local changes in the natural or biological variability. Therefore, the S/TRF representation should incorporate the local structure by construction. The residual $Y(s,t)$ is also calculated locally by means of linear combinations of the S/TRF values of $X(s,t)$ in a neighborhood.

The general S/TRF- ν/μ Eq. (1.2) can be discretized as follows

$$\sum_{i=1}^{m} \sum_{j=1}^{p} q(s,t;s_i,t_j)X(s_i,t_j) = Y(s,t), \tag{1}$$

where the local weights $q(s,t;s_i,t_j)$ represent the discretized detrending operator $Q[\cdot]$ and the $X(s_i,t_j)$ represent the values of the random field $X(s,t)$ at the space-time locations

(s_i, t_j). In the following, the residual field is in certain places denoted by Y_q in order to emphasize the dependence on the detrending operator.

REMARK 1: If the orders of continuity are uniform over the domain the weights are also uniform, and they can be denoted by q_{ij}, where the indices i and j determine the position of the neighbors relative to any location (s,t). This notation is used for brevity below, but we warn the reader that in practice the orders of continuity vary within the domain, and the weights q_{ij} depend on the observation point (s,t).

The operator $Q[\cdot]$ eliminates the polynomial trends of the continuous S/TRF $X(s,t)$. Linear combination of the S/TRF values $X(s_i,t_j)$ weighted by the q_{ij} has the same effect on the discretized system. We illustrate this property by means of the following example.

EXAMPLE 1: We evaluate the weights q_{ij} for a S/TRF with continuity orders $\nu = 1 / \mu = 1$ in $R^1 \times T$. The detrending operator $Q[\cdot]$ has the following continuum representation

$$Q \equiv \partial^4 / \partial s^2 \, \partial t^2 . \tag{2}$$

The discretized difference operator to $O(\delta s^2 \, \delta t^2)$ accuracy is given by (Abramowitz and Stegun, 1970)

$$\frac{\partial^4 X_{0,0}}{\partial s^2 \, \partial t^2} \cong \frac{X_{1,1} + X_{-1,1} + X_{1,-1} + X_{-1,-1} - 2X_{1,0} - 2X_{-1,0} - 2X_{0,1} - 2X_{0,-1} + 4X_{0,0}}{\delta s^2 \, \delta t^2}, \tag{3}$$

where $X_{k,l} \equiv X(s + k\,\delta s, t + l\,\delta t)$ and δs, δt denote the space-time lattice spacing. In view of the approximation (3) the weights are given by the following

$$\left.\begin{array}{l} q_{1,1} = q_{-1,1} = q_{1,-1} = q_{-1,-1} = 1 \\ q_{1,0} = q_{-1,0} = q_{0,-1} = q_{0,1} = -2 \\ q_{0,0} = 4 \end{array}\right\}. \tag{4}$$

By definition, the weights (4) eliminate the linear polynomials in space and time. Hence, they satisfy the following sum rules

$$\sum_{j=-1}^{1} q_{ij}\, j^{\zeta} = \sum_{j=-1}^{1} q_{ij}\, i^{\rho} = \sum_{i=-1}^{1} \sum_{j=-1}^{1} q_{ij}\, (i^{\rho}\, j^{\zeta}) = 0, \quad \rho, \zeta = 0,1. \tag{5}$$

In order to see how the sum rules are obtained, consider the monomials $p_{\rho/\zeta}(s,t)=s^{\rho}\,t^{\zeta}$ such that $0\leq\rho,\zeta\leq 1$. The effect of the detrending operator on these monomials can be expressed as follows

$$Q[p_{\rho/\zeta}(s,t)]\cong\sum_{i=-1}^{1}\sum_{j=-1}^{1}q_{ij}\,(s+i\delta s)^{\rho}\,(t+j\delta t)^{\zeta}+O(\delta s^2\,\delta t^2). \tag{6}$$

If we solve the equation $Q[p_{\rho/\zeta}(s,t)]=0$ up to $O(\delta s^2\,\delta t^2)$ for all the combinations $0\leq\rho,\zeta\leq 1$ the following relations are obtained

$$Q[p_{0/0}(s,t)]=\sum_{i=-1}^{1}\sum_{j=-1}^{1}q_{ij}=0. \tag{7}$$

$$Q[p_{1/0}(s,t)]=\sum_{i=-1}^{1}\sum_{j=-1}^{1}q_{ij}(s+i\delta s)=0, \tag{8}$$

$$Q[p_{0/1}(s,t)]=\sum_{i=-1}^{1}\sum_{j=-1}^{1}q_{ij}(t+j\delta t)=0, \tag{9}$$

$$Q[p_{1/1}(s,t)]=\sum_{i=-1}^{1}\sum_{j=-1}^{1}q_{ij}(s+i\delta s)(t+j\delta t)=0. \tag{10}$$

Eqs. (7)-(10) lead to the sum rules in Eq. (5). Similar sum rules are also satisfied for higher-order derivatives. □

In numerical studies the discretized derivatives involve linear combinations of the field's values at points enclosed within a space-time neighborhood Λ. The size of this neighborhood depends on the continuity orders, since it should include a sufficient number of points for an accurate estimation of the derivatives. Let $(\mathbf{s}+\boldsymbol{\delta}\mathbf{s}_i,t+\delta t_j)$, where $i=0,\ldots,m(\nu)$, $j=0,\ldots,p(\mu)$ denote the number points in the local neighborhood Λ around $(\mathbf{s},t)$, and $\boldsymbol{\delta}\mathbf{s}_0=\delta t_0=0$.

Definition 1: *A discrete S/TRF residual $Y_q(\mathbf{s},t)$ defined by means of*

$$Y_q(\mathbf{s},t)=\sum_{i=0}^{m(\nu)}\sum_{j=0}^{p(\mu)}q_{ij}\,X(\mathbf{s}+\boldsymbol{\delta}\mathbf{s}_i,t+\delta t_j)\;, \tag{11}$$

is called a spatiotemporal increment of order ν/μ (S/TI- ν/μ) if the following is true

$$\sum_{i=0}^{m(\nu)} \sum_{j=0}^{p(\mu)} q_{ij}\, g_\alpha(\mathbf{s} + \boldsymbol{\delta}\mathbf{s}_i, t + \delta t_j) = 0, \tag{12}$$

for all monomials with $\alpha = 1, \ldots, N_n(\nu/\mu)$.

The S/TI- ν/μ is useful for practical purposes, since S/TI- ν/μ is involved in estimation (mapping) methods, as we discuss further in Chapter VI.

Definition 2: *Consider an S/TRF* $X(\mathbf{s},t)$ *for which an S/TI-* ν/μ $Y_q(\mathbf{s},t)$ *can defined as in Definition 1. If the* $Y_q(\mathbf{s},t)$ *is a homogeneous-stationary S/TRF, the* $X(\mathbf{s},t)$ *is called a discrete S/TRF* ν/μ.

The discrete domain Definition 2 motivates the term: "S/TRF with homogeneous increments of order ν in space and stationary increments of order μ in time" to describe a discrete S/TRF- ν/μ. This is a generalization of the S/TRF with homogeneous-stationary 0th-order increments --which corresponds to ν/μ=0/0 in this scheme.

EXAMPLE 2: Let $X(s,t)$ be an S/TRF-1/1 in $R^1 \times T$. Then $\partial^4 X(s,t)/\partial s^2\, \partial t^2 = Y(s,t)$, where $Y(s,t)$ is a homogeneous-stationary residual field. As we saw in Example 1 above the discrete representation of $Q[X]$ is given by Eq. (3). The Eqs. (7) through (10) establish that the weights q_{ij} satisfy the conditions in Eq. (12). Hence, the residual $Y(s,t)$ is an S/TI-1/1. Moreover, since $Y(s,t)$ is homogeneous-stationary, $X(s,t)$ is an S/TRF-1/1. □

Eq. (11) provides a general model for constructing S/TRF residuals of the process $X(\mathbf{s},t)$ on a lattice. By fitting this model to the data at a scale Λ, information is obtained about the space-time dynamics at this scale. A practical approach for constructing the S/TI- ν/μ is suggested by

EXAMPLE 3: In order to estimate the value of a natural process at an unmeasured point $(\mathbf{s}_k, t_\ell)$, a linear superposition of measured values at neighboring points is used (a detailed exposition is presented in Chapter VI). Let $X(\mathbf{s},t)$ be an S/TRF- ν/μ modelling sulfate deposition concentration with known values in a neighborhood around $(\mathbf{s}_k, t_\ell)$ that includes measured points $(\mathbf{s} + \boldsymbol{\delta}\mathbf{s}_i, t + \delta t_j)$, $i = 1, \ldots, m(\nu)$ and $j = 1, \ldots, p(\mu)$. The linear expression

$$\hat{X}(\mathbf{s}_k, t_\ell) = \sum_{i=1}^{m(\nu)} \sum_{j=1}^{p(\mu)} q_{ij}\, X(\mathbf{s}_k + \boldsymbol{\delta}\mathbf{s}_i, t_\ell + \delta t_j) \tag{13}$$

is used to estimate the value of the random field at (s_k, t_ℓ). The weights q_{ij} should be chosen so as to guarantee an unbiased estimator

$$\overline{\hat{X}(s_k,t_\ell) - X(s_k,t_\ell)} = 0. \tag{14}$$

The mean of the S/TRF- ν/μ is given by the polynomial

$$\overline{X(s_k,t_\ell)} = \sum_{\alpha=1}^{N_n(\nu/\mu)} a_\alpha \, g_\alpha(s_k,t_\ell). \tag{15}$$

In light of Eqs. (14) and (15) it can be shown (Christakos, 1992) that the weights satisfy the condition (12), and thus the residual S/TRF

$$Y_q(s_k,t_\ell) = \hat{X}(s_k,t_\ell) - X(s_k,t_\ell) \tag{16}$$

is an S/TI- ν/μ. □

6. SPECTRAL REPRESENTATIONS

Representations of S/TRF in frequency space are obtained by means of Fourier transforms (FT) using spectral densities of residual correlation functions. The FT of polynomial functions do not exist in the ordinary sense --because polynomial functions are not bounded-- but generalized FT can be defined and evaluated (e.g., Christakos 1984a, 1992). The FT of polynomials are used in permissibility conditions for generalized covariance functions according to Bochner's theorem (see, also, below). Spectral representations of S/TRF are particularly useful, because they take advantage of the homogeneity-stationarity properties of the residual field. Even if the ordinary FT of the S/TRF $X(\boldsymbol{s},t)$ does not exist, spectral densities for increments of order ν/μ can be defined in the ordinary sense. Thus, for homogeneous and stationary fields the spectral density is used for the centered covariance, while for fields with homogeneous-stationary increments, it is used for the semi-variogram.

Definition 1: *The ordinary FT of a function $f(\boldsymbol{s},t)$ in the frequency space $(\omega, \boldsymbol{\kappa})$ is defined by means of the integral*

$$\tilde{f}(\boldsymbol{\kappa},\omega) = \int d\boldsymbol{s} \int dt \, e^{i\boldsymbol{\kappa}\cdot\boldsymbol{s} - i\omega t} f(\boldsymbol{s},t), \tag{1}$$

and the inverse FT (IFT) is given by means of the integral

$$f(s,t) = \int_{\kappa}\int_{\omega} \tilde{f}(\kappa,\omega)\, e^{-(i\kappa\cdot s - i\omega t)}, \tag{2}$$

where κ is the spatial frequency (wavevector), ω is the temporal frequency.

REMARK 1: In the following for convenience we also use $w = (\kappa_1,\ldots,\kappa_n,\omega)$ to denote both the spatial and temporal frequencies in unified notation, and $\breve{w} = (\kappa_1,\ldots,\kappa_n,-\omega)$. Then, $\int_w = \int_\kappa \int_\omega$, and $\kappa \cdot s - \omega t = \breve{w} \cdot p$.

For an S/TRF- ν/μ a spectral representation can be used for its increments.

EXAMPLE 1: In the case $\nu/\mu = 0/0$ (homogeneous-stationary increments) the following representation is possible

$$X(s,t) = X(\mathbf{0},0) + c_1 t + d_1 \cdot s + \int_w \tilde{\Delta}_x(w), \tag{3}$$

where $\tilde{\Delta}_x(w)$ is a random density function that satisfies the conditions

$$\overline{\tilde{\Delta}_x(w)} = 0, \tag{4}$$

$$\overline{\tilde{\Delta}_x^{\dagger}(w_1)\tilde{\Delta}_x(w_2)} = (2\pi)^{n+1}\,\delta(w_1 + w_2)\left|\tilde{\Delta}_x(w_1)\right|^2, \tag{5}$$

where $\tilde{\Delta}_x^{\dagger}$ denotes the complex conjugate. For real-valued S/TRF $X(s,t)$ the density function is symmetric $\tilde{\Delta}_x^{\dagger}(w) = \tilde{\Delta}_x(-w)$. If the Q-operator is given by

$$Q = \mathrm{a}_0\, \partial/\partial t + \sum_{i=1}^{n} \mathrm{a}_i\, \partial/\partial s_i, \tag{6}$$

the $\nu = 0/\mu = 0$ increment is

$$Y(s,t) = \overline{Y} + \int_w e^{i\breve{w}\cdot p}(i\mathbf{a}\cdot\breve{w})\tilde{\Delta}_x(w), \tag{8}$$

where $\mathbf{a} = (\mathrm{a}_0,\ldots,\mathrm{a}_\mathrm{n})$ and $\overline{Y} = c_1 + \mathbf{a}\cdot d_1$. □

7. GENERALIZED SPATIOTEMPORAL RANDOM FIELDS

Generalized random fields represent a natural extension of ordinary S/TRF, based on the work of Schwartz on distribution theory (Schwartz, 1950 and 1951). The theory of generalized *spatial* random fields was spearheaded by Itô (1954) and Gel'fand (1955) with important contributions by Yaglom (1957) and Matheron (1973). Generalized *spatiotemporal* random fields were introduced by Christakos (1991 and 1992).

Generalized S/TRFs are defined in terms of *linear* functionals. A functional is a rule that assigns a numerical value to a mathematical expression (e.g., the definite Riemann integral of a function over a finite support represents a functional). Generalized S/TRF have important applications in environmental health science for two reasons: Firstly, they provide the mathematical framework for rigorous formulations of coarse graining. Coarse grained S/TRF can be used to represent averaging effects in exposure measurements due to finite instrument bandwidth, or effects of numerical averaging. Secondly, they lead to well-defined representations for S/TRF that do not have well-defined point values. Intermittent processes, e.g., rainfall, are discontinuous and do not have well-defined values at all space-time points. However, coarse grained values can be obtained for intermittent processes by averaging over specific space-time windows. The resulting processes, generally, depend on the scale of the coarse graining window. A standard example of a generalized function defined in terms of a functional instead of its point values is the delta function. Some functionals are presented below, in a stochastic context.

EXAMPLE 1: Assume that the S/TRF $X(\boldsymbol{s},t)$ has integrable and differentiable sample functions (realizations). The following are simple functionals

$$X[\textstyle\int](\Omega) = \int_{\Omega} d\boldsymbol{p}\, X(\boldsymbol{p}), \tag{1}$$

$$X[\partial/\partial t](\boldsymbol{p}_k) = \partial X(\boldsymbol{s}_k,t)/\partial t\big|_{t=t_k}. \tag{2}$$

If the spatiotemporal dependence of the coarse graining functional operator is evident, the notation can be shortened to $X[\partial/\partial t]$ and $X[\int]$. □

In environmental health modelling, coarse graining functionals involve integrals of S/TRF with test functions $q(\boldsymbol{s},t)$ that belong to a space $\mathcal{D}$. S/TRF functionals are denoted by $X[q]$ and satisfy the linearity conditions $X[q_1+q_2] = X[q_1] + X[q_2]$ and $X[\alpha q] = \alpha X[q]$.

A practical space $\mathcal{D}$ contains functions that are (a) continuous, (b) integrable and (c) infinitely differentiable. It is also customary to require that the test functions and all their derivatives vanish outside a certain interval in $n+1-$dimensional space, the support, denoted by $\Omega(q)$ [Schwartz 1950, 1951; Itô 1954; Gel'fand, 1955]. Test functions with finite support provide appropriate models for the observation effect (Christakos, 1992).

EXAMPLE 2: A test function in $\mathcal{D}$ is given by the following

$$q_1(\boldsymbol{r},\tau)=\begin{cases}\exp(-\dfrac{r^2}{R^2-r^2}-\dfrac{\tau^2}{T^2-\tau^2}), & \text{for } r<R \text{ and } |\tau|<T \\ 0, & \text{for } r\geq R \text{ and } |\tau|\geq T\end{cases} \quad (3)$$

All the derivatives of $q_1(\boldsymbol{r},\tau)$ are continuous at the boundary of the support. In Fig. 4 we

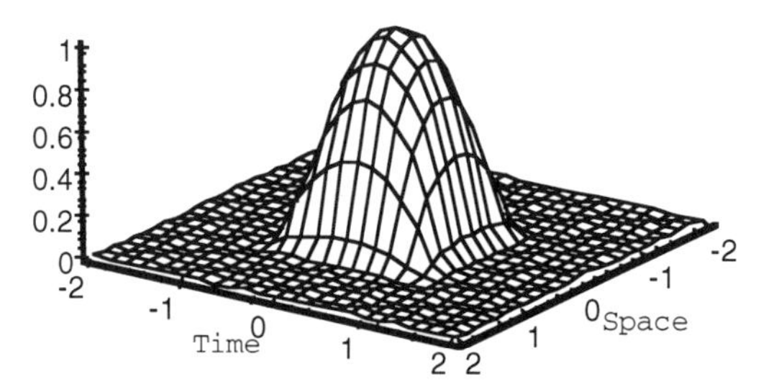

FIGURE 4: Plot of the test function $q_1(\boldsymbol{r},\tau)$ vs. $r=|\boldsymbol{r}|$ and τ for $R=1$ and $T=1$ [in appropriate units].

plot the spherically symmetric test function $q_1(\boldsymbol{r},\tau)$ in space-time. □

The space $\mathcal{D}'$ of infinitely differentiable functions which, together with their derivatives, tend to zero faster than $1/|\boldsymbol{r}|$ $(1/t)$ as $|\boldsymbol{r}|\to\infty$ $(t\to\infty)$ is also useful. In the space $\mathcal{D}'$ the finite extent of the support is traded for mathematical convenience.

EXAMPLE 3: A test function in $\mathcal{D}'$ is given by the Gaussian

$$q_2(\boldsymbol{r},\tau)=\exp(-r^2/R^2-\tau^2/T^2), \quad (4)$$

which has an infinite support (Fig. 5). However, it can be used to model a finite support

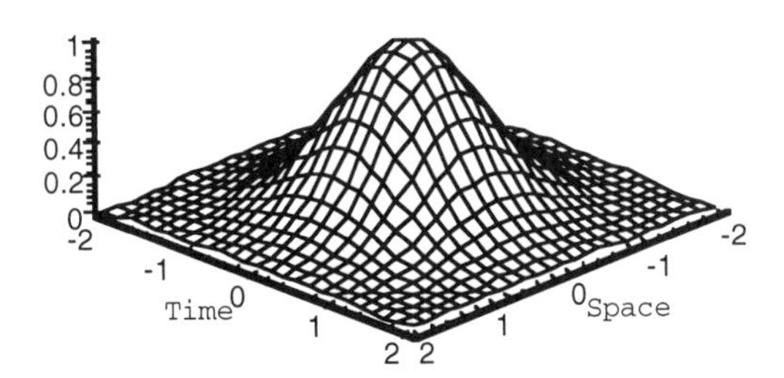

FIGURE 5: Plot of the test function $q_2(\boldsymbol{r},\tau)$ versus $r=|\boldsymbol{r}|$ and τ for $R=1$ and $T=1$ [in appropriate units].

filter, because it decays fast outside the range determined by R and T. □

REMARK 1: Separable test functions can be expressed as the product of a spatial and a temporal component, i.e., $q(\boldsymbol{p})=q_s(\boldsymbol{s})q_t(t)$. If the measurements are essentially localized in time the test function $q_t(t)=\delta(t-t_0)$ can be used.

If the S/TRF has point values $X(\boldsymbol{p})$, generalized S/TRF $X[q]$ can be defined by means of the following expression

$$X[q]=\int_{\Omega(q)}d\boldsymbol{p}\,X(\boldsymbol{p})q(\boldsymbol{p}), \tag{5}$$

which corresponds to an average of $X(\boldsymbol{p})$ over the support $\Omega(q)$. A different linear functional is defined by means of the convolution integral

$$X[q](\boldsymbol{p})=\int_{\Omega(q)}d\boldsymbol{p}'\,X(\boldsymbol{p}')q(\boldsymbol{p}-\boldsymbol{p}'). \tag{6}$$

Functional (6) is a function of the space-time point which represents a nonlocal average of the S/TRF $X(\boldsymbol{p})$ over a window determined by the test function. Expressions such as (5) and (6) provide mathematical representations of the coarse graining involved in measurements of random fields. The actual point values of the S/TRF are inaccessible to the observer, because the measurement process involves some averaging of the random field over the characteristic scales of the measuring apparatus (e.g., Cushman, 1984; Yaglom, 1987). The coarse graining process can significantly modify the properties of random fields (observation effect). If the apparatus function is known, the point values of the S/TRF are determined by the deconvolution of Eqs. (5) or (6), which can be realized by means of maximum entropy techniques. Such methods have been used successfully for enhanced information recovery by spectrum deconvolution in atomic spectroscopy (Davies *et al.*, 1991; von der Linden, 1995; Fisher *et al.*, 1997). In the context of environmental studies, the observation effect has been pointed out in theoretical investigations (Cushman, 1984; Baveye and Sposito, 1984).

The probability distribution and the stochastic moments of generalized S/TRF $X[q]$ are defined in a straightforward manner: the point values χ_i of the S/TRF are replaced by the functionals $X[q_i]$. The mean value functional of a generalized S/TRF is given by $m_x[q] = \overline{X[q]}$, and the covariance by $C_x[q_1, q_2] = \overline{X^{\dagger}(q_1)X(q_2)}$ --in this case the subscripts denote space-time points. The properties and classifications that were defined earlier for S/TRF can be extended to generalized S/TRF, as well.

In view of the definition of the function spaces $\mathcal{D}$ and $\mathcal{D}'$ given above, the partial derivatives of the test functions also belong to the same spaces. The partial derivatives of the generalized S/TRF $X[q]$ exist and are defined by means of the following functional

$$\partial X[q]/\partial p_i = -X[\partial q/\partial p_i], \tag{7}$$

Hence, according to the above, the derivatives of the generalized S/TRF can be evaluated by means of the derivatives of the test function.

The derivative of an ordinary S/TRF may not exist in the ordinary sense, but it may still be defined in terms of generalized S/TRF. The derivatives of generalized S/TRF in $\mathcal{D}$ always exist as generalized random fields. The partial derivatives of generalized S/TRF of any order are obtained by a straightforward extension of Eq. (7), viz.,

$$\partial^m X[q]/\partial p_i^m = (-1^m) X[\partial^m q(\boldsymbol{p})/\partial p_i^m]. \tag{8}$$

Generalized S/TRF share the stochastic symmetries of ordinary S/TRF. Thus, homogeneous and stationary generalized S/TRF are defined by means of the second order moment functionals (in the weak sense) or the pdf (strict sense). Generalized S/TRF with higher continuity orders demonstrate the stochastic symmetries in appropriately defined generalized increments. The latter are generated by difference operators that remove trends from the initial S/TRF thus producing a homogeneous and stationary residual.

Various natural and other complex processes can be characterized using generalized S/TRF, including river discharge records, financial time series (e.g., Mandelbrot, 1981) and fluid turbulence dynamics (Kraichnan, 1961). Most analytical studies of generalized S/TRF processes involve S/TRF with homogeneous and stationary increments, also called locally homogeneous and stationary S/TRF. In the following example we study a S/TRF with a generalized time derivative.

EXAMPLE 4: A random process $X(t)$ is obtained if the variability of an S/TRF $X(s,t)$ is due mainly to temporal behavior, or if the spatial fluctuations are eliminated by averaging. Assume that $X(t)$ has stationary increments, i.e., the $Y_\tau(t) = X(t+\tau) - X(t)$ is stationary for all τ values. A common example is *Brownian motion*, also known as *Wiener process* (Wiener, 1930), which provides a physical model for small scale dynamics that lead to apparently noisy behavior at larger scales (e.g., Kubo *et al.*, 1991). It has many applications in natural sciences (e.g., Feder, 1987), economics (Mandelbrot, 1982), etc. Brownian motion increments are stationary and uncorrelated. Realizations of Brownian motion $X(t)$ in the normalized time interval $[0,1]$ is obtained by means of the infinite series

$$X(t) = \frac{\sqrt{2}}{\pi} \sum_{n=1}^{\infty} Z_n \frac{\sin[(n-1/2)\pi t]}{(n-1/2)}. \tag{9}$$

In the above, Z_n is a $N(0,1)$ (standard Gaussian) random variable. The series converges because $(n-1/2)$ in the denominator which cuts off high order terms. In Fig. 6 we show a

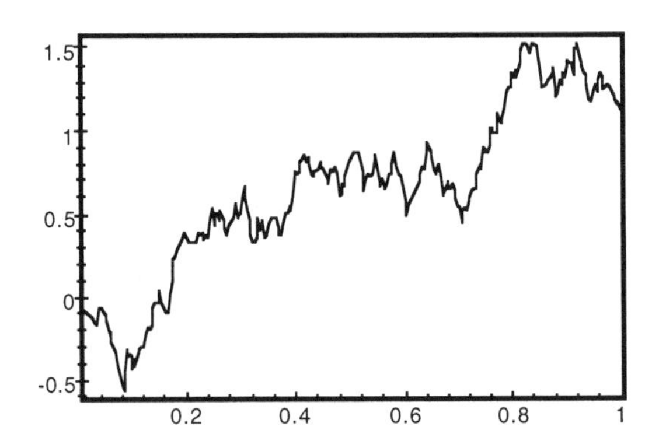

FIGURE 6: Realization of the Brownian motion $X(t)$ for $t \in [0,1]$.

realization of $X(t)$ obtained by truncating the series at $n = 1024$. For many applications it is useful to estimate the velocity of Brownian motion. However, the sample paths (realizations) are almost everywhere continuous but nondifferentiable functions of time (e.g., Feller, 1966); hence, the derivative $X'(t)$ does not exist. This is shown by taking the derivatives of the terms in the series for $X(t)$, which leads to the following series

$$Y(t) = \sqrt{2} \sum_{n=1}^{\infty} Z_n \cos[(n - 1/2)\pi t]. \qquad (10)$$

The $Y(t)$ diverges, because the fast fluctuations (high orders) are undamped. This is evidenced in approximations of $Y(t)$ that truncate the series after a finite number of terms. The plot in Fig. 7 was obtained by truncating the series $Y(t)$ after 1024 terms and exhibits

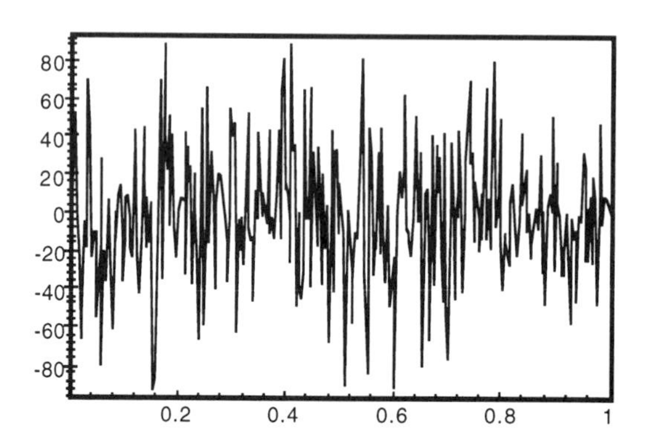

FIGURE 7: Realization of the truncated diverging series $Y(t)$.

erratic behavior with large fluctuations. The problem is solved by evaluating the derivative of the generalized random process $X[q] = \int_{-\infty}^{\infty} dt' \, X(t-t') q(t')$, instead of the point process $X(t)$. The test function q filters out the fast fluctuations that cause the divergence. Thus, the derivative $Y[q] = X'[q]$ exists and is stationary. If the Gaussian test function

$$q(t) = (\sqrt{2\pi}\sigma)^{-1} \exp(-t^2/2\sigma^2) \qquad (11)$$

is used, the generalized derivative $X'[q]$ is represented by means of the series

$$X'[q] = \sqrt{2} \sum_{n=1}^{\infty} Z_n \cos[(n - 1/2)\pi t] \exp[-(n - \tfrac{1}{2})^2 \pi^2 \sigma^2/2]. \qquad (12)$$

The high order terms in $X'[q]$ are damped by the exponential which ensures convergence of the series, and hence the derivative is well-defined. An approximation of the generalized derivative for a filter width $\sigma = 0.01$ obtained by truncating the series for $X'[q]$ after 1024 terms is shown in Fig. 8. □

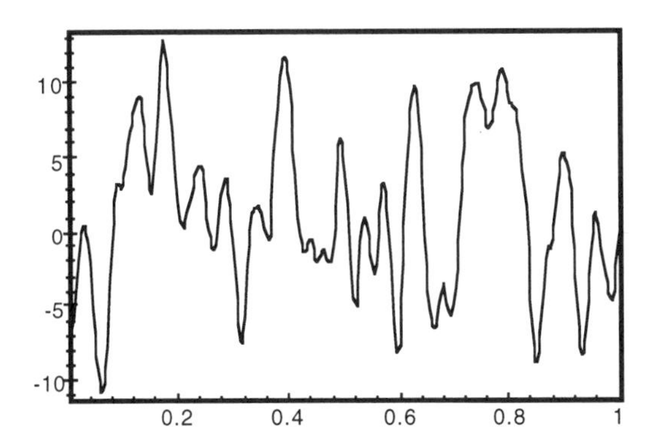

FIGURE 8: Realization of the generalized derivative $X'[q]$.

8. PERMISSIBILITY CRITERIA

The covariances of a real-valued S/TRF- ν/μ $X(s,t)$ and its residual field $Y(s,t)$ are related via the following expression

$$C_y(\boldsymbol{r},\tau) = (QQ')\,C_x(s,t;s',t'), \tag{1}$$

where $\boldsymbol{r} = \boldsymbol{s} - \boldsymbol{s}'$ and $\tau = t - t'$; the operators Q and Q' act at the points (s,t) and (s',t') respectively. In addition, it is essential to mention that the covariance of $X(s,t)$ satisfies the *decomposition relationship* (Christakos, 1992)

$$C_x(s,t;s',t') = k_x(\boldsymbol{r},\tau) + p_{\nu/\mu}(s,t)\,p_{\nu/\mu}(s',t'), \tag{2}$$

where $k_x(\boldsymbol{p}-\boldsymbol{p}')$=$k_x(\boldsymbol{r},\tau)$ is the generalized spatiotemporal covariance, the order of which depends only on the spatial and temporal lags $\boldsymbol{r} = \boldsymbol{s} - \boldsymbol{s}'$ and $\tau = t - t'$; the Q-operators filter out the polynomials $p_{\nu/\mu}(s,t)$ and $p_{\nu/\mu}(s',t')$.

As we saw in Chapter III, for homogeneous-stationary S/TRF the permissibility conditions are based on Bochner's theorem. In the case of nonhomogeneous-nonstationary S/TRF, the general permissibility criterion that the covariance function $c_x(\boldsymbol{p},\boldsymbol{p}')$ be non-negative definite still holds. Bochner's theorem can not be used, however, because the FT is not well-defined. In this case, the permissibility conditions can be expressed in terms of the *generalized spatiotemporal covariance* $k_x(\boldsymbol{p}-\boldsymbol{p}')=k_x(\boldsymbol{r},\tau)$, $\boldsymbol{r}=\boldsymbol{s}-\boldsymbol{s}'$ and $\tau=t-t'$, which includes all the homogeneous-stationary terms of $c_x(\boldsymbol{p},\boldsymbol{p}')$ [Christakos, 1984a, 1992]. Certain of these terms are not integrable, hence their FT does not exist in the ordinary sense. Fortunately, it is possible to define *generalized FT* (*GFT*) by means of test functions (Gel'fand and Shilov, 1964). The Bochner-Schwartz theorem provides conditions in terms of the GFT, which a non-negative definite generalized function must satisfy.

8.1 Analysis in the Continuous Spatiotemporal Domain - The Generalized Spatiotemporal Covariance

To obtain the basic analytical results needed for our environmental health applications, let $Q^{\dagger}=\sum_{|\mathbf{p}|=\nu+1}\alpha^{\dagger}_{\mathbf{p}/\mu}\,\partial^{|\mathbf{p}|+\mu+1}/\partial s^{|\mathbf{p}|}\,\partial t^{\mu+1}$ be the complex conjugate operator of the spatiotemporal differential operator Q of order ν/μ defined in Eq. (1.3). As before, the corresponding S/TRF-ν/μ $X(\boldsymbol{s},t)$ satisfies $Q[X(\boldsymbol{s},t)]=Y(\boldsymbol{s},t)$, where $Y(\boldsymbol{s},t)$ is a homogeneous-stationary residual. Multiplying the last equation by $Q^{\dagger}[X(\boldsymbol{s}',t')]=Y(\boldsymbol{s}',t')$ and then taking the expected value of the product, the following expression is obtained

$$U_Q\,k_x(\boldsymbol{r},\tau)=c_y(\boldsymbol{r},\tau). \tag{3}$$

where $U_Q=QQ^{\dagger}$ is a linear space-time differential operator, and $c_y(\boldsymbol{r},\tau)$ is the covariance of $Y(\boldsymbol{s},t)$. In deriving Eq. (3), the fact that the Q-operators filter out the polynomials $p_{\nu/\mu}(\boldsymbol{s},t)$ and $p_{\nu/\mu}(\boldsymbol{s}',t')$ has been taken into account. If equation $Q[X(\boldsymbol{s},t)]=Y(\boldsymbol{s},t)$ is solved, the generalized spatiotemporal covariance of $X(\boldsymbol{s},t)$ is determined by the residual and is permissible by construction. However, an explicit solution of this equation is not always possible. Instead, it is often easier to solve Eq. (3) that expresses the generalized spatiotemporal covariance $k_x(\boldsymbol{r},\tau)$ in terms of the residual covariance $c_y(\boldsymbol{r},\tau)$. This approach leads to a number of constant parameters (e.g., coefficients of polynomial terms) that are not determined by the solution. Permissibility criteria are then required in order to ensure that $k_x(\boldsymbol{r},\tau)$ is a permissible generalized spatiotemporal covariance. The need to derive permissibility criteria independent of the solution of the equations above, leads to the following definition.

Definition 1: *A generalized function* $k_x(\boldsymbol{r},\tau)$ *is called conditionally non-negative definite if* $U_Q\,k_x(\boldsymbol{r},\tau)$ *is a non-negative definite function.*

It can be shown that a non-negative definite function is also conditionally non-negative definite (Gel'fand and Vilenkin, 1964; p.175). The converse, however, is not always true. Moreover, the following theorem can be proven for a spatiotemporal generalized covariance.

Theorem 1: *A function $k_x(\boldsymbol{r},\tau)$ can be considered a permissible generalized spatiotemporal covariance if and only if it is conditionally non-negative definite* (Christakos, 1991; 1992).

From Definition 1 and Theorem 1, the required permissibility criteria of the generalized spatiotemporal covariance can be derived using the GFT, as follows. The generalized covariance $k_x(\boldsymbol{r},\tau)$ can be viewed as a generalized function defined on the spaces $\mathcal{D}$ or $\mathcal{D}'$ by means of a linear functional (k_x, q) as follows

$$(k_x, q) = \int_{\Omega(q)} d\boldsymbol{p}\, q(\boldsymbol{p}) k_x(\boldsymbol{p}), \tag{4}$$

where $k_x(\boldsymbol{p}) = k_x(\boldsymbol{r},\tau)$ and $q(\boldsymbol{p})$ is a test function in one of the above spaces [for simplicity, we assume that $k_x(\boldsymbol{r},\tau)$ is a real function]. The GFT $\tilde{k}_x(\boldsymbol{\kappa},\omega)$ of $k_x(\boldsymbol{r},\tau)$ is defined by means of the following expression

$$(k_x, q) = (\tilde{k}_x, \tilde{q}); \tag{5}$$

the functional $(\tilde{k}_x, \tilde{q})$ denotes the frequency space integral

$$(\tilde{k}_x, \tilde{q}) = \int_w \tilde{q}(\boldsymbol{w}) \tilde{k}_x(\boldsymbol{w}), \tag{6}$$

where $\tilde{q}(\boldsymbol{w}) = \tilde{q}(\boldsymbol{\kappa},\omega)$ is the ordinary FT of the test function, and $\tilde{k}_x(\boldsymbol{w}) = \tilde{k}_x(\boldsymbol{\kappa},\omega)$ is the GFT of $k_x(\boldsymbol{r},\tau)$. Explicit expressions for GFTs are given in Gel'fand and Shilov (1964; pp. 359-367).

EXAMPLE 1: The GFT of the polynomial $p_{\rho/\zeta}(\boldsymbol{r},\tau) = r^{\rho}\, \tau^{\zeta}$, where $r = |\boldsymbol{r}|$ in n spatial dimensions, is

$$\tilde{p}_{\rho/\zeta}(\boldsymbol{\kappa},\omega) = \tilde{p}_{\rho(1)}(\boldsymbol{\kappa})\, \tilde{p}_{\zeta(2)}(\omega), \tag{7}$$

where

$$\tilde{p}_{\rho(1)}(\boldsymbol{\kappa}) = 2^{\rho+n}\,\pi^{n/2}\,\Gamma(\tfrac{\rho+n}{2})\big/[\Gamma(-\tfrac{\rho}{2})\,\kappa^{\rho+n}], \tag{8}$$

with $\kappa = |\boldsymbol{\kappa}|$, and

$$\tilde{p}_{\zeta(2)}(\omega) = 2^{\zeta+1}\,\pi^{1/2}\,\Gamma(\tfrac{\zeta+1}{2})\big/[\Gamma(-\tfrac{\zeta}{2})\,\omega^{\zeta+1}]. \tag{9}$$

The value of the Gamma function for negative values of its argument is obtained from the recursive expression $\Gamma(\alpha+1) = \alpha\,\Gamma(\alpha)$; the latter leads to

$$\Gamma(-\tfrac{\rho}{2}) = \begin{cases} (-1)^{\ell}/\ell!, & \rho = 2\ell \\ (-1)^{\ell+1}\,2^{\ell+1}\,\pi^{1/2}/(2\ell+1)!!, & \rho = 2\ell+1 \end{cases}, \tag{10}$$

which is a useful formula in calculations involving GFTs. □

If the ordinary FT of the function $k_x(\boldsymbol{r},\tau)$ exists, it is identical to the GFT. In addition, the properties of the FT of partial derivatives are also satisfied for the GFTs of the function $k_x(\boldsymbol{r},\tau)$. For example

$$(\partial k_x/\partial r_j, q) = i(\kappa_j\,\tilde{k}_x, \tilde{q}), \tag{11}$$

and

$$(\partial k_x/\partial\tau, q) = -i(\omega\,\tilde{k}_x, \tilde{q}). \tag{12}$$

Similarly, for the operator U_Q we have

$$(U_Q\,k_x, q) = (\tilde{U}_Q\,\tilde{k}_x, \tilde{q}), \tag{13}$$

where $\tilde{U}_Q$ is the U_Q equivalent in Fourier space. In general, the operator $\tilde{U}_Q$ is obtained from U_Q by means of the transformations $\partial/\partial r_j \to i\kappa_j$ and $\partial/\partial\tau \to -i\omega$.

EXAMPLE 2: If the operator U_Q is given by $U_Q = \nabla^{2\nu+2}\,\partial^2/\partial\tau^{2\mu+2}$, the $\tilde{U}_Q$ is given by

$$\tilde{U}_Q = \kappa^{2\nu+2}\, \omega^{2\mu+2}, \tag{14}$$

where $\kappa = |\boldsymbol{\kappa}|$. □

Bochner-Schwartz Theorem: *A generalized function $f_x(\boldsymbol{r},\tau)$ is non-negative definite if and only if it has a GFT --defined by means of Eq. (4) above-- where $\tilde{f}_x(\boldsymbol{\kappa},\omega)$ is a real-valued, non-negative spectral density such that the tempered integral*

$$\iint d\boldsymbol{\kappa}\, d\omega \frac{\tilde{f}_x(\boldsymbol{\kappa},\omega)}{(1+\kappa^2)^{p_1}\,(1+\omega^2)^{p_2}}. \tag{15}$$

converges for some numbers $p_1 \geq 0, p_2 \geq 0$.

The Bochner-Schwartz theorem can be used to formulate permissibility criteria for the generalized covariance as follows.

Permissibility Criterion: *A function $k_x(\boldsymbol{r},\tau)$ is conditionally non-negative definite if $\tilde{k}_x(\boldsymbol{\kappa},\omega)$ is a real-valued and non-negative function such that the tempered integral*

$$\iint d\boldsymbol{\kappa}\, d\omega \frac{\tilde{U}_Q\, \tilde{k}_x(\boldsymbol{\kappa},\omega)}{(1+\kappa^2)^{p_1}\,(1+\omega^2)^{p_2}}. \tag{16}$$

converges for some numbers $p_1 \geq 0, p_2 \geq 0$.

Note that the conditions involved in the permissibility criterion are satisfied if:

I. $\tilde{k}_x(\boldsymbol{\kappa},\omega) \geq 0$.

II. $\tilde{U}_Q\, \tilde{k}_x(\boldsymbol{\kappa},\omega)$ does not have any non-integrable singularities around $\kappa = |\boldsymbol{\kappa}| = 0,\ \omega = 0$.

III. As $\kappa,\ \omega \to \infty$, the function $\tilde{U}_Q\, \tilde{k}_x(\boldsymbol{\kappa},\omega)$ increases slower than $\kappa^{2p_1}\, \omega^{2p_2}$ for some numbers $p_1 \geq 0, p_2 \geq 0$.

The expression $\tilde{U}_Q = \kappa^{2\nu+2}\, \omega^{2\mu+2}$ is commonly used for the detrending operator, and the permissibility conditions are described by the following theorem.

Theorem 2 (Christakos, 1991): *The conditions for $k_x(\boldsymbol{r},\tau)$ to be a generalized spatiotemporal covariance of order ν/μ are*

$$\kappa^{2\nu+2+(n-1)}\,\omega^{2\mu+2}\,\tilde{k}_x(\boldsymbol{\kappa},\omega)\geq 0, \tag{17}$$

(where the κ^{n-1} is included because the integration is in R^n, i.e., $\boldsymbol{d\kappa}=d\Omega_n\,d\kappa\,\kappa^{n-1}$) and

$$\lim_{r\to\infty}k_x(\boldsymbol{r},\tau)/r^{2\nu+2}=\lim_{\tau\to\infty}k_x(\boldsymbol{r},\tau)/\tau^{2\mu+2}=0. \tag{18}$$

REMARK 1: The polynomial terms arise naturally in generalized spatiotemporal covariance models: they represent trends or fluctuation correlations. The distinction between trends and fluctuations is important: Brownian motion is a zero mean random process, but its variance increases linearly with the support size.

8.2 Some Examples in the Discrete Space-Time Domain

Since numerical applications are analyzed in a discrete space-time domain, it will be instructive to consider a few examples.

EXAMPLE 3: Let $X(s,t)$ be an S/TRF-0/0 in $R^1\times T$. Then, $\partial^2X(s,t)/\partial s\,\partial t=Y(s,t)$, where $Y(s,t)$ is a homogeneous-stationary residual. Using forward finite differences to approximate the derivative we can write

$$Y(s,t)=\partial^2X(s,t)/\partial s\,\partial t\approx(X_{1,1}-X_{0,1}-X_{1,0}+X_{0,0})/\delta s\,\delta t. \tag{19}$$

In view of Eq. (5.1) and Eq. (19) above the weights q_{ij} are given by

$$q_{00}=q_{11}=1,\;\; q_{01}=q_{10}=-1. \tag{20}$$

It is easy to show that the above weights eliminate the constant terms, e.g.,

$$\sum_{i=0}^{1}\sum_{j=0}^{1}q_{ij}\,s_i^0\,t_j^0=\sum_{i=0}^{1}\sum_{j=0}^{1}q_{ij}=0. \tag{21}$$

Consider now the separable function $k_x(r_{ij}, \tau_{lm}) = r_{ij}\, \tau_{lm}$, where $r_{ij} = |s_i - s_j|$ and $\tau_{lm} = |t_l - t_m|$. A necessary condition for k_x to be a 0/0 generalized covariance is that $\overline{Y_q^2} \geq 0$, where $Y_q = Q[X]$. This condition is expressed as

$$\sum_{i=0}^{1} \sum_{j=0}^{1} \sum_{l=0}^{1} \sum_{m=0}^{1} q_{il}\, q_{jm} r_{ij}\, \tau_{lm} \geq 0, \tag{22}$$

for all the points (s_i, t_l), (s_j, t_m) and all the weights q_{il}, and q_{jm} that satisfy the condition of Eq. (21). The double sum inside the brackets in Eq. (22) can be expressed as follows

$$\sum_{i=0}^{1} \sum_{j=0}^{1} q_{il}\, q_{jm} r_{ij} = (q_{0l}\, q_{1m} + q_{1l}\, q_{0m})\, \delta s; \tag{23}$$

in light of Eq. (23) the left hand side of the inequality (22) becomes

$$\delta s \sum_{l=0}^{1} \sum_{m=0}^{1} \tau_{lm} (q_{0l}\, q_{1m} + q_{1l}\, q_{0m}) = 4\delta s\, \delta t \geq 0 \tag{24}$$

which proves that the necessary condition is satisfied. □

EXAMPLE 4: In the case of a homogeneous-stationary S/TRF or an S/TRF-0/0, the space-time semi-variogram

$$\gamma_x(\boldsymbol{p}_i - \boldsymbol{p}_j) = \tfrac{1}{2} Var[X(\boldsymbol{p}_i) - X(\boldsymbol{p}_j)] \tag{25}$$

may be used instead of the generalized covariance. Assume that $X(\boldsymbol{p})$ is an S/TRF-0/0. In the discrete domain, the coefficients q_i must satisfy

$$q_i\, r_i^{\rho}\, t_i^{\zeta} = 0, \;\; \rho, \zeta = 0, 1. \tag{26}$$

Note that only one subscript is sufficient since a single space-time label is used. Then, one can define the residual S/TRF

$$Y_q = \sum_i q_i\, X(\boldsymbol{p}_i) = \sum_i q_i [X(\boldsymbol{p}_i) - X(\boldsymbol{0})]. \tag{27}$$

In view of Eq. (26) and the fact that $X(\boldsymbol{p})$ is an S/TRF-0/0, it follows that $\overline{Y_q} = 0$. The variance of the residual is equal to the weighted combination of the increment covariance

$$Var[\sum_i q_i X(\boldsymbol{p}_i)] = \sum_i \sum_j q_i q_j \overline{[X(\boldsymbol{p}_i) - X(\boldsymbol{0})][X(\boldsymbol{p}_j) - X(\boldsymbol{0})]}. \tag{28}$$

The definition (25) allows us to express the increment covariance in Eq. (28) as

$$\overline{[X(\boldsymbol{p}_i) - X(\boldsymbol{0})][X(\boldsymbol{p}_j) - X(\boldsymbol{0})]} = \gamma_x(\boldsymbol{p}_i) + \gamma_x(\boldsymbol{p}_j) - \gamma_x(\boldsymbol{p}_i - \boldsymbol{p}_j) \tag{29}$$

Hence, using Eq. (29) the following expression for the variance of the residual is obtained

$$Var[Y_q] = \sum_j q_j \sum_i q_i \, \gamma_x(\boldsymbol{p}_i) + \sum_i q_i \sum_i q_i \, \gamma_x(\boldsymbol{p}_j) - \sum_i q_i \sum_i q_i \, \gamma_x(\boldsymbol{p}_i - \boldsymbol{p}_j). \tag{30}$$

In light of Eqs. (25) and (26) the first two terms in the right hand side of Eq. (30) are zero, thus leading to the following expression for the variance of the residual

$$Var[Y_q] = -\sum_i \sum_j q_i q_j \, \gamma_x(\boldsymbol{p}_i - \boldsymbol{p}_j) \tag{31}$$

Hence, the function $-\gamma_x(\boldsymbol{p}_i - \boldsymbol{p}_j)$ is a space-time generalized covariance of order 0/0, and this is also true for any function $c_o - \gamma_x(\boldsymbol{p}_i - \boldsymbol{p}_j)$, where c_o is a constant. □

8.3 Permissibility Conditions for Polynomial Covariances

Let us consider the permissibility conditions for polynomial generalized covariances of S/TRF with continuity orders ν/μ, which can be expressed as follows

$$k_x(\boldsymbol{r}, \tau) = \sum_{\rho=0}^{2\nu+1} \sum_{\zeta=0}^{2\mu+1} (-1)^{s(\rho)+s(\zeta)} \, a_{\rho/\zeta} \, r^{\rho} \, \tau^{\zeta}; \tag{32}$$

the $s(\rho)$ and $s(\zeta)$ are sign functions given by

$$s(x) = \tfrac{1}{2}(x - \delta_{x,2l+1}), \tag{33}$$

where $x = \rho, \zeta$. The sign functions compensate the negative signs introduced by the GFT. Hence, the GFT of all the monomials involved in Eq. (32) are positive, and the right hand side of (32) satisfies by construction the permissibility condition (17). Therefore, it is only necessary to check that the permissibility condition (18) is also satisfied. In many cases the polynomial generalized covariance can be expressed as a product of separable space and time polynomials as follows

$$k_x(\boldsymbol{r},\tau) = k_{x(1)}(\boldsymbol{r})k_{x(2)}(\tau) = [\sum_{\rho=0}^{2\nu+1}(-1)^{s(\rho)}\alpha_\rho r^\rho][\sum_{\zeta=0}^{2\mu+1}(-1)^{s(\zeta)}b_\zeta \tau^\zeta], \tag{34}$$

where the polynomial coefficients are related via $a_{\rho/\zeta} = \alpha_\rho b_\zeta$. If both the space and time components are permissible covariances, their product is also a permissible covariance. In light of Eqs. (8), (9) and (17) above, the permissibility conditions are expressed as

$$\tilde{k}_{x(1)}(\boldsymbol{\kappa}) = \sum_{\rho=0}^{2\nu+1} c_\rho \alpha_\rho \kappa^{2\nu+1-\rho} \geq 0 \quad \text{for all } \kappa \geq 0, \tag{35}$$

and

$$\tilde{k}_{x(2)}(\omega) = \sum_{\zeta=0}^{2\mu+1} d_\zeta b_\zeta \omega^{2\mu+1-\zeta} \geq 0 \quad \text{for all } \omega \geq 0 \tag{36}$$

where the coefficients c_ρ and d_ζ of the GFT are given by

$$c_\rho = (-1)^{s(\rho)} 2^{\rho+n} \pi^{n/2} \Gamma(\rho/2 + n/2)\Gamma^{-1}(-\rho/2), \tag{37}$$

$$d_\zeta = (-1)^{s(\zeta)} 2^{\zeta+1} \pi^{1/2} \Gamma(\zeta/2 + 1/2)\Gamma^{-1}(-\zeta/2). \tag{38}$$

The permissibility conditions are satisfied if the coefficients of the leading powers of the polynomials $\tilde{k}_{x(1)}(\boldsymbol{\kappa})$ and $\tilde{k}_{x(2)}(\omega)$ are positive and their roots are either complex conjugate or real and negative.

EXAMPLE 5: Polynomial covariances that involve only odd powers are commonly used in applications. We examine the permissibility criteria for such models in the special cases of purely spatial-purely temporal models. Consider the generalized covariance

$$k_x(r) = \sum_{\rho=0}^{\nu} (-1)^\rho \alpha_\rho r^{2\rho+1} \quad . \tag{39}$$

Note that the polynomial in Eq. (39) represents a temporal generalized covariance if r is replaced by τ and ν by μ. The permissibility condition is expressed as

$$k_x(r) = \sum_{\rho=0}^{\nu} c_\rho \alpha_\rho \kappa^{2(\nu-\rho)} \geq 0 \tag{40}$$

for all $\kappa \geq 0$, where the GFT coefficients are given by means of the expression

$$c_\rho = 2^{\rho+n}\,\pi^{(n-1)/2}\,\Gamma[(2\rho+n+1)/2](2\rho+1)!! \tag{41}$$

We have evaluated the coefficients c_ρ for $\rho = 0,1,2,3$ and $n = 1,2,3$. Their values are shown in Table 3. The cases $\nu = 0$ and $\nu = 1$ are trivial: Permissibility conditions require

TABLE 3: Values of the GFT coefficients

$n=1$	$n=2$	$n=3$
$c_0 = 2$	$c_0 = 2\pi$	$c_0 = 8\pi$
$c_1 = 12$	$c_1 = 18\pi$	$c_1 = 96\pi$
$c_2 = 240$	$c_2 = 450\pi$	$c_2 = 2880\pi$
$c_3 = 10080$	$c_3 = 22050\pi$	$c_3 = 161280\pi$

that $\alpha_0 \geq 0$ ($\nu = 0$) and $\alpha_0, \alpha_1 \geq 0$ ($\nu = 1$). For $\nu = 2$ these conditions lead to

$$\tilde{k}_x(\kappa) = c_0(\alpha_0\,\kappa^2 + c_1\,\alpha_1\,\kappa/c_0 + c_2\,\alpha_2/c_0) \geq 0, \tag{42}$$

where the constants c_i depend on the dimension n, and they are given in Table 3. Since all the c_i are positive, for the inequality (42) to hold for all κ the following must be true

$$\alpha_0 \geq 0. \tag{43}$$

The roots of the polynomial $\tilde{k}_x(\kappa)$ in (42) are given by

$$\kappa_{1,2} = [-c_1\,\alpha_1 \pm \sqrt{c_1^2\,\alpha_1^2 - 4\,\alpha_0\,\alpha_2\,c_2\,c_0}\,]/2\,\alpha_0\,c_0 \tag{44}$$

The two roots are complex conjugate if $\alpha_0, \alpha_2 \geq 0$ and $|\alpha_1| \leq 2\sqrt{\alpha_0\,\alpha_2\,c_0\,c_2}/c_1$. The roots are real and negative if $\alpha_0, \alpha_2 \geq 0$ and $\alpha_1 \geq 2\sqrt{\alpha_0\,\alpha_2\,c_0\,c_2}/c_1$. Hence, the permissibility conditions are met when the coefficients satisfy

$$\begin{cases} \alpha_0, \alpha_2 \geq 0 \\ \alpha_1 \geq 2\sqrt{\alpha_0\,\alpha_2\,c_0\,c_2}/c_1 \end{cases}. \tag{45}$$

In Fig. 9 we plot the GFT $\tilde{k}_x(\kappa)$ for four values of the parameter α_1. The $\tilde{k}_x(\kappa)$ becomes

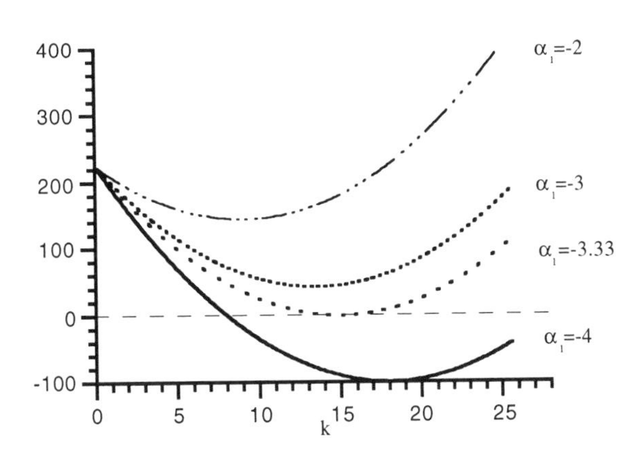

FIGURE 9: Plot of the two-dimensional GFT $\tilde{k}_x(\kappa)/c_0$ vs. κ for $n=2$ and $\alpha_0=\alpha_2=1$. Curves for four values of the parameter α_1 are shown.

negative when $\alpha_1 < -10/3$, in agreement with Eq. (45). □

8.4 Permissibility Conditions for Vector S/TRF

The scalar S/TRF analysis can be extended in the case of vector S/TRF

Definition 2: *A vector S/TRF is defined as*

$$\boldsymbol{X}(\boldsymbol{p}) = [X_1(\boldsymbol{p}),\ldots,X_M(\boldsymbol{p})]^T, \tag{46}$$

where $X_\ell(\boldsymbol{p})$ *(* $\ell = 1,\ldots,M$ *) are scalar S/TRFs.*

Most of the properties of the vector S/TRF can be derived from those of its component S/TRFs. Indeed, the vector random field $\boldsymbol{X}(\boldsymbol{p})$ is said to be spatially nonhomogeneous and temporally nonstationary, if its vector mean value $\overline{\boldsymbol{X}}(\boldsymbol{p})$ is a function of the space-time point $\boldsymbol{p}$, and its centered matrix covariance

$$\boldsymbol{c}_x(\boldsymbol{p},\boldsymbol{p}') = \overline{[\boldsymbol{X}(\boldsymbol{p}) - \overline{\boldsymbol{X}}(\boldsymbol{p})][\boldsymbol{X}(\boldsymbol{p}') - \overline{\boldsymbol{X}}(\boldsymbol{p}')]^T} \tag{47}$$

is a function of $\boldsymbol{p}$ and $\boldsymbol{p}'$. Homogeneity-stationarity (in the weak sense) occur when the mean $\overline{X}(\boldsymbol{p})$ is constant and the spatial covariance $c_x(\boldsymbol{p},\boldsymbol{p}')$ depends only on the space-time lags $\boldsymbol{r}$ and τ. A more general class of vector S/TRFs can be defined as follows.

Definition 3: *A vector S/TRF-* $\boldsymbol{\nu}/\boldsymbol{\mu}$ $X(\boldsymbol{p})$, $\boldsymbol{\nu}=[\nu_1 \ldots \nu_M]^T$ *and* $\boldsymbol{\mu}=[\mu_1 \ldots \mu_M]^T$, *consists of* M *scalar S/TRF-* ν_ℓ/μ_ℓ, $X_\ell(\boldsymbol{p})$, $\ell=1,\ldots,M$.

Higher orders ν_ℓ and μ_ℓ impose fewer restrictions on the data, allowing the application of the random field concepts to a wider range of environmental health processes than classical statistical models permit.

Given the vector S/TRF- $\boldsymbol{\nu}/\boldsymbol{\mu}$ $X(\boldsymbol{p})$ of dimension M, one can define the matrix $\boldsymbol{K}_x(\boldsymbol{r},\tau)$ with elements the generalized covariances and cross-covariances $k_{ij}(\boldsymbol{r},\tau)$, $i,j=1,\ldots,M$. In particular, the $k_{ii}(\boldsymbol{r},\tau)$ denote the generalized covariances of the S/TRF components X_i, and $k_{ij}(\boldsymbol{r},\tau)$, with $i \neq j$ the generalized cross-covariances. The latter are defined as follows: Consider detrending operators Q_i such that $Q_i X_i(\boldsymbol{s},t)=Y_i(\boldsymbol{s},t)$, where the $Y_i(\boldsymbol{s},t)$ are homogeneous-stationary residuals. The $k_{ij}(\boldsymbol{r},\tau)$ are the solutions of the PDEs

$$U_{Q(ij)}\, k_{ij}(\boldsymbol{r},\tau)=c_{ij}(\boldsymbol{r},\tau), \tag{48}$$

where $U_{Q(ij)}=Q_i^{\dagger}Q_j$ and $c_{ij}(\boldsymbol{r},\tau)=\overline{Y_i(\boldsymbol{s},t)Y_j(\boldsymbol{s},t)}-\overline{Y_i(\boldsymbol{s},t)}\,\overline{Y_j(\boldsymbol{s},t)}$. The permissibility criterion can be formulated as an extension of the Bochner-Schwartz theorem in terms of the FT matrix covariance $\tilde{\boldsymbol{K}}_x(\boldsymbol{\kappa},\omega)$ that has elements $\tilde{k}_{ij}(\boldsymbol{\kappa},\omega)$.

Definition 4: *The Hermitian matrix* $\tilde{\boldsymbol{K}}_x(\boldsymbol{w})$ *is non-negative definite if every characteristic root of the polynomial* $p(\lambda)=Det\left[\lambda \boldsymbol{I}-\tilde{\boldsymbol{K}}_x(\boldsymbol{w})\right]$ *is non-negative for all* $\boldsymbol{w}$.

A consequence of this definition is that both the trace and the determinant of $\tilde{\boldsymbol{K}}_x(\boldsymbol{w})$ are positive for all $\boldsymbol{w}$. Also, it can be shown that the matrix $\tilde{\boldsymbol{K}}_x(\boldsymbol{w})$ is non-negative definite if and only if all the principal submatrices $\tilde{\boldsymbol{K}}_x[\alpha|\alpha](\boldsymbol{w})$ are non-negative definite (Marcus and Minc, 1992). Principal submatrices are obtained from the initial matrix by excluding rows and columns designated by the set α, which contains $0 \leq j \leq M-1$ elements. If, e.g., $\alpha=\{1,2\}$ the principal submatrix is obtained by excluding the first two rows and columns.

Permissibility Criterion: *The matrix* $\boldsymbol{K}_x(\boldsymbol{r},\tau)$ *is a permissible covariance if the matrix* $\tilde{\boldsymbol{K}}_x(\boldsymbol{w})$ *is Hermitian non-negative definite, and if there are numbers* $p_1 \geq 0, p_2 \geq 0$ *such that the following tempered integrals converge for all* $i,j=1,\ldots,M$.

$$\int_w \frac{\tilde{U}_{Q(ij)}\,\tilde{k}_{ij}(\boldsymbol{w})}{(1+\kappa^2)^{p_1}\,(1+\omega^2)^{p_2}}. \tag{49}$$

EXAMPLE 1: Let us consider a vector spatial random field. Let $\boldsymbol{K}_x(\boldsymbol{r})$ be its matrix of isotropic generalized covariances and cross-covariances for $M=2$, i.e.,

$$\boldsymbol{K}_x(r)=\begin{bmatrix} k_1(r) & k_{12}(r) \\ k_{21}(r) & k_2(r) \end{bmatrix}, \tag{50}$$

where polynomial models are used, i.e.,

$$\left.\begin{aligned} k_j(r) &= \sum_{\rho=1}^{v} (-1)^{\rho+1}\, c_{j,\rho}\, r^{2\rho+1} \quad ,j=1,2 \\ k_{ij}(r) &= \sum_{\rho=1}^{v} (-1)^{\rho+1}\, c_{ij,\rho}\, r^{2\rho+1} \quad ,i\neq j=1,2 \end{aligned}\right\}. \tag{51}$$

Depending on the order v_ℓ of each random field, some generalized covariance and cross-covariance representations are shown in Table 4. Reduced models (with some coefficients set equal to zero) are also allowed. If $k_1(r)$ and $k_2(r)$ are polynomials of degree ρ, $k_{12}(r)$

TABLE 4: Spatial generalized covariance and cross-covariance models

v	$k_j(r)$	$k_{12}(r)$	$k_{21}(r)$
0	$-c_{j,0}\,r$	$-c_{12,0}\,r$	$-c_{21,0}\,r$
1	$-c_{j,0}\,r+c_{j,1}r^3$	$-c_{12,0}\,r+c_{12,1}r^3$	$-c_{21,0}\,r+c_{21,1}r^3$
2	$-c_{j,0}\,r+c_{j,1}r^3-c_{j,2}r^5$	$-c_{12,0}\,r+c_{12,1}r^3-c_{12,2}r^5$	$-c_{21,0}\,r+c_{21,1}r^3-c_{21,2}r^5$
	j=1 and 2		

and $k_{21}(r)$ cannot be of order higher than ρ. The corresponding representations for $k_{12}(r)$ are also shown in Table 5, where the coefficients $c_{j,\rho}$ and $c_{ij,\rho}$ (i, j=1,2 and ρ=0,1 and 2) must satisfy the constraints described in Table 5 (Cassiani and Christakos, 1998) where n denotes the number of dimensions. □

TABLE 5: Permissibility criteria for the generalized covariances of Table 4

ν	Constraints
0	$c_{1,0}, c_{2,0} \geq 0$ and $c_{12,0}\, c_{21,0} \leq c_{1,0}\, c_{2,0}$
1	$c_{1,0}, c_{2,0}, c_{1,1}, c_{2,1} \geq 0,\ c_{12,0}\, c_{21,0} \leq c_{1,0}\, c_{2,0},\ c_{12,1}\, c_{21,1} \leq c_{1,1}\, c_{2,1}$, and $c_{12,0}\, c_{21,1} + c_{12,1}\, c_{21,0} - c_{1,0}\, c_{2,1} - c_{1,1}\, c_{2,0})^2 \leq 4(c_{12,0}c_{21,0} - c_{1,0}\, c_{2,0})(c_{12,1}c_{21,1} - c_{1,1}\, c_{2,1})$
2	$c_{1,0}, c_{2,0}, c_{1,2}, c_{2,2} \geq 0,\ c_{1,1} \geq -\sqrt{\frac{20(n+3)}{3(n+1)} c_{1,0}\, c_{1,2}},\ c_{2,1} \geq -\sqrt{\frac{20(n+3)}{3(n+1)} c_{2,0}\, c_{2,2}}$ $c_{1,1} + c_{2,1} \geq -\sqrt{\frac{20(n+3)}{3(n+1)} (c_{1,0} + c_{2,0})(c_{1,2} + c_{2,2})}\ (n = 1,2,3)$ $\sum_{p=0}^{4} a_p x^{4-p} \leq 0$ for all $x \geq 0$ (a_p depend on the spatial dimension n)

9. GENERALIZED COVARIANCE MODELS DERIVED FROM PDEs

Here it is essential to recall that Eq. (8.3) is useful in obtaining generalized spatiotemporal covariance models. The Q-operators transform under space and time inversion as $Q_{-\boldsymbol{p}}^{(\nu+1/\mu+1)} = (-1)^{\nu+\mu+2}\, Q_{\boldsymbol{p}}^{(\nu+1/\mu+1)}$ (the superscript indicates the continuity orders, the subscript the space-time point). Hence, U_Q can be expressed as $U_Q = (-1)^{\nu+\mu}\, L^{(\nu+1/\mu+1)}$, where $L^{(\nu+1/\mu+1)}$ is a linear differential operator of order $\nu+1$ in space and $\mu+1$ in time. The solutions of the PDE (8.3) for $k_x(\boldsymbol{r},\tau)$ involve integrals of the homogeneous-stationary covariance $c_y(\boldsymbol{r},\tau)$. Generalized covariance models can be obtained using various detrending operators U_Q. For practical applications, detrending operators that allow explicit solutions of Eq. (8.3) are preferred. With this purpose in mind, we define the following covariance detrending operator

$$U_Q = (-1)^{\nu+\mu} (\partial^{2\mu+2}/\partial\tau^{2\mu+2})(\nabla^2)^{\nu+1}, \tag{1}$$

where ∇^2 denotes the Laplacian; in a Cartesian coordinate system $\nabla^2 = \sum_{i=1}^{n} \partial^2/\partial r_i^2$. In the case of spatially isotropic residual covariance it is useful to express the Laplacian in spherical coordinates as follows: $\nabla^2 = \partial^2/\partial r^2 + 2r^{-1}\, \partial/\partial r$.

REMARK 1: Note that $[\nabla^2]^{\nu+1} \neq [\sum_{i=1}^{n} \partial^{\nu+1}/\partial r_i^{\nu+1}]^2$ except for $\nu = -1, 1$. Hence, the generalized covariance *is not* the covariance of the S/TRF $X(\boldsymbol{s},t)$ that satisfies the PDE

$$(\partial^{\mu+1}/\partial t^{\mu+1})(\sum_{i=1}^{n} \partial^{\nu+1}/\partial s_i^{\nu+1}) X(s,t) = Y(s,t). \tag{2}$$

Using the expression of Eq. (1) for U_Q, generalized covariance models can be obtained from the solution of the following inhomogeneous PDE

$$(-1)^{\nu+\mu} (\nabla^2)^{\nu+1} \partial^{2\mu+2} k_x(\boldsymbol{r},\tau) / \partial \tau^{2\mu+2} = c_y(\boldsymbol{r},\tau). \tag{3}$$

The complete solution of Eq. (3) is a superposition of the *particular,* $k_{x(p)}(\boldsymbol{r},\tau)$, and the homogeneous equation solutions (Bender and Orszag, 1978). The *homogeneous* solution is given by a polynomial of degree $2\nu + 1/2\mu + 1$. The particular solution is obtained explicitly using Green's function techniques. Integration of Eq. (3) over time leads to

$$(\nabla^2)^{\nu+1} k_{x(p)}(\boldsymbol{r},\tau) = (-1)^{\nu} c_z(\boldsymbol{r},\tau), \tag{4}$$

where the centered covariance of the S/TRF $Z(s,t)$, defined in Eq. (4.4), is obtained by

$$c_z(\boldsymbol{r},\tau) = \int_{-\infty}^{\infty} d\tau' \, G_1(\tau - \tau') c_y(\boldsymbol{r},\tau'). \tag{5}$$

The temporal kernel (Green's function) $G_1(\tau - \tau')$ is given by the following expression

$$G_1(\tau - \tau') = (-1)^{\mu} (\tau - \tau')^{2\mu+1} \theta(\tau - \tau') / (2\mu + 1)!, \tag{6}$$

where $\theta(\tau - \tau')$ is the step function, i.e., $\theta(\tau - \tau') = 1$ if $\tau \geq \tau'$ and $\theta(\tau - \tau') = 0$ if $\tau < \tau'$. The particular solution for the generalized covariance is given by

$$k_{x(p)}(\boldsymbol{r},\tau) = \int \boldsymbol{dr}' \, G_2(\boldsymbol{r} - \boldsymbol{r}') c_z(\boldsymbol{r}',\tau). \tag{7}$$

The spatial kernel (Green's function) $G_2(\boldsymbol{r} - \boldsymbol{r}')$ is defined as follows

$$G_2(\boldsymbol{r} - \boldsymbol{r}') = \begin{cases} \dfrac{(-1)^{\nu}}{(2\nu+1)!} (r - r')^{2\nu+1} \theta(r - r'), & n = 1 \\ \dfrac{1}{2^{2\nu+1} \pi (\nu!)^2} |\boldsymbol{r} - \boldsymbol{r}'|^{2\nu} \log |\boldsymbol{r} - \boldsymbol{r}'|, & n = 2, \\ \dfrac{(-1)^{\nu+1} \Gamma(1/2 - \nu)}{2^{2\nu+2} \pi^{3/2} \nu!} |\boldsymbol{r} - \boldsymbol{r}'|^{2\nu-1}, & n = 3 \end{cases} \tag{8}$$

where $\theta(r-r')$ denotes the unit step function, $\theta(r-r')=1$ if $r \geq r'$ and $=0$ if $r<r'$. The Gamma function $\Gamma(1/2-\nu)$ is evaluated using $\Gamma(\alpha-1)=(\alpha-1)\Gamma(\alpha)$ and $\Gamma(1/2)=\sqrt{\pi}$. Note that the 1-D spatial kernel is given by the same functional expression as the time kernel. In 2-D and 3-D the kernel is an even function of the spatial lag, i.e., $G_2(\boldsymbol{r}-\boldsymbol{r}')=G_2(\boldsymbol{r}'-\boldsymbol{r})$. The generalized covariances in 1-D are not symmetric under reflection of the lag vector. Finally, the complete solution for the generalized covariance is

$$k_x(\boldsymbol{r},\tau)=k_{x(p)}(\boldsymbol{r},\tau)+p_{2\nu+1}(\boldsymbol{r})\,p_{2\mu+1}(\tau), \tag{9}$$

where

$$k_{x(p)}(\boldsymbol{r},\tau)=\int_{-\infty}^{\tau}d\tau'\int d\boldsymbol{r}'\,K(\boldsymbol{r}-\boldsymbol{r}',\tau-\tau')c_y(\boldsymbol{r}',\tau'). \tag{10}$$

The spatiotemporal kernel $K(\boldsymbol{r}-\boldsymbol{r}',\tau-\tau')$ is the product of the space and time Green's functions, i.e.,

$$K(\boldsymbol{r}-\boldsymbol{r}',\tau-\tau')=G_1(\tau-\tau')G_2(\boldsymbol{r}-\boldsymbol{r}'). \tag{11}$$

Eqs. (6), (8), and (9)-(11), summarized in Table 6, are the fundamental expressions used in the evaluation of the generalized spatiotemporal covariance from the residual covariance.

REMARK 2: Note that the coefficients of the polynomial terms in Eq. (9) are not determined from the solution of the covariance detrending Eq. (8.3). An explicit expression for the covariance involves solving Eq. (8.3). However, solutions of the Q-

TABLE 6: Fundamental Equations of the Generalized Spatiotemporal Covariance

$$k_x(\boldsymbol{r},\tau)=k_{x(p)}(\boldsymbol{r},\tau)+p_{2\nu+1}(\boldsymbol{r})\,p_{2\mu+1}(\tau)$$

$$k_{x(p)}(\boldsymbol{r},\tau)=\int_{-\infty}^{\tau}d\tau'\int d\boldsymbol{r}'\,K(\boldsymbol{r}-\boldsymbol{r}',\tau-\tau')c_y(\boldsymbol{r}',\tau')$$

$$K(\boldsymbol{r}-\boldsymbol{r}',\tau-\tau')=G_1(\tau-\tau')G_2(\boldsymbol{r}-\boldsymbol{r}')$$

$$G_1(\tau-\tau')=(-1)^{\mu}\,(\tau-\tau')^{2\mu+1}\,\theta(\tau-\tau')/(2\mu+1)!$$

$$G_2(\boldsymbol{r}-\boldsymbol{r}')=\begin{cases}\dfrac{1}{2^{2\nu+1}\,\pi(\nu!)^2}|\boldsymbol{r}-\boldsymbol{r}'|^{2\nu}\log|\boldsymbol{r}-\boldsymbol{r}'|, & n=2\\[2ex] \dfrac{(-1)^{\nu+1}\,\Gamma(1/2-\nu)}{2^{2\nu+2}\,\pi^{3/2}\,\nu!}|\boldsymbol{r}-\boldsymbol{r}'|^{2\nu-1}, & n=3\end{cases}$$

equations are not easily obtained for $n \geq 2$ and $\nu, \mu \geq 0$. In contrast, explicit solutions for the generalized covariance can be obtained from the covariance detrending equation, as we showed above. The price to be paid for bypassing the Q-equation is the indeterminacy of the polynomial coefficients. The latter are constrained by the permissibility conditions for the generalized covariance.

10. SPACE-TIME SEPARABILITY

As we saw in previous sections, it is a common modelling approach to consider correlations that have separable spatial and temporal components. This is particularly useful when the correlations do not follow from a physical model but are inferred on the basis of existing data. The assumption of separability restricts the class of permissible covariance functions. Ideally, one should verify that the investigated environmental health process is consistent with this assumption. However, this is not usually possible; thus, empirical covariance functions that satisfy the permissibility criteria and approximate reasonably well the data statistics are used in practice. Separable models are flexible and, thus, very useful for modelling natural fields and health effects.

Consider the case of the separable residual covariance models $c_y(\boldsymbol{r}, \tau)$

$$c_y(\boldsymbol{r}, \tau) = \sigma^2 \rho_{y(1)}(\tau) \rho_{y(2)}(\boldsymbol{r}), \tag{1}$$

where σ^2 is a constant variance, $\rho_{y(1)}(\tau)$ is a purely temporal and $\rho_{y(2)}(\boldsymbol{r})$ a purely spatial correlation function. The generalized spatiotemporal covariance $k_x(\boldsymbol{r}, \tau)$ is also separable

$$k_x(\boldsymbol{r}, \tau) = \sigma^2 k_{x(1)}(\tau) k_{x(2)}(\boldsymbol{r}), \tag{2}$$

where the components $k_{x(1)}(\tau)$ and $k_{x(2)}(\boldsymbol{r})$ are given by

$$\left.\begin{aligned} k_{x(1)}(\tau) &= \int_{-\infty}^{\tau} d\tau' \, G_1(\tau - \tau') \rho_{y(1)}(\tau') + p_{2\mu+1}(\tau) \\ k_{x(2)}(\boldsymbol{r}) &= \int d\boldsymbol{r}' \, G_2(\boldsymbol{r} - \boldsymbol{r}') \rho_{y(2)}(\boldsymbol{r}') + p_{2\nu+1}(\boldsymbol{r}) \end{aligned}\right\}, \tag{3}$$

where the spatial integral extends over the support of the environmental health process. The spatial component of the generalized covariance's particular solution is given by the n-dimensional integral $\int d\boldsymbol{r}' \, G_2(\boldsymbol{r} - \boldsymbol{r}') \rho_{y(2)}(\boldsymbol{r}')$. If the residual covariance is spatially isotropic, the particular part of the generalized covariance is also isotropic. For a 3-D isotropic covariance the spatial integral in Eq. (3) can be replaced by the 1-D integral

$$\int d\boldsymbol{r}' \, G_2(\boldsymbol{r}-\boldsymbol{r}') \rho_{y(2)}(\boldsymbol{r}') = \int_0^\infty d\tau' \, G_2^*(r,r') \rho_{y(2)}(r'), \tag{4}$$

where $r = |\boldsymbol{r}|$, $r' = |\boldsymbol{r}'|$, and the kernel $G_2^*(r,r')$ is given by the following integral over the surface of the n-dimensional unit sphere

$$G_2^*(r,r') = (r')^{n-1} \int d\Omega_n \, G_2(\boldsymbol{r}-\boldsymbol{r}'). \tag{5}$$

Evaluation of the integral in 3-D leads to the following expression for the kernel $G_2^*(r,r')$

$$G_2^*(r,r') = \frac{(-1)^{2\nu+1}}{2^{\nu+1}\, \nu! [\prod_{i=1}^{\nu+1} (2i-1)]} (\frac{r'}{r}) [(r+r')^{2\nu+1} - |r-r'|^{2\nu+1}]. \tag{6}$$

Anisotropic residual covariances can be reduced to isotropic functions by means of a rescaling transformation.

11. NEW GENERALIZED SPATIOTEMPORAL COVARIANCE MODELS

The class of generalized spatiotemporal covariances is very rich (in fact, it is considerably richer than the class of ordinary covariances). Here we examine in more detail a few popular space-time generalized covariance models and we propose some new ones. The most widely used model is the polynomial.

11.1 The Polynomial Model

If the residual random field $Y(s,t)$ is white noise, the generalized covariance of the field $X(s,t)$ is a space-time polynomial; namely, if

$$c_y(\boldsymbol{r},\tau) = \sigma^2 \, \delta(\boldsymbol{r}) \delta(\tau), \tag{7}$$

then it follows from Eqs. (8.3) that the generalized covariance is given by

$$k_x(\boldsymbol{r},\tau) = \sigma^2 [G_1(\boldsymbol{r}) + p_{2\nu+1}(\boldsymbol{r})][G_2(\tau) + p_{2\mu+1}(\tau)]. \tag{8}$$

In 1-D and in 3-D the above is a polynomial of degree $2\nu+1/2\mu+1$, see the Eq. (9.8). In 2-D the particular solution of the generalized covariance includes the *logarithmic term* $G_1(\boldsymbol{r}) \propto |\boldsymbol{r}|^{2\nu} \log|\boldsymbol{r}|$. To our knowledge, the origin of the logarithmic term has not been explicitly demonstrated before, even though it has been used empirically for smoothing in generalized covariance models (e.g., Christakos and Thesing, 1993). The function $G_1(\boldsymbol{r})$ is isotropic by construction, i.e., $G_1(\boldsymbol{r}) = G_1(r = |\boldsymbol{r}|)$, but the polynomial term $p_{2\nu+1}(\boldsymbol{r})$ does not have to be isotropic. Nonetheless, the isotropic hypothesis is often used in environmental health studies. In addition, space-time nugget terms are added to the generalized covariance.

In light of the above, the generalized spatiotemporal polynomial covariance can be expressed as follows

$$k_x(|\boldsymbol{r}|,\tau) = \alpha_0\,\delta(r)\delta(\tau) + \delta(r)\sum_{\zeta=0}^{\mu} \mathrm{a}_\zeta(-1)^{\zeta+1}\tau^{2\zeta+1} + \delta(\tau)\sum_{\rho=0}^{\nu} \mathrm{b}_\rho(-1)^{\rho+1} r^{2\rho+1}$$
$$+\sum_{\rho=0}^{\nu}\sum_{\zeta=0}^{\mu} d_{\rho/\zeta}(-1)^{\rho+\zeta} r^{2\rho+1}\tau^{2\zeta+1} + \delta_{n,2}\, r^{2\nu}\log r \sum_{\zeta=0}^{\mu}(-1)^{\zeta} c_\zeta\, \tau^{2\zeta+1}, \quad (9)$$

where the coefficients α_0, a_ζ, b_ρ c_ζ and $d_{\rho/\zeta}$ must satisfy certain relationships derived from the permissibility conditions. The first 3 terms in Eq. (9) represent space-time nuggets (discontinuities at the space-time origin), and the 4th term is purely polynomial. The last term which is logarithmic in the space lag is obtained only in 2-D. The polynomial generalized covariance model is useful for environmental health processes that have white noise residuals, but due to its simplicity it has been widely used in geostatistics. The assumption, however, of white noise residual is restrictive; more flexible models are obtained using residuals with finite range correlations.

11.2 The Exponential Model

This is a more general model than the polynomial, which is obtained from a residual S/TRF with exponentially decaying correlations. The simplest case involves a spatially isotropic residual with separable space-time covariance,

$$c_y(\boldsymbol{r},\tau) = \sigma^2 \exp(-|\boldsymbol{r}|/\xi - |\tau|/\lambda)\ . \quad (10)$$

The form of the generalized spatiotemporal covariance associated with Eq. (10) depends on the spatial dimensionality.

In view of separability, the generalized covariance in $R^1 \times T$ is given by

$$k_{x(p)}(r,\tau) = (-1)^{\mu+\nu}\,\sigma^2\,\xi^{2\nu+2}\,\lambda^{2\mu+2}\,X_\nu(r/\xi)\,\Psi_\mu(\tau/\lambda). \tag{11}$$

The functions $X_\nu(r/\xi)$ and $\Psi_\mu(\tau/\lambda)$ depend on the dimensionless distance r/ξ and time τ/λ. The $X_\nu(r/\xi)$ is expressed as a combination of exponential and polynomial terms,

$$X_\nu(r/\xi) = \exp(-|r|/\xi) + \theta(r/\xi)\sum_{i=0}^{2\nu+1}[(r/\xi)^i - (-r/\xi)^i]/i!\,. \tag{12}$$

Analogous expressions are obtained for the space functions $\Psi_\nu(x)$, i.e.,

$$\Psi_\mu(\tau/\lambda) = \exp(-|\tau|/\lambda) + \theta(\tau/\lambda)\sum_{i=0}^{2\mu+1}[(\tau/\lambda)^i - (-\tau/\lambda)^i]/i!. \tag{13}$$

EXAMPLE 1: In Table 7 below, we show the functions $\Psi_\mu(x)$, $x = \tau/\lambda$, for three values

TABLE 7: The Functions $\Psi_\mu(x)$, $x = \tau/\lambda$, in $R^1 \times T$

	$\Psi_\mu(x)$
$\mu = 0$	$\Psi_0(x) = \exp(-\lvert x\rvert) + 2x\,\theta(x)$
$\mu = 1$	$\Psi_1(x) = \exp(-\lvert x\rvert) + \theta(x)(2x + x^3/3)$
$\mu = 2$	$\Psi_2(x) = \exp(-\lvert x\rvert) + \theta(x)(2x + x^3/3 + x^5/60)$

of the time continuity order $\mu = 0,1,2$. In Fig. 10 we plot $\Psi_\mu(x)$ vs. the dimensionless lag x. For negative lags the $\Psi_\mu(x)$ increase exponentially and do not depend on the continuity order. For positive lags, the $\Psi_\mu(x)$ are modified by the polynomial terms. The $\Psi_\mu(x)$ are not symmetric under the reflection operation $\tau \to -\tau$. □

Generalized spatiotemporal exponential covariance models are obtained by combinations of the space functions $X_\nu(r/\xi)$ and the time functions $\Psi_\mu(\tau/\lambda)$ following Eq. (11). These covariances contain no undetermined coefficients and satisfy the permissibility criteria, since they are generated directly by integration of a permissible residual covariance. The models can be augmented by adding the homogeneous solution -- i.e. polynomials of the form $p_{2\nu+1}(r)p_{2\mu+1}(\tau)$. The polynomial coefficients must satisfy relations specified by the permissibility conditions for the total generalized covariance.

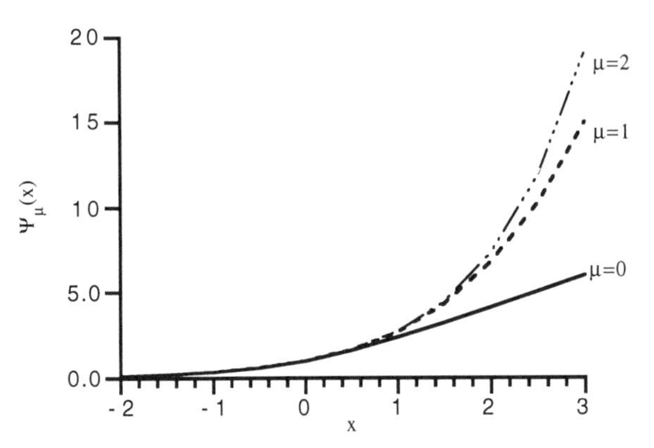

FIGURE 10: Plot of the function $\Psi_\mu(x)$ vs. the dimensionless lag $x = \tau/\lambda$ in $R^1 \times T$.

EXAMPLE 2: The exponential generalized covariance given below (Christakos, 1992; p.131) contains products of polynomial and exponential terms,

$$k_x(r,\tau) = (-1)^{\nu+\mu}\, \xi^{2\nu+2}\, \lambda^{2\mu+2} \left[\frac{\Gamma(2\nu+2,-r/\xi)\,\Gamma(2\mu+2,-\tau/\lambda)}{\Gamma(2\nu+2)\,\Gamma(2\mu+2)} \right] e^{-r/\xi-\tau/\lambda}, \tag{14}$$

and it is obtained using an asymmetric residual covariance that vanishes for negative lags. □

A symmetric residual covariance results in the simpler covariance expressions obtained above, in which the polynomial and the exponential terms are decoupled. In view of the fact that real-valued covariances must honor the property $c(\tau) = c(-\tau)$ the symmetric residual covariance model is a more physical choice.

In $R^3 \times T$, the particular solution of the generalized covariance is given by

$$k_{x(p)}(\boldsymbol{r},\tau) = (-1)^{\mu+\nu+1}\, \sigma^2\, \xi^{2\nu+2}\, \lambda^{2\mu+2}\, X_\nu(r/\xi)\, \Psi_\mu(\tau/\lambda), \tag{15}$$

The time functions $\Psi_\mu(\tau/\lambda)$ are given by Eq. (13) above. In view of Eqs. (10.4) and (10.6) the spatial functions $X_\nu(r/\xi)$ are given by the following ($x = r/\xi$),

$$X_\nu(x) = \frac{(-1)^\nu}{2^{\nu+1}\, \nu! \prod_{i=1}^{\nu+1}(2i-1)}\, x^{-1} \int_0^\infty dy\, y \exp(-y) [(x+y)^{2\nu+1} - |x-y|^{2\nu+1}]. \tag{16}$$

EXAMPLE 3: The integral in Eq. (16) can be evaluated explicitly. The results for $\nu = 0,1,2$ are shown in Table 8. We plot the three functions $X_\nu(x)$ in Fig. 11. In contrast to the 1-D

TABLE 8: The Functions $X_\nu(x)$, $x = r/\xi$, in $R^3 \times T$

	$X_\nu(x)$
$\nu = 0$	$X_0(x) = 2x^{-1}(1-e^{-x}) - e^{-x}$
$\nu = 1$	$X_1(x) = -4x^{-1}(1-e^{-x}) + e^{-x} - x$
$\nu = 2$	$X_2(x) = 6x^{-1}(1-e^{-x}) - e^{-x} + 2x(1+x^3/24)$

case, the values of the 3-D functions at zero lag depend on the continuity order. For $\nu = 0$, the $X_0(x)$ lacks a polynomial term and it decreases monotonically, in contrast with the $X_\nu(x)$ for $\nu > 0$ which increase monotonically due to the polynomial terms. The results for $\nu = 0,1,2$ can be summarized as follows

$$X_\nu(x) = (-1)^\nu [\alpha_\nu x^{-1}(1-e^{-x}) - e^{-x}] + p_\nu(x), \tag{17}$$

where α_ν are suitable coefficients. A Taylor expansion shows that $X_\nu(x)$ are not singular at zero lag, as one may suspect from the term $1/x$. Instead, the zero lag value is given by $X_\nu(0) = (-1)^\nu(\alpha_\nu - 1)$. □

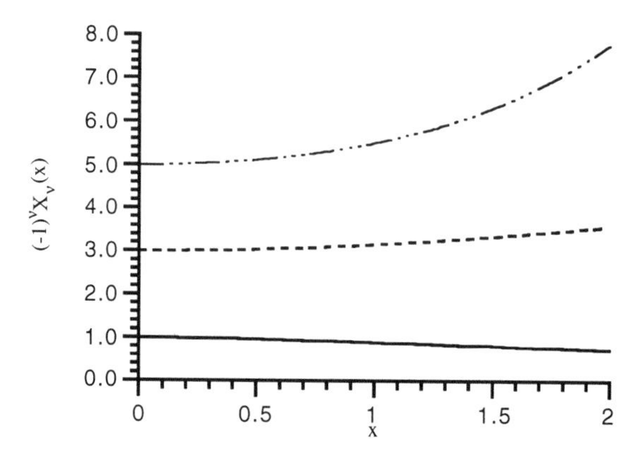

FIGURE 11: Plot of the function $(-1)^\nu X_\nu(x)$ vs. the dimensionless lag $x = r/\xi$ in $R^3 \times T$.

Generalized covariance functions in 3-D are also permissible in 2-D (Christakos, 1992). This is a useful property, since explicit construction of models in 2-D is difficult due to the logarithmic term in the spatial kernel. To date S/TRF maps are generated on spatially flat grids due to computational considerations. However, 3-D models can provide more information of effects of geographic features, and they will be used more widely in the future as the computer technology improves.

11.3 The Gaussian model

The Gaussian residual covariance is given by

$$c_y(\mathbf{r},\tau) = \sigma^2 \exp(-|\mathbf{r}|^2/\xi^2 - \tau^2/\lambda^2) \ . \tag{18}$$

The generalized covariance in $R^3 \times T$ is given by Eq. (12) above, where the spatial function $X_\nu(x)$. $x = r/\xi$, now represents the integral

$$X_\nu(x) = \frac{(-1)^\nu}{2^{\nu+1}\,\nu!\prod_{i=1}^{\nu+1}(2i-1)} x^{-1}\int_0^\infty dy\, y e^{-y^2}\,[(x+y)^{2\nu+1} - |x-y|^{2\nu+1}]. \tag{19}$$

The time functions $\Psi_\mu(x)$, $x = \tau/\lambda$, are given by the following integral,

$$\Psi_\mu(x) = [\int_{-\infty}^{x} dy\, e^{-y^2}(x-y)^{2\mu+1}]/(2\mu+1)!. \tag{20}$$

EXAMPLE 4: Explicit evaluation of $X_\nu(x)$ for $\nu = 0,1,2$ leads to the results in Table 9

TABLE 9: The Functions $X_\nu(x)$, $x = r/\xi$, in $R^3 \times T$

	$X_\nu(x)$
$\nu = 0$	$X_0(x) = \sqrt{\pi}\, erf(x)/4\,x$
$\nu = 1$	$X_1(x) = -[\sqrt{\pi}\, erf(x)(1+2x^2)/(16\,x) + e^{-x^2}/8]$
$\nu = 2$	$X_2(x) = \sqrt{\pi}\, erf(x)(3+12x^2+4x^4)/(384\,x) + (10+4x^2)e^{-x^2}/384$

below. The time functions $\Psi_\mu(x)$, for $\mu = 0,1,2$ lead to the results in Table 10 below. □

TABLE 10: The Functions $\Psi_\mu(x)$, $x = \tau/\lambda$, in $R^3 \times T$

	$\Psi_\mu(x)$
$\mu = 0$	$\Psi_0(x) = \{\sqrt{\pi}[1 + erf(x)]x + e^{-x^2}\}/2$
$\mu = 1$	$\Psi_1(x) = \{\sqrt{\pi}[1 + erf(x)](2x^3 + 3x) + 2(1 + x^2)e^{-x^2}\}/4$
$\mu = 2$	$\Psi_1(x) = \{\sqrt{\pi}[1 + erf(x)](4x^5 + 20x^3 + 15x) + (8 + 18x^2 + 4x^4)e^{-x^2}\}/8$

After this strong dose of technical analysis, we think that it is suitable to close the chapter with an a real world data set.

EXAMPLE 5: The breast cancer $\nu - \mu$ map of Fig. 3 was produced assuming a generalized covariance model of the form of Eq. (9). Observation scales Λ of varying spatiotemporal size (D in km and T in days) were used at different parts of Eastern U.S. The size of Λ is usually selected so that the average ozone estimation error at a set of control points within each Λ is minimized (at control points the measured ozone values are assumed unknown, and they are subsequently estimated using a S/TRF mapping method; see Chapter VI). The size of Λ provides an estimate of the space-time *zone of influence*. Data within the zone of influence of a control point are significant for the estimation of the ozone value at this point. The zone of influence concept has some similarities with the notion of correlation length discussed in Rao *et al.* (1995). The zone of influence has the advantage of involving composite spatial-temporal influences, while the analysis of Rao *et al.* (1995) focused on the residuals of purely temporal ozone series neglecting composite space-time variations.

Chapter V: Spatiotemporal Mapping of Environmental Health Processes - the BME Approach

"Knowledge comes from the synthesis of observations and concepts.
A priori knowledge comes purely from reasoning, independent
of observations. A posteriori knowledge comes from observations.
Without the senses we cannot become aware of any object,
but without understanding we should form no conception of it."

I. Kant

1. ABOUT SPATIOTEMPORAL MAPS

For a very large number of environmental health problems, the required outcome of the analysis is one or more spatiotemporal maps. These maps may convey visual information regarding the distribution of variables in space-time (e.g., spring water solute contents; breast cancer incidence), usually obtained from data sets. Maps may also incorporate other broad-based knowledge (general and case-specific knowledge, analytic hypotheses, models of physical laws, boundary and initial conditions, etc.; §I.5). While the first viewpoint is more descriptive, the second one is more explanatory. The *Bayesian maximum entropy* (*BME*) approach discussed in this Chapter favors a perspective that combines both: a spatiotemporal map representing the evolution of an environmental variable in space-time should be the outcome of an analysis that incorporates the set of observations available in space-time as well as other useful knowledge bases.

In practical applications, techniques for predicting exposure and health effect values are many times necessary either because measurements cannot be made of a particular situation or the assessment is being done of a future scenario. Another important advantage of these techniques is that they are much cheaper than monitoring programs.

EXAMPLE 1: An important component in environmental health management is the assessment of the consequences of a policy decision. Spatiotemporal mapping techniques can provide the decision-maker with various possible scenario-maps regarding the effects of future environmental changes, etc.. □

Mapping techniques that can be used to construct spatiotemporal maps accurately and efficiently are discussed in detail in the following sections, as well as in subsequent Chapters. To whet your appetite, a few maps are presented below.

EXAMPLE 2: In Fig. 1 we see: (a) a map of ozone concentrations over Eastern U.S..

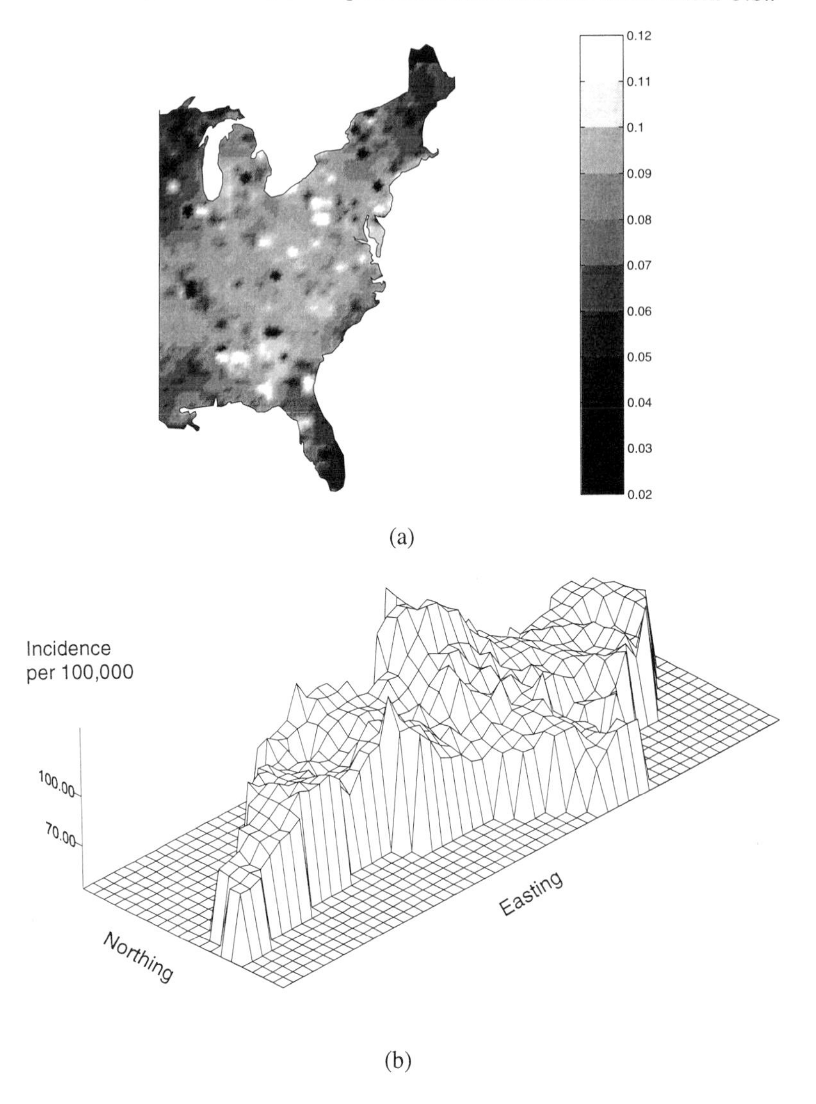

FIGURE 1 : Maps of: (a) ozone concentration (in *ppm*) over Eastern U.S. on July 15th, 1995; (b) breast cancer incidence (per 100,000 cases) in the state of North Carolina during the year 1991.

(Christakos and Vyas, 1998a), and (b) a map of breast cancer incidence in North Carolina (Christakos and Lai, 1997). The former is an indispensable tool in ascertaining compliance with ambient air quality standards; the latter can help elucidate causal mechanisms, explain disease occurrences at a certain scale, and offer guidance in health management and administration. □

The S/TRF theory of the previous Chapters will provide the necessary theoretical foundation for the development and implementation of the space-time mapping techniques in this and the following chapters. Indeed, rigorous environmental health studies can use the mathematical power of S/TRF techniques for generating space-time exposure maps. Such maps can be used for dividing the domain of interest into zones of high and low health risk according to health-based regulatory standards and exposure limits. Thus, S/TRF mapping techniques provide valuable information for etiologic implications, hazard identification, and risk management purposes. Furthermore, by comparing the results obtained using different combinations of incidence and exposure data, S/TRF models can lead to useful criteria of disease-exposure association. Other possible applications are discussed in the following sections.

2. SPACE-TIME MAPPING FUNDAMENTALS

2.1 The Basic Epistemic Framework of the BME Approach

Generally, a mapping approach has three parts: (i) the *knowledge bases* $\mathcal{K}$ available; (ii) the *estimator* $\hat{X}$, which denotes the mathematical formulation used to approximate the actual (but unknown) environmental health process X; and (iii) the *estimates* $\hat{\chi}$ of the actual values χ generated from the estimator $\hat{X}$ usually on a regular grid in space-time. These grid values constitute a *map*.

As we saw in §I.5, there are various knowledge bases available in environmental health science including (a) *general* knowledge (justified beliefs relative to the mapping situation overall, general laws of science, structured patterns and assumptions, previous experience independent of any case-specific observations, etc.); and (b) *case-specific* knowledge (measurements and perceptual evidence, empirical propositions, expertise with the specific situation, incomplete evidence, etc..). The case-specific data set χ_{data} is usually divided into two groups: Hard data χ_{hard} (i.e., obtained from real-time observation devices), and soft data χ_{soft} (interval values, probability statements, expert's assessments, etc.).

What should the important features of the space-time estimator $\hat{X}$ be? The view proposed by Christakos (1990a and b, 1992) and discussed in detail in this chapter is expressed by the following

Basic idea: *Just as any other product of scientific reasoning, space-time maps should be derived in terms of a sound epistemic framework or paradigm.*

The *holistic* system of environmental health science discussed in the preceding Chapters has two components: an epistemic component related to our understanding of a process, and an ontological component which refers to physical reality. The BME mapping of natural and health processes capitalizes on the epistemic character of the holistic system. In particular, the epistemic framework or paradigm considered in this Chapter is based on considerations that most philosophers of science will find acceptable. This is a paradigm -- essentially foreshadowed by the fundamental principles of §I.2-- which distinguishes between three main stages of knowledge-gaining, processing and interpretation as follows:

(i) The *prior* stage. Space-time mapping does not work in an intellectual vacuum. Instead it always start with a basic set of assumptions and general knowledge.

(ii) The *pre-posterior* (or *meta-prior*) stage. Case-specific knowledge is collected and analyzed.

(iii) The *posterior* stage. Results from (a) and knowledge from (b) should be processed to produce the required space-time map.

Reflecting upon the above paradigm, Christakos (1990a and b, 1992) proposed the *Bayesian maximum entropy* (BME) mapping approach that satisfies the four theses of Table 1. The crux of these theses is that a space-time mapping approach should be both

TABLE 1: Four Theses Underlying the BME Concept [from Christakos, 1990a]

Thesis 1 : At the prior stage, an inverse relation holds between information and probability: The more informative is a theory, a belief etc., the less probable it is to occur.

Thesis 2 : High prior information regarding the map. In the stochastic context this involves the maximization of the expected (or potential) value of an appropriate information measure given general knowledge.

Thesis 3 : Case-specific knowledge of the pre-posterior stage and prior probability law are introduced into knowledge processing rules to produce the posterior map.

Thesis 4 : High posterior probability (or low posterior information) for the map.

informative and cogent. More specifically, the BME way of thinking aims at a proper balance between *informativeness* (prior information maximization --Thesis 2), and *cogency* (posterior probability maximization --Thesis 4). Both requirements involve probabilities, but are conditional probabilities relative to different knowledge bases at each stage. At the prior stage the information we seek to maximize is conditioned to general knowledge. Theses 1 and 2 express standard epistemic rules: the more unspecified and general a certain statement is, the more alternatives it includes. Hence, it is more probable but also less informative. Conversely, the information increases with the number of alternatives excluded by the statement. At the posterior stage the probability we seek to maximize is conditioned to the case-specific knowledge of Thesis 3.

2.2 Other Mapping Approaches

A different approach to space-time mapping is in terms of traditional optimization criteria. Since mapping produces estimates in space-time rather than the actual values of the process of interest, an optimization criterion must be decided which these estimates should satisfy. One approach is to select a *loss function* that represents the accuracy of estimation with respect to the actual values, and then to generate estimates that minimize the expected value of the loss function. This introduces an optimization criterion based on the loss function, for which there are several options. The most common is the squared difference of the estimate and the actual value. Its expected value leads to the well-known *minimum mean square error* (*MMSE*) criterion. The latter, which is essentially a special case of BME analysis, is presented in Chapter VI.

EXAMPLE 1: The maps of Fig. 1 above were derived by means of an MMSE method. □

Yet another mapping approach is to obtain the estimates as the solutions of PDEs governing the evolution of the environmental health process in space-time; these solutions should satisfy the BC/IC available.

EXAMPLE 2: The contour maps in Fig. 2 show (a) the wetting phase (water) constant pressure contours, and (b) the non-wetting phase (oil) constant pressure contours within the effective flow. These maps are the solutions of the PDE governing two-phase flow (the pressure is measured in meters of equivalent water column). Note, however, that knowledge in terms of PDEs can be also incorporated into the BME analysis (see §3 below). □

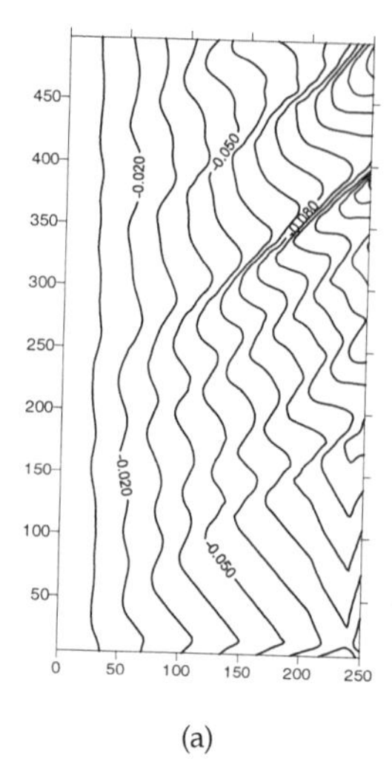

(a)

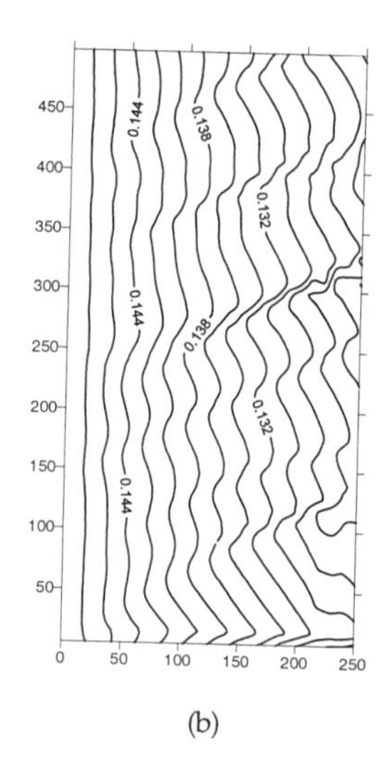

(b)

FIGURE 2: Maps of (a) wetting phase constant pressure contours; (b) non-wetting phase constant pressure contours for two-phase flow in a heterogeneous medium (Christakos *et al.*, 1998a).

2.3 Operational Concepts of Scale

In §I.6 various concepts of scale were reviewed. For mapping purposes, operational concepts of scale are necessary that characterize available measurements, suggest new types of measurements, are in agreement with the inherent scales of natural processes, and lead to better models. Two such operational definitions of scale are as follows:

Observation scale (or neighborhood): *It is a subset of the support scale that includes the data used in determining the model parameters and obtaining predictions by means of this model.*

EXAMPLE 3: In $R^2 \times T$, an observation neighborhood $\Lambda = D$ km / T mo involves data within a space-time cylinder of diameter equal to D km and height equal to T months (Fig. 3). The center of the cylinder is the reference point, e.g., an unsampled point where a prediction is sought based on the data at the other points within the neighborhood. □

Mapping scale (or resolution): *It is expressed by the resolution of the numerical grid.*

EXAMPLE 4: In $R^2 \times T$ the unit grid cell is defined as $\Delta s_1 \times \Delta s_2 \times \Delta t$, where Δs_1 and

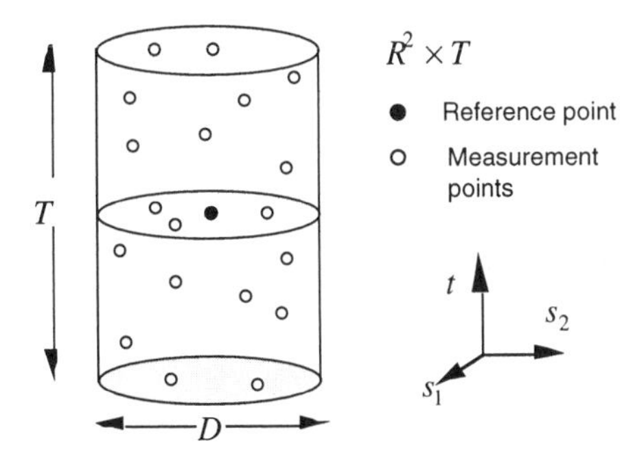

FIGURE 3: An illustration of the spatiotemporal observation scale (neighborhood); D is the spatial diameter and T is the time period considered.

Δs_2 are the lattice spacing in the two spatial directions, and Δt is the time step. □

Using finer mapping resolution leads to better resolution of the space-time variability. On the other hand, the computational effort required for an increase in resolution may be significant; in addition, one has to be concerned whether the added variability introduced by increased resolution is, indeed, a property of the system and not an artifact of the simulation method.

3 . MATHEMATICAL FORMULATION OF THE BME APPROACH

Technically, the term "space-time mapping" includes: spatial interpolation (for estimation at points that lie within the region containing the data points), spatial extrapolation (for estimation at points that lie outside the region of available data), and temporal prediction (for extrapolation to future time instants).

Let us consider the S/TRF $X(\boldsymbol{p}) = X(\boldsymbol{s},t)$ that represents an environmental health process, in general. As we saw in §I.3, there are two basic approaches of determining a point in space-time: one in terms of the pair $(\boldsymbol{s}_i, t_j)$ and one in terms of the vector $\boldsymbol{p}_i = (\boldsymbol{s},t)_i$. Each one of these approaches has its usefulness in the space-time context. A realization of the S/TRF $X(\boldsymbol{p})$ at the space-time points $\boldsymbol{p}_{\text{map}} = [\boldsymbol{p}_1 \cdots \boldsymbol{p}_m\, \boldsymbol{p}_{k_1} \cdots \boldsymbol{p}_{k_\rho}]^T$ is denoted by the set of values $\boldsymbol{\chi}_{\text{map}} = [\chi_1 \cdots \chi_m\, \chi_{k_1} \cdots \chi_{k_\rho}]^T$. Because in many mapping applications the lack of a sufficient number of hard data $\boldsymbol{\chi}_{\text{hard}}$ is the main limiting factor, incorporating other sources of case-specific knowledge (e.g., soft data $\boldsymbol{\chi}_{\text{soft}}$) in the space-time mapping procedure can be very beneficial. Certain forms of hard and soft data were already discussed in §I.5.

The multi-point, space-time mapping problem can be formulated as follows (Fig. 4):

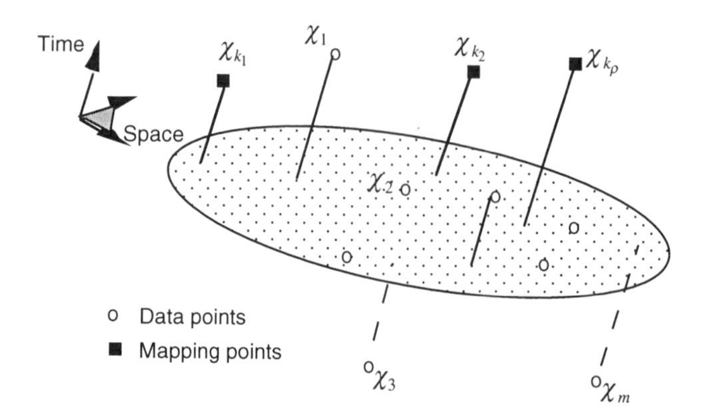

FIGURE 4: An illustration of the space-time mapping problem

Multi-point spatiotemporal mapping problem: *Consider general and case-specific knowledge bases $\mathcal{K}$ consisting of a set of hard and soft data of an environmental health process $X(\boldsymbol{p})$ at the space-time points $\boldsymbol{p}_i$ (i=1,...,m), i.e. [see, §I.5]*

$$\boldsymbol{\chi}_{\text{data}} = \boldsymbol{\chi}_{\text{hard}} \cup \boldsymbol{\chi}_{\text{soft}} = [\chi_1,\ldots,\chi_m]^T. \tag{1}$$

We seek to construct an S/TRF estimator $\hat{X}(\boldsymbol{p})$ that provides estimates $\hat{\boldsymbol{\chi}}_{\boldsymbol{k}} = [\hat{\chi}_{k_1} \cdots \hat{\chi}_{k_\rho}]^T$ of the actual (but unknown) values χ_{k_ℓ} of $X(\boldsymbol{p})$ at points $\boldsymbol{p}_{\boldsymbol{k}} = [\boldsymbol{p}_{k_1} \cdots \boldsymbol{p}_{k_\rho}]^T$ ($k_\ell \neq i$, $\ell = 1,\ldots,\rho$).

In most applications the mapping points $\boldsymbol{p}_{\boldsymbol{k}}$ lie on the nodes of a space-time grid, and the environmental health process values at the nodes are estimated based on the estimator $\hat{X}(\boldsymbol{p}_{\boldsymbol{k}})$ which is, in general, a random functional of the data $X(\boldsymbol{p}_i)$ within local neighborhoods $\Lambda_{\boldsymbol{k}}$ around $\boldsymbol{p}_{\boldsymbol{k}}$, viz.,

$$\hat{X}(\boldsymbol{p}_{\boldsymbol{k}}) = \mathcal{G}[\mathcal{K}, X(\boldsymbol{p}_i), \boldsymbol{p}_i \in \Lambda_{\boldsymbol{k}}], \tag{2}$$

where the functional $\mathcal{G}[\cdot]$ has an arbitrary form. In practical mapping situations, one seeks one realization of the field $\hat{X}(\boldsymbol{p}_k)$ at points $\boldsymbol{p}_k$; hence, Eq. (2) may be written in terms of S/TRF-realizations as

$$\hat{\boldsymbol{\chi}}_k = \mathcal{G}[\mathcal{K}, \boldsymbol{\chi}_{\text{data}}, \boldsymbol{p}_i \in \Lambda_k]. \tag{3}$$

Making clear the distinction between the random (2) and the deterministic (3) forms will avoid confusion in the derivation of the mapping equations below. In view of the above, the estimator is defined in terms of $\mathcal{G}[\cdot]$. MMSE mapping methods usually make an assumption about the form of $\mathcal{G}[\cdot]$ (e.g., linear combination of data; Chapter VI). In BME analysis the $\mathcal{G}[\cdot]$ follows directly from the available information, and no assumptions are required.

REMARK 1: While MMSE mapping (Chapter VI) provides an estimate $\hat{\chi}_k$ of χ_k at a single-point $\boldsymbol{p}_k$, multi-point BME mapping provides estimates $\hat{\chi}_{k_j}$ of χ_{k_j} at several points $\boldsymbol{p}_{k_j}$ $(j=1,...,\rho)$ simultaneously. Multi-point mapping, provided that it is practically feasible, is a considerable improvement over single-point mapping, because it involves a multivariate pdf, which is more informative than the univariate pdf used in single-point mapping. Indeed, while multi-point mapping returns several correlated estimates at different locations, single-point mapping returns one estimate independently of the process values at neighboring nodes.

To see deeper into the meaning of the epistemological framework underlying BME, some mathematical analysis is necessary below. In addition, several examples will be also included that aim to elucidate issues that are, perhaps, hidden in the general mathematical formulation. BME allows a variety of choices for the space-time estimates $\hat{\boldsymbol{\chi}}_k = [\hat{\chi}_{k_1} \cdots \hat{\chi}_{k_\rho}]^T$ at points $\boldsymbol{p}_k = [\boldsymbol{p}_{k_1} \cdots \boldsymbol{p}_{k_\rho}]^T$.

A reasonable estimate choice $\boldsymbol{\chi}_k = \hat{\boldsymbol{\chi}}_k$, is one that maximizes the posterior pdf (Thesis 4 of Table 1).

BME mapping concept: *The space-time mapping problem consists of finding estimates* $\hat{X}_k = \hat{X}(\boldsymbol{p}_k)$ *of a process* $X(\boldsymbol{p})$ *at space-time points* $\boldsymbol{p}_k = [\boldsymbol{p}_{k_1} \cdots \boldsymbol{p}_{k_\rho}]^T$, *given data (hard or soft) at points* $\boldsymbol{p}_{\text{data}} = [\boldsymbol{p}_1 \cdots \boldsymbol{p}_m]^T$ $(i=1,2,...,m \neq k_\ell,\ \ell = 1,...,\rho)$ *such that the conditions of Table 2 below are satisfied. The* $\hat{X}(\boldsymbol{p}_k)$ *is called the BME mode estimator.*

TABLE 2: Conditions of the BME Space-time Mode Estimator

Prior information maximization : The expected information

$$\overline{InfX_{\text{map}}}(\boldsymbol{p}_{\text{map}}) = -\int d\boldsymbol{\chi}_{\text{map}} f_x(\boldsymbol{\chi}_{\text{map}};\boldsymbol{p}_{\text{map}})\,\ell og\, f_x(\boldsymbol{\chi}_{\text{map}};\boldsymbol{p}_{\text{map}}) \qquad (4)$$

is maximized with respect to the prior pdf $f_x(\boldsymbol{\chi}_{\text{map}};\boldsymbol{p}_{\text{map}})$, subject to general knowledge expressed in terms of the statistics

$$\overline{g_\alpha}(\boldsymbol{p}_{\text{map}}) = \int d\boldsymbol{\chi}_{\text{map}}\, f_x(\boldsymbol{\chi}_{\text{map}};\boldsymbol{p}_{\text{map}})\, g_\alpha(\boldsymbol{\chi}_{\text{map}}), \qquad \alpha = 0,1,\ldots,N_c; \qquad (5)$$

for α=0, $g_0 = 1$, and the respective statistic is

$$\overline{g_0}(\boldsymbol{p}_{\text{map}}) = \int d\boldsymbol{\chi}_{\text{map}}\, f_x(\boldsymbol{\chi}_{\text{map}};\boldsymbol{p}_{\text{map}}) = 1, \qquad (6)$$

i.e., $\overline{g_0}$ is a normalization constant.

Pre-posterior incorporation of data : Case-specific knowledge expressed by the data vector $\boldsymbol{\chi}_{\text{data}} = \boldsymbol{\chi}_{\text{hard}} \cup \boldsymbol{\chi}_{\text{soft}} = [\chi_1,\ldots,\chi_m]^T$ is considered.

Posterior probability maximization : The case-specific knowledge is incorporated by means of the knowledge processing rule

$$f_x(\boldsymbol{\chi}_k|\boldsymbol{\chi}_{\text{data}}) = A^{-1}\, \Psi[\boldsymbol{\chi}_{\text{soft}}, f_x(\boldsymbol{\chi}_{\text{map}};\boldsymbol{p}_{\text{map}})], \qquad (7)$$

where $A = \Psi[\boldsymbol{\chi}_{\text{soft}}, f_x(\boldsymbol{\chi}_{\text{data}};\boldsymbol{p}_{\text{map}})]$ and $\Psi[\cdot]$ is an operator. Eq. (7) gives the posterior pdf, which is maximized with respect to $\boldsymbol{\chi}_k$, thus obtaining the desired estimates $\hat{\boldsymbol{\chi}}_k$, i.e.,

$$\max_{\chi_k = \hat{\chi}_k} f_x(\boldsymbol{\chi}_k|\boldsymbol{\chi}_{\text{data}}). \qquad (8)$$

Another, equivalent, way of choosing an estimate at the posterior stage is to aim at a choice $\boldsymbol{\chi}_k = \hat{\boldsymbol{\chi}}_k$ that minimizes the *information gain*

$$Inf(\boldsymbol{\chi}_k|\boldsymbol{\chi}_{\text{data}}) = Inf(\boldsymbol{\chi}_{\text{map}}) - Inf(\boldsymbol{\chi}_{\text{data}}). \qquad (9)$$

Eq. (9) provides an estimate $\hat{\boldsymbol{\chi}}_k$ that adds the minimum possible amount of information to that already extracted from the data $\boldsymbol{\chi}_{\text{data}}$. The prior pdf is, generally, a function of the space-time points $\boldsymbol{p}_{map}$, i.e., $f_x(\boldsymbol{\chi}_{\text{map}};\boldsymbol{p}_{map})$. In the following the space-time vector $\boldsymbol{p}_{map}$ will occasionally be dropped for simplicity, and the notation $f_x(\boldsymbol{\chi}_{\text{map}})$ will be used instead.

Satisfaction of the conditions outlined in Table 2, leads to the general BME space-time mapping approach summarized in Table 3. In light of Table 3, the main steps of BME

TABLE 3: The BME Space-time Mapping Approach

$$\left.\begin{aligned}&\overline{g_\alpha}(\boldsymbol{p}_{map}) = \int d\boldsymbol{\chi}_{\text{map}}\, g_\alpha(\boldsymbol{\chi}_{\text{map}}) \exp \Im[\boldsymbol{\chi}_{\text{map}};\boldsymbol{p}_{\text{map}}], \quad \alpha = 0,1,\ldots,N_c \\ &\Im[\boldsymbol{\chi}_{\text{map}};\boldsymbol{p}_{\text{map}}] = \textstyle\sum_{\alpha=0}^{N_c} \mu_\alpha(\boldsymbol{p}_{\text{map}})\, g_\alpha(\boldsymbol{\chi}_{\text{map}})\end{aligned}\right\} \tag{10}$$

$$\frac{\partial}{\partial \chi_{k_\ell}} \Psi\{\boldsymbol{\chi}_{\text{soft}}, \exp \Im[\boldsymbol{\chi}_{\text{map}};\boldsymbol{p}_{\text{map}}]\}_{\boldsymbol{\chi}_k = \hat{\boldsymbol{\chi}}_k} = 0, \qquad \ell = 1,\ldots,\rho \tag{11}$$

$$\left.\begin{aligned}&f_x(\boldsymbol{\chi}_k | \boldsymbol{\chi}_{\text{data}}) = A^{-1}\, \Psi\{\boldsymbol{\chi}_{\text{soft}}, Z^{-1} \exp \Im[\boldsymbol{\chi}_{\text{map}};\boldsymbol{p}_{\text{map}}]\} \\ &A = \Psi\{\boldsymbol{\chi}_{\text{soft}}, \exp \Im[\boldsymbol{\chi}_{\text{data}};\boldsymbol{p}_{\text{data}}]\} \\ &Z^{-1} = \exp[\mu_0(\boldsymbol{p}_{\text{map}})]\end{aligned}\right\} \tag{12}$$

space-time mapping are as follows:

(i) Eqs. (10) are solved for the *Lagrange multipliers* $\mu_\alpha(\boldsymbol{p}_{\text{map}})$.

(ii) The $\mu_\alpha(\boldsymbol{p}_{\text{map}})$ are substituted into Eqs. (11), which are then solved for the unknown estimates $\boldsymbol{\chi}_k = \hat{\boldsymbol{\chi}}_k$, with $\hat{\boldsymbol{\chi}}_k = [\hat{\chi}_{k_1} \ldots \hat{\chi}_{k_\rho}]^T$.

(iii) The posterior pdf is obtained from Eq. (12). To ensure that the solution maximizes the pdf, the Hessian of the pdf at $\boldsymbol{\chi}_k = \hat{\boldsymbol{\chi}}_k$, must be nonpositive-definite.

REMARK 2: Scientific explanation and prediction are to some extent parallel processes in space-time mapping. In this context, the posterior pdf (12) is useful not only in making predictions but, also, because it pertains to the goal of capturing significant *generalizations* which are important in the explanatory description of the natural phenomenon represented by the map. Based on a small number of porosity values (see Fig. 5 below), e.g., one can generate a map that involves a much larger set of porosity values. The statement "a porosity map is $\hat{\boldsymbol{\chi}}_k = [\underbrace{9.5\ 13.7\ \ldots\ 10.6}_{\rho=5{,}000 \text{ points}}]^T$ --the values represent percentages-- with a probability 0.7", is a generalization that provides more information than the small initial data set.

The Lagrange multipliers $\mu_\alpha(\boldsymbol{p}_{\text{map}})$, $\alpha = 0,1,\ldots,N_c$, in Eqs. (10) depend on the general knowledge, and the operator $\Psi[\cdot]$ in Eqs. (11) depends on the case-specific knowledge.

These knowledge dependencies are best explained in terms of a few examples (without loss of generality, the single-point case $\rho = 1$ is considered).

EXAMPLE 1: Assume that the general knowledge available is expressed in terms of the following g_α-functions at point $\boldsymbol{p}$

$$\left.\begin{array}{l} \overline{g_0}(\boldsymbol{p}) = 1 \\ \overline{g_\alpha}(\boldsymbol{p}) = \overline{X^\alpha}(\boldsymbol{p}) = \int d\chi\, \chi^\alpha f_x(\chi;\boldsymbol{p}),\ \alpha = 1, \ldots, N_c \end{array}\right\}. \tag{13}$$

Then, maximization of the expected information (4) gives the prior pdf

$$f_x(\chi;\boldsymbol{p}) = \exp[\textstyle\sum_{\alpha=0}^{N_c} \mu_\alpha(\boldsymbol{p})\chi^\alpha], \tag{14}$$

where the Lagrange multipliers are found by substituting Eq. (14) into Eq. (13) and solving for $\mu_\alpha(\boldsymbol{p})$ at each point $\boldsymbol{p}$. □

EXAMPLE 2: Assume that the general knowledge is represented by the PDE governing groundwater flow

$$\nabla^2 X + \nabla Y \cdot \nabla X = 0, \tag{15}$$

where $X = H(\boldsymbol{s})$ represents the hydraulic head and $Y = \ell og K(\boldsymbol{s})$ the log-conductivity. The general law is expressed in stochastic terms by means of the expectations

$$\nabla^2 \overline{X} + \overline{\nabla Y \cdot \nabla X} = 0, \tag{16}$$

or

$$\int\int d\chi\, d\psi\, \chi \nabla^2 f_{xy}(\chi, \psi;\boldsymbol{p}) + \int\int d\chi\, d\psi\, (\chi\psi) \nabla \cdot \nabla f_{xy}(\chi, \psi;\boldsymbol{p},\boldsymbol{p}')\big|_{\boldsymbol{p}=\boldsymbol{p}'} = 0, \tag{17}$$

where $\nabla \cdot \nabla = \sum_{i=1}^{n} \partial^2 / \partial p_i\, \partial p_i'$. Hence, the g_α-functions at point $\boldsymbol{p}$ should include $\overline{g_1}(\boldsymbol{p},\boldsymbol{p}') = \overline{X(\boldsymbol{p})Y(\boldsymbol{p}')}$. Maximization of the expected information (4) gives the prior pdf

$$f_x(\chi, \psi;\boldsymbol{p},\boldsymbol{p}') = \exp[\mu_0(\boldsymbol{p},\boldsymbol{p}') + \mu_1(\boldsymbol{p},\boldsymbol{p}')\chi\psi]. \tag{18}$$

By substituting Eq. (18) into Eq. (17) we get

$$\int\int d\chi\, d\psi\, \chi[\nabla^2\mu_0(\boldsymbol{p},\boldsymbol{p}')+\chi\psi\nabla^2\mu_1(\boldsymbol{p},\boldsymbol{p}')]\exp[\mu_0(\boldsymbol{p},\boldsymbol{p}')+\mu_1(\boldsymbol{p},\boldsymbol{p}')\chi\psi]+\int\int d\chi\, d\psi(\chi\psi)$$
$$[\nabla\cdot\nabla\mu_0(\boldsymbol{p},\boldsymbol{p}')+\chi\psi\nabla\cdot\nabla\mu_1(\boldsymbol{p},\boldsymbol{p}')]\exp[\mu_0(\boldsymbol{p},\boldsymbol{p}')+\mu_1(\boldsymbol{p},\boldsymbol{p}')\chi\psi]\|_{\boldsymbol{p}=\boldsymbol{p}'}=0, \qquad (19)$$

which together with the normalization equation

$$\int d\chi\, d\psi\, \exp[\mu_0(\boldsymbol{p},\boldsymbol{p}')+\mu_1(\boldsymbol{p},\boldsymbol{p}')\chi\psi]=1, \qquad (20)$$

must be solved with respect to $\mu_\alpha(\boldsymbol{p},\boldsymbol{p}')$, $\alpha=0,1$. In this formulation, the mean $\overline{g_1}(\boldsymbol{p})$ may be given implicitly as the solution of Eq. (16). □

EXAMPLE 3: Assume that the case-specific knowledge consists of the hard data of Eq. (I.5.1) and the soft (interval) data of Eq. (I.5.2). The posterior operator $\Psi[\cdot]$ is then written as (Christakos and Li, 1998)

$$\Psi\{\boldsymbol{\chi}_{\text{soft}},\boldsymbol{\chi}_{\text{map}};\boldsymbol{p}_{\text{map}}]\}=\int_{\boldsymbol{I}} d\boldsymbol{\chi}_{\text{soft}}\, f_x(\boldsymbol{\chi}_{\text{map}};\boldsymbol{p}_{\text{map}}), \qquad (21)$$

where $\boldsymbol{I}$ denotes the domain of $\boldsymbol{\chi}_{\text{soft}}$. □

EXAMPLE 4: Assume that the case-specific knowledge includes the hard data of Eq. (I.5.1) and the probabilistic data of Eq. (I.5.3). The posterior operator $\Psi[\cdot]$ is then written as (Christakos, 1998a)

$$\Psi\{\boldsymbol{\chi}_{\text{soft}},\boldsymbol{\chi}_{\text{map}};\boldsymbol{p}_{\text{map}}]\}=\int_{\boldsymbol{I}} dF_x(\boldsymbol{\chi}_{\text{soft}})\, f_x(\boldsymbol{\chi}_{\text{map}}), \qquad (22)$$

where $\boldsymbol{I}$ denotes the domain of $\boldsymbol{\chi}_{\text{soft}}$. □

Since the posterior pdf is available through the BME approach, in addition to the mode estimate, several other choices are possible. For example, one may choose as an estimate the median,

$$\hat{\boldsymbol{\chi}}_k = F_x^{-1}(0.5), \qquad (23)$$

or the conditional mean

$$\hat{\boldsymbol{\chi}}_k = \overline{X(\boldsymbol{p}_k)|\boldsymbol{\chi}_{\text{data}}}. \qquad (24)$$

Percentiles, quantiles, etc. may be also defined from the posterior pdf. Note that in the case of a symmetric pdf the estimates (22) and (24) coincide, etc..

The determination of the space-time mapping accuracy depends on the shape of the posterior pdf $f_x(\boldsymbol{\chi}_k | \boldsymbol{\chi}_{\text{data}})$. If the posterior pdf has a single maximum one can distinguish between symmetric and asymmetric cases. If the posterior pdf is symmetric around the maximum, then the maximum coincides with the mean. For many single maximum pdf's, a measure of the accuracy of the BME estimate $\hat{\chi}_k$ may be obtained by means of the standard deviation of the posterior pdf at points $\boldsymbol{p}_k$, i.e.,

$$\sigma_x(\boldsymbol{p}_k) = StDev[f_x(\boldsymbol{\chi}_k | \boldsymbol{\chi}_{\text{data}})]. \tag{25}$$

Eq. (25) offers an excellent measure of mapping accuracy in cases of symmetric or approximately symmetric pdfs. In the case of an asymmetric posterior pdf, the best estimate and the standard deviation do not necessarily offer a satisfactory description of the situation and we need to look at the full distribution as provided by the $f_x(\boldsymbol{\chi}_k | \boldsymbol{\chi}_{\text{data}})$ at points $\boldsymbol{p}_k$. However, in many situations simpler alternative descriptions provide useful results.

EXAMPLE 5: Consider the single-point estimation with $\rho = 1$. For an asymmetric pdf with a single maximum, a reasonable measure of accuracy of the BME estimate is provided by the *confidence interval* $w_{x,\eta}(\boldsymbol{p}_k)$ determined as follows: Given an appropriate value η for the probability $P[\hat{\chi}_k - a \leq X(\boldsymbol{p}_k) \leq \hat{\chi}_k + b]$, the values a and b must be determined so that

$$\eta = P[\hat{\chi}_k - a \leq X(\boldsymbol{p}_k) \leq \hat{\chi}_k + b] = q^{-1} \int_{\hat{\chi}_k - a}^{\hat{\chi}_k + b} d\chi_k \, f_x(\chi_k | \boldsymbol{\chi}_{\text{data}}), \tag{26}$$

where q is a normalization constant. Then, the confidence interval is given by

$$w_{x,\eta}(\boldsymbol{p}_k) = a + b, \tag{27}$$

and the $X(\boldsymbol{p}_k)$ lies in the interval $[\hat{\chi}_k - a, \hat{\chi}_k + b]$ with probability η. □

In the case of several maxima, the estimation uncertainty should be considered separately for each maximum. A complete posterior pdf may be necessary in order to obtain a satisfactory picture of the existing uncertainties.

4. SOME ANALYTICAL RESULTS

Depending on the knowledge bases available (hard vs. soft data, etc.) and the goals of mapping (multi-point vs. single-point, etc.) a variety of analytical formulations of BME are possible. Due to space limitations, theoretical formulations for only a few cases are considered below. However, there are infinite possibilities, limited only by the availability of data and physical knowledge. This theoretical richness is a real powerful feature of BME analysis. While no proofs are given for most of the analytical results presented, proper referencing is made to the existing literature.

In all cases considered below, we assume that hard data exist at points $\boldsymbol{p}_i$ ($i = 1,2,...,m_h$) as well as soft data of various possible forms (such as those discussed in §I.5) at points $\boldsymbol{p}_i$ ($i = m_h + 1,...,m$).

4.1 Multi-Point BME

Proofs of the following propositions can be found in Christakos (1998a, b and c) and Christakos *et al.* (1998b).

Proposition 1: *The general knowledge is the non-centered ordinary covariance. Then, the multi-point BME estimate* $\boldsymbol{\chi}_k = [\chi_{k_1} \cdots \chi_{k_\rho}]^T$ *is the solution of the set of equations*

$$\sum_{i=1}^{m_h} C^{-1}_{ik_\ell} \chi_i + \sum_{i=m_h+1}^{m} C^{-1}_{ik_\ell} \overline{x_i}(\hat{\boldsymbol{\chi}}_k) + \sum_{i=k_1}^{k_\rho} C^{-1}_{ik_\ell} \hat{\chi}_i = 0, \quad \ell = 1,...,\rho. \tag{1}$$

$C^{-1}_{ik_\ell}$ *is the* ik_ℓ*-th element of the inverse matrix* $\boldsymbol{C}^{-1}_{\text{map}}$*, where* $\boldsymbol{C}_{\text{map}}$ *is the matrix of the non-centered covariances* C_{ik_ℓ} *between the points* $\boldsymbol{p}_i$ ($i = 1,...,m$) *and* $\boldsymbol{p}_{k_\ell}$ ($\ell = 1,...,\rho$)*; and*

$$\overline{x_i}(\hat{\boldsymbol{\chi}}_k) = \frac{B\int_D d\Xi(\boldsymbol{\chi}_{\text{soft}})\chi_i \exp\Im[\boldsymbol{\chi}_{\text{map}}]_{\boldsymbol{\chi}_k = \hat{\boldsymbol{\chi}}_k}}{B\int_D d\Xi(\boldsymbol{\chi}_{\text{soft}}) \exp\Im[\boldsymbol{\chi}_{\text{map}}]_{\boldsymbol{\chi}_k = \hat{\boldsymbol{\chi}}_k}}, \tag{2}$$

where the B*,* $\boldsymbol{D}$ *and* $\Xi(\cdot)$ *determine the form of the* Ψ*-posterior operator.*

The form of the Ψ-posterior operator depends on the case-specific knowledge available. In particular: if the soft data are of the form of Eq. (I.5.2), then $B = A^{-1}$, $\boldsymbol{D} = \boldsymbol{I}$

and $\Xi(\boldsymbol{\chi}_{\text{soft}}) = \boldsymbol{\chi}_{\text{soft}}$; if the soft data are of the form of Eq. (I.5.3), then $B = A^{-1}$, $\boldsymbol{D} = \boldsymbol{I}$ and $\Xi(\boldsymbol{\chi}_{\text{soft}}) = F_x(\boldsymbol{\chi}_{\text{soft}})$; etc..

4.2 Single-Point BME

Proposition 2: *The general knowledge is the centered ordinary covariance. Then, the BME estimate* $\hat{\chi}_k = \chi_{\text{mode}}$ *is the solution of equation*

$$\sum_{i=1}^{m_h} c_{ik}^{-1}(\chi_i - \overline{x_i}) + \sum_{i=m_h+1}^{m} c_{ik}^{-1}[\overline{x_i}(\hat{\chi}_k) - \overline{x_i}] + c_{kk}^{-1}(\hat{\chi}_k - \overline{x_k}) = 0, \tag{3}$$

where c_{ik}^{-1} *is the* ik*-th element of the inverse matrix* $\boldsymbol{c}_{\text{map}}^{-1}$, *where* $\boldsymbol{c}_{\text{map}}$ *is the matrix of the centered covariances* c_{ik} *between the points* $\boldsymbol{p}_i$ ($i = 1,...,m$) *and* $\boldsymbol{p}_k$,

$$\overline{x_i}(\hat{\chi}_k) - m_i = \frac{B\int_D d\Xi(\boldsymbol{\chi}_{\text{soft}})(\chi_i - m_i)\exp\Im[\boldsymbol{\chi}_{\text{map}}]_{\chi_k=\hat{\chi}_k}}{B\int_D d\Xi(\boldsymbol{\chi}_{\text{soft}})\exp\Im[\boldsymbol{\chi}_{\text{map}}]_{\chi_k=\hat{\chi}_k}}. \tag{4}$$

and the B, $\boldsymbol{D}$, $\Xi(\cdot)$ *were defined in §3.1 above.*

EXAMPLE 1: In this study we will use porosity data from the West Lyon field in West-Central Kansas (Olea, 1997). The geomorphology of the area includes Mississippian (Lower Carboniferous) sediments deposited in the shallow epicontinental seas that covered much of the North America in the Late Paleozoic. The porosity data (in %) were collected over an area of approximately 2.5×4.5 *miles*^{2}. A total of 76 data were available. The analysis of the porosity field is two-fold. In the first part we will allow BME to use all 76 hard data. The general knowledge consists of the mean and the semi-variogram (Fig. 5) of the porosity data set. Based on these data, BME produces the porosity map of Fig. 6. Note that this BME map is the same as the simple kriging (SK; Chapter VI) map obtained using the same data, mean and semi-variogram (a detailed comparison of BME vs. minimum mean square error techniques such as SK is given in Chapter VI that follows). In the second part of the analysis we used only 56 hard data, while at the remaining 20 points we assumed soft interval data with a width of 1 unit and centered at the measured porosity value (assumed unknown). BME analysis accounts for both hard and soft data, and produces the map of Fig. 7; note that despite the uncertainty introduced by the soft data, it closely resembles the spatial structure of the previous map (Fig. 6). □

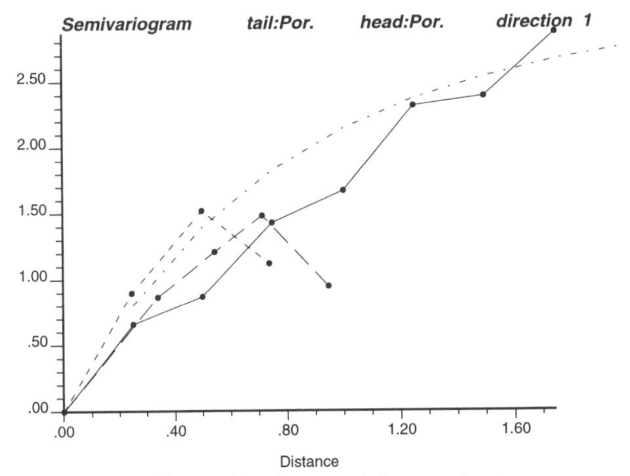

FIGURE 5: The semi-variogram of the porosity data

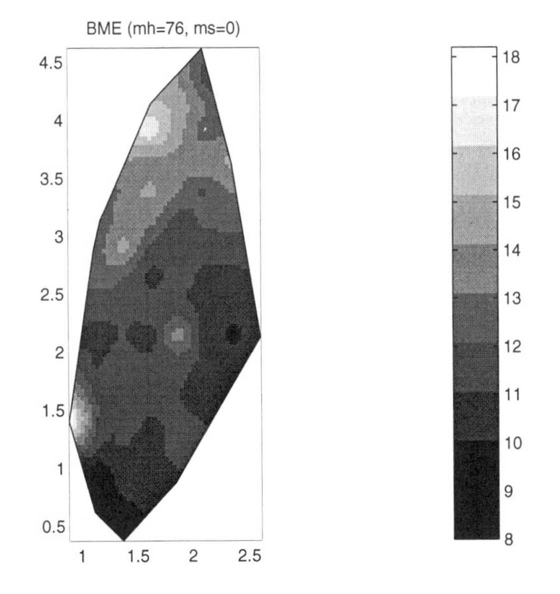

FIGURE 6: The BME porosity map using 76 hard porosity data (in %).

Proposition 3: *The general knowledge is the semi-variogram. Then, the BME estimate $\hat{\chi}_k$ is the solution of equation*

$$\hat{\chi}_k \sum_{i=1}^{m} \gamma_{ik}^{-1} - \sum_{i=1}^{m_h} \gamma_{ik}^{-1} \chi_i - \sum_{i=m_h+1}^{m} \gamma_{ik}^{-1} \overline{x_i}(\hat{\chi}_k) = 0\,; \tag{5}$$

γ_{ik}^{-1} *is the ij-th element of the inverse matrix* $\boldsymbol{\gamma}_{\text{map}}^{-1}$, *where*

$$\boldsymbol{\gamma}_{\text{map}}: \; \gamma_{\text{map},ij} = \gamma_{ik}\, \delta_{i,j}(1-\delta_{i,m+1}) \tag{6}$$

is the semi-variogram matrix between the points $\boldsymbol{p}_i$ $(i=1,...,m)$ *and* $\boldsymbol{p}_k$, $\gamma_{ik} = \frac{1}{2}\overline{(\chi_i - \chi_k)^2}$, *and* $\overline{x_i}(\hat{\chi}_k)$ *is of the form of Eq. (4) above with*

$$\Im[\boldsymbol{\chi}_{\text{map}}] = -\tfrac{1}{4}\sum_{i=1}^{m} \gamma_{ik}^{-1}(\chi_i - \chi_k)^2. \tag{7}$$

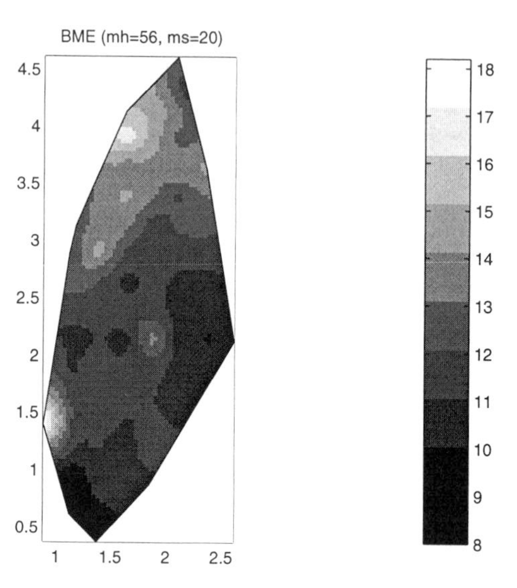

FIGURE 7: The BME porosity map using 56 hard and 20 soft data (in %).

Proposition 4: *Assume that we are dealing with an S/TRF-* ν/μ *(Chapter IV) in which case the general knowledge is the generalized spatiotemporal covariance of order* ν/μ *between any pair of points in* $\boldsymbol{p}_{\text{map}} = \{\boldsymbol{p}_i;\; i=1, ..., m, k\}$. *The BME estimate* $\hat{\chi}_k$ *is the solution of equation*

$$\frac{\partial}{\partial \chi_k}\Psi[\boldsymbol{\chi}_{\text{soft}}, f_x(\boldsymbol{\chi}_{\text{map}})]_{\chi_k=\hat{\chi}_k} = B\int_D d\Xi(\boldsymbol{\chi}_{\text{soft}})\{\exp[\Im(\boldsymbol{\chi}_{\text{map}})]\frac{\partial\Im(\boldsymbol{\chi}_{\text{map}})}{\partial\chi_k}\}_{\chi_k=\hat{\chi}_k} = 0 \tag{8}$$

where

$$\Im[\boldsymbol{\chi}_{\text{map}}] = -\tfrac{1}{2}\boldsymbol{Q}^T(\boldsymbol{\chi}_{\text{map}})\boldsymbol{c}_Q^{-1}(\boldsymbol{k}_{\text{map}})\,\boldsymbol{Q}(\boldsymbol{\chi}_{\text{map}}), \tag{9}$$

where $\boldsymbol{Q}$ *is the vector operator associated with the S/TRF-* ν/μ, *and* $\boldsymbol{k}_{\text{map}}$ *is the generalized covariance matrix between the space-time points* $\boldsymbol{p}_i$ *such that*

$$\boldsymbol{c}_Q(\boldsymbol{k}_{\text{map}}) = \overline{\boldsymbol{Q}(\boldsymbol{\chi}_{\text{map}})\boldsymbol{Q}^T(\boldsymbol{\chi}_{\text{map}})}. \tag{10}$$

Other applications may be found in Christakos (1992, 1998a, b and c), Christakos and Li (1998), Serre *et al.* (1998) and Christakos *et al.* (1998b).

5. OTHER FORMULATIONS OF THE BME APPROACH

5.1 Functional BME

For notational simplicity, single-point estimation ($\rho = 1$) will be considered. Suppose that we are seeking an estimate of the functional

$$X_v(\boldsymbol{p}_k) = F[X(\boldsymbol{p}), v], \tag{1}$$

where $F[\cdot,\cdot]$ is an arbitrary function, given data (hard and/or soft) χ_{data} at points $\boldsymbol{p}_i$ ($i = 1,\ldots,m$). The Lagrange multipliers μ_α are solutions of the system of equations

$$\overline{g_\alpha} = Z^{-1}\int d\boldsymbol{\chi}_{\text{data}}\int d\chi_v\, g_\alpha(\boldsymbol{\chi}_{\text{data}}, \chi_v)\exp\Im, \quad \alpha = 0,1,\ldots,N_c; \tag{2}$$

the BME equation is given by

$$\frac{\partial}{\partial \chi_v}\Psi[\boldsymbol{\chi}_{\text{soft}},\, Z^{-1}\exp\Im]\Big|_{\chi_v=\hat{\chi}_v} = 0, \tag{3}$$

and the posterior pdf is

$$f_x(\chi_v|\boldsymbol{\chi}_{\text{data}}) = A^{-1}\,\Psi[\boldsymbol{\chi}_{\text{soft}},\, Z^{-1}\exp\Im], \tag{4}$$

where $\Im = \sum_{\alpha=1}^{N_c}\mu_\alpha\, g_\alpha(\boldsymbol{\chi}_{\text{data}}, \chi_v)$ and A is a normalization constant. The form of the Ψ-posterior operator in Eqs. (3) and (4) depends on the case-specific knowledge available.

EXAMPLE 1: Assume that the case-specific knowledge consists of the hard data (I.5.1) and the soft (interval) data (I.5.2). Then, Eqs. (3) and (4) reduce to

$$\sum_{\alpha=1}^{N_c} \mu_\alpha \int_I d\boldsymbol{\chi}_{\text{soft}} \left[\frac{dg_\alpha(\boldsymbol{\chi}_{\text{data}}, \chi_v)}{d\chi_v} \exp \Im\right]_{\chi_v = \hat{\chi}_v} = 0, \tag{5}$$

$$f_x(\chi_v | \boldsymbol{\chi}_{\text{data}}) = (AZ)^{-1} \int_I d\boldsymbol{\chi}_{\text{soft}} \exp \Im. \tag{6}$$

The g_α are now functions of the point and the functional random fields involved (they may include point, functional as well as point-functional covariances, etc.). □

BME analysis may provide a method for studying change-of-scale problems. Let us illustrate this useful feature by means of the following

EXAMPLE 2: Let χ_{eff} be the effective parameter of a medium. Assume that data (hard and soft) χ_j ($j = 1, 2, \ldots, m$) are distributed on a grid whose grid-block length is representative of the original data sample size. The original grid blocks are coalesced into larger blocks, e.g., $\ell << m$ neighboring blocks are coalesced into one block. The grid block effective parameters of the ℓ-block configurations are calculated in some way, say

$$\chi_{eff} = F[\boldsymbol{\chi}_{\text{data}}^{(\ell)}] \tag{7}$$

where $\boldsymbol{\chi}_{\text{data}}^{(\ell)} = [\chi_1 \ldots \chi_\ell]^T$, and $F[\cdot]$ is a known transformation (e.g., in the case of χ_{eff}=effective permeability of a medium, $F[\cdot]$ could be some kind of equivalent resistor network formula). Assuming that the case-specific knowledge consists of the hard data of Eq. (I.5.1) and the soft (interval) data of Eq. (I.5.2), the BME equations is written as

$$\sum_{\alpha=1}^{N_c} \mu_\alpha \int_I d\boldsymbol{\chi}_{\text{soft}}^{(\ell)} \left[\frac{dg_\alpha(\boldsymbol{\chi}_{\text{data}}^{(\ell)}, \chi_{eff})}{d\chi_{eff}} \exp \Im\right]_{\chi_{eff} = \hat{\chi}_{eff}} = 0, \tag{8}$$

and

$$\overline{g_\alpha} = Z^{-1} \int d\boldsymbol{\chi}_{\text{data}}^{(\ell)} \int d\chi_{eff}\, g_\alpha(\boldsymbol{\chi}_{\text{data}}^{(\ell)}, \chi_{eff}) \exp \Im, \quad \alpha = 0, 1, \ldots, N_c, \tag{9}$$

where $\Im = \sum_{\alpha'=1}^{N_c} \mu_{\alpha'} g_{\alpha'}(\boldsymbol{\chi}_{\text{data}}^{(\ell)}, \chi_{eff})$, and the $\overline{g_\alpha} = \overline{g_\alpha}(\boldsymbol{\chi}_{\text{data}}^{(\ell)}, \chi_{eff}) = \overline{g_\alpha}(\boldsymbol{\chi}_{\text{data}}^{(\ell)}, F[\boldsymbol{\chi}_{\text{data}}^{(\ell)}])$ should be calculated from the data. This includes the statistics of the point values χ_j, as well as the statistics of the effective parameter χ_{eff}. For example, the mean $\overline{g_1} = \overline{x_{eff}}$ is easily found by averaging the χ_{eff}-values assigned by equation

$$X_v(\boldsymbol{p}_k) = v^{-1} \int_{v(\boldsymbol{p}_k)} d\boldsymbol{u}\, X(\boldsymbol{s}_k - \boldsymbol{u}, t) \tag{10}$$

at every ℓ-block configuration. Various other statistical moments of χ_{eff} can be found in a similar fashion. The solution of Eqs. (8) and (9) determines the Lagrange multipliers μ_α and the $\hat{\chi}_{eff}$, which will be the effective parameter of the original heterogeneous medium. Moreover, the posterior pdf of χ_{eff} is given by

$$f_x(\chi_{eff} | \boldsymbol{\chi}_{\text{data}}^{(\ell)}) = (AZ)^{-1} \int_I d\boldsymbol{\chi}_{\text{soft}}^{(\ell)} \exp \Im. \tag{11}$$

If necessary, the procedure can be repeated along the lines of a renormalization approach (e.g., Creswick *et al.*, 1992). I.e., the χ_{eff}-values obtained from Eq. (7) for the ℓ-block configurations are assigned to a new, coarser grid; then, the F-transformation is applied to these χ_{eff}-values, leading to a new effective estimate $\hat{\chi}_{eff}(1)$ and a new posterior pdf with a smaller variance. The procedure is repeated several (say, n) times until the posterior pdf obtained does not show a significant change. The corresponding $\hat{\chi}_{eff}(n)$-value will be the effective parameter of the medium. □

5.2 Vector BME

Assume that the prime environmental health process $X(\boldsymbol{p})$ is related to the $N-1$ secondary processes $\boldsymbol{W} = [W_2(\boldsymbol{p}), \ldots, W_N(\boldsymbol{p})]^T$. We seek to estimate $X(\boldsymbol{p})$ at point $\boldsymbol{p}_k$. In this case, the Lagrange multipliers μ_α are the solution of the system of equations

$$\overline{g_\alpha} = Z^{-1} \int d\boldsymbol{\chi}_{\text{map}} \int d\boldsymbol{\Omega}_{\text{soft}}\, g_\alpha(\boldsymbol{\chi}_{\text{map}}, \boldsymbol{\Omega}_{\text{data}}) \exp \Im, \quad \alpha = 0, 1, \ldots, N_c, \tag{7}$$

where $\boldsymbol{\Omega}_{\text{data}} = [\boldsymbol{\omega}_{2,\text{data}}, \ldots, \boldsymbol{\omega}_{N,\text{data}}]^T$ and $\boldsymbol{\Omega}_{\text{soft}} = [\boldsymbol{\omega}_{2,\text{soft}}, \ldots, \boldsymbol{\omega}_{N,\text{soft}}]^T$ are data of the secondary processes $\boldsymbol{W}$, and $\Im = \sum_{\alpha'=1}^{N_c} \mu_{\alpha'} g_{\alpha'}(\boldsymbol{\chi}_{\text{map}}, \boldsymbol{\Omega}_{\text{data}})$; the BME estimation equation is given by

$$\frac{\partial}{\partial \chi_k} \Psi[\boldsymbol{\chi}_{\text{soft}}, \boldsymbol{\Omega}_{\text{soft}}, Z^{-1} \exp \Im] \Big|_{\chi_k = \hat{\chi}_k} = 0, \tag{8}$$

and the posterior pdf is

$$f_x(\chi_k|\boldsymbol{\chi}_{\text{data}}, \boldsymbol{\Omega}_{\text{data}}) = A^{-1}\,\Psi[\boldsymbol{\chi}_{\text{soft}}, \boldsymbol{\Omega}_{\text{soft}}, \; Z^{-1}\exp\Im]. \tag{9}$$

The g_α are important functions that incorporate whatever is known about the two fields. This knowledge may include spatiotemporal statistics (covariances, cross-covariances, etc.), physical laws, etc..

6. BME vs. MMSE ESTIMATES

The BME theory is formulated in a way that preserves most of the references of earlier theories, which are its limiting cases. As we shall see in more detail in Chapter VI (§VI.7 and 8), for Gaussian fields the MMSE estimator obtained using a set of hard data and up to second order statistics coincides with the BME estimate derived using the same hard data and prior statistics (in both cases the estimator is a linear combination of the data). Moreover, popular linear MMSE estimators, such as kriging, are limiting cases of the BME estimator.

Chapter VI: SPATIOTEMPORAL MMSE MAPPING

"The real truths are those that can be invented."
K. Kraus

1. INTRODUCTION

This chapter treats spatiotemporal *minimum mean square error* ($MMSE$) mapping techniques in considerable detail. Due to their popularity in many of scientific fields, we decided to include a separate chapter on MMSE techniques, despite the fact that in principle these techniques are special cases of the BME analysis. As we already discussed in the previous chapter, mapping techniques provide the tools for generating accurate predictive maps and for deductive analysis. Mapping requires important decisions regarding a few crucial issues, such as (i) the type of data to be included, (ii) the mapping objectives and (iii) the form of the estimator. The MMSE mapping techniques considered in this chapter are designed to incorporate mainly hard data. The other two issues are addressed as follows.

Since mapping methods generate space-time estimates rather than the actual values of the mapped process, an optimization criterion is required. There are several ways to implement an optimization criterion. One way is to define a *loss function* L expressing the deviation of the estimate from the actual values. Then, the estimate is the value that minimizes the expectation $\overline{L}$ of the loss function. There are several options for the loss function, the most widely used being the square of the difference between the estimate and the actual value. The expected value $\overline{L}$ of this loss function leads to the well-known MMSE criterion on which the estimators presented in this chapter are based.

The choice of estimators $\hat{X}$ used to produce the MMSE maps is usually restricted within the limits of a specific class of functions for reasons of mathematical convenience and computational efficiency. In general, linear forms are easier to use than arbitrary (non-linear) ones. The price paid for linearity is the generation of "sub-optimal" estimators and lower quality mapping. One should not ignore the fact, however, that there exist a few special cases in which linear estimators are optimal.

2. MATHEMATICAL FORMULATION

Within the MMSE context the space-time mapping problem can be formulated as follows:

Space-time mapping problem: *Consider a set of hard data χ_i of a process $X(\boldsymbol{p})$ at the space-time points $\boldsymbol{p}_i$ (i=1,...,m) represented by the vector*

$$\chi_{\text{hard}} = [\chi_1, \ldots, \chi_m]^T. \tag{1}$$

We seek to construct an S/TRF estimator $\hat{X}(\boldsymbol{p})$ that provides estimates of the actual (but unknown) values of the S/TRF $X(\boldsymbol{p})$ at points $\boldsymbol{p}_k$ ($k \neq i$), located on a regular space-time grid.

Generally, the estimator $\hat{X}(\boldsymbol{p}_k)$ will be a random functional of the available data $X(\boldsymbol{p}_i)$ within a local neighborhood Λ_k around $\boldsymbol{p}_k$, viz.,

$$\hat{X}(\boldsymbol{p}_k) = \mathcal{G}[X(\boldsymbol{p}_i), \boldsymbol{p}_i \in \Lambda_k], \tag{2}$$

where the functional $\mathcal{G}[\cdot]$ will be determined on the basis of the MMSE estimation process. In practical mapping applications, we seek one realization of the field $\hat{X}(\boldsymbol{p}_k)$ at each point $\boldsymbol{p}_k$; hence, Eq. (2) may be written in terms of S/TRF-realizations as

$$\hat{\chi}_k = \mathcal{G}[\boldsymbol{\chi}_{\text{hard}}, \boldsymbol{p}_i \in \Lambda_k]. \tag{3}$$

Making clear the distinction between the random (2) and the deterministic (3) forms will prevent confusion in the derivation of the mapping equations below. Note that unlike Eqs. (VI.3.2) and (VI.3.3), the above equations do not depend on knowledge bases $\mathcal{K}$ other than the available hard data.

The choice of the S/TRF estimator $\hat{X}$ can be made in a variety of ways. One major group of estimators involves the minimization of a single loss function with respect to the estimator. More specifically, let $L[\hat{X}(\boldsymbol{p}_k), X(\boldsymbol{p}_k)]$ be a loss function expressing the "cost" of using the estimator $\hat{X}(\boldsymbol{p}_k)$ instead of the real process is $X(\boldsymbol{p}_k)$. The estimate should minimize the expected value of the loss function, namely,

$$\overline{L} = \int d\chi_{\text{hard}}\, d\chi_k\, L[\hat{\chi}_k, \chi_k] f_x(\chi_{\text{data}}, \chi_k), \tag{4}$$

where all the components of the vector χ_{hard} as well as χ_k are assumed to be integrated in the above. By minimizing Eq. (4) with respect to $\mathcal{G}$ taking into account Eq. (3), one obtains the following *fundamental integral equation* of space-time mapping

$$\int d\chi_k \frac{\partial L\{\mathcal{G}[\boldsymbol{\chi}_{\text{hard}}], \chi_k\}}{\partial \mathcal{G}} f_x(\boldsymbol{\chi}_{\text{hard}}, \chi_k) = 0. \tag{5}$$

To proceed further we need to choose a loss function L. This choice depends on a variety of factors including the available information/data and the mapping objectives. It can be shown (Christakos, 1992) that if L is the absolute error (where the error is defined as the difference between the estimator and the actual process), Eq. (5) leads to the median estimator. Also, if L is the square error, Eq. (4) defines the *minimum mean square error (MMSE)* criterion and Eq. (5) provides the conditional mean estimator

$$\hat{X}(\boldsymbol{p}_k) = \overline{X(\boldsymbol{p}_k)|\chi_{\text{hard}}}. \tag{6}$$

In this chapter we will consider MMSE estimators. Moreover, due to practical considerations related to the calculation of the statistics (means, covariances) involved in the derivation of the estimator $\hat{X}(\boldsymbol{p}_k)$, additional assumptions are needed regarding the functional form of $\mathcal{G}$. Linear $\mathcal{G}$-forms will be examined below.

3. LINEAR MMSE SPACE-TIME MAPPING

3.1 Mathematical Formulation

We assume that measurements of the field are available at discrete locations $\boldsymbol{s}_i$ $(i = 1, 2, \ldots, m)$ and times t_j $(j = 1, 2, \ldots, p)$. In practice, measurements may not be available for all time instants at every location; due to this reason the $\boldsymbol{p}$-notation is preferred. Hence, we assume that there are measurements at the space-time points $\boldsymbol{p}_i$, $i = 1, \ldots, N$ where $N = m \times p$. We seek to establish a mapping approach that provides linear estimates $\hat{X}(\boldsymbol{p}_k)$ at several unmeasured points $\boldsymbol{p}_k \neq \boldsymbol{p}_i$ located on a regular space-time grid, i.e.,

$$\hat{X}(\boldsymbol{p}_k) = \sum_{i=1}^{N'} \xi_i X(\boldsymbol{p}_i), \tag{1}$$

where ξ_i is the weight associated with the datum at $\boldsymbol{p}_i$; The linear sum includes measurements at points in the neighborhood of $\boldsymbol{p}_k$, hence $N' \leq N$. Such an approach is indeed suggested by the

Biased MMSE concept: *Find estimates $\hat{X}(\boldsymbol{p}_k)$ of a process $X(\boldsymbol{p}_k)$ at space-time points $\boldsymbol{p}_k$, given hard data at points $\boldsymbol{p}_i$ ($i=1,2,\ldots,N \neq k$), such that the conditions of Table 1 are satisfied.*

TABLE 1: Conditions of the Linear MMSE Space-Time Estimator

Linearity : The estimator is a linear function of the data, i.e., Eq. (1) is valid.

Loss-function : This is the square of the difference between the estimator and the actual (but unknown) S/TRF, i.e.,

$$L[\hat{X}, X] = [\hat{X}(\boldsymbol{p}_k) - X(\boldsymbol{p}_k)]^2. \tag{2}$$

Optimality : The expected value of the loss function (mean square estimation error)

$$\sigma_x^2(\boldsymbol{p}_k) = \overline{L[\hat{X}, X]} = \overline{[\hat{X}(\boldsymbol{p}_k) - X(\boldsymbol{p}_k)]^2}, \tag{3}$$

is minimized with respect to the weights ξ_i.

Let us express Eq. (1) in the vector form

$$\hat{X}(\boldsymbol{p}_k) = \boldsymbol{\xi}^T \boldsymbol{X}, \tag{4}$$

where $\boldsymbol{\xi}$ represents the vector of weights

$$\boldsymbol{\xi} = [\xi_1 \ldots \xi_{N'}]^T, \tag{5}$$

and $\boldsymbol{X}$ represents the S/TRF vector at the measured points

$$\boldsymbol{X} = [X(\boldsymbol{p}_1) \ldots X(\boldsymbol{p}_{N'})]^T. \tag{6}$$

REMARK 1: For future reference, note that Eq. (4) can be written in terms of realizations as

$$\hat{\chi}_k = \boldsymbol{\xi}^T \boldsymbol{\chi}_{\text{hard}} \tag{7}$$

where $\boldsymbol{\chi}_{\text{hard}} = [\chi_1 \ldots \chi_{N'}]^T$; see, also, Eq. (2.3) above.

By substituting Eq. (4) into Eq. (3) and minimizing the resulting expression with respect to the weights ξ_i we obtain the linear system of equations

$$\mathbf{C}_{\mathbf{x}}\,\xi = \mathbf{C}^*_{\mathbf{x}} \tag{8}$$

where

$$\mathbf{C}_{\mathbf{x}} = \overline{\boldsymbol{X}\boldsymbol{X}^T} \tag{9}$$

represents the covariance matrix between the data points with elements $C_{ij} = C_x(\boldsymbol{p}_i;\boldsymbol{p}_j)$, $i, j = 1,\ldots,N'$; and $\mathbf{C}^*_{\mathbf{x}}$ the vector of covariances between the data points and the estimation point

$$\mathbf{C}^*_{\mathbf{x}} = \overline{X(\boldsymbol{p}_k)\boldsymbol{X}} = [C_x(\boldsymbol{p}_1;\boldsymbol{p}_k) \;\ldots\; C_x(\boldsymbol{p}_{N'};\boldsymbol{p}_k)]^T \tag{10}$$

The solution of Eq. (8) gives $\xi = \mathbf{C}^{-1}_{\mathbf{x}}\,\mathbf{C}^*_{\mathbf{x}}$, and the corresponding estimation error variance (3) can be written as

$$\begin{aligned}\sigma^2_x(\boldsymbol{p}_k) &= \overline{X(\boldsymbol{p}_k)^2} + \xi^T\,\overline{\boldsymbol{X}\boldsymbol{X}^T}\,\xi - 2\xi^T\,\overline{X(\boldsymbol{p}_k)\boldsymbol{X}} = \overline{X(\boldsymbol{p}_k)^2} + \xi^T\,\mathbf{C}_{\mathbf{x}}\,\xi - 2\,\xi^T\mathbf{C}^*_{\mathbf{x}} \\ &= \overline{X(\boldsymbol{p}_k)^2} - \mathbf{C}^{*T}_{\mathbf{x}}\,\mathbf{C}^{-1}_{\mathbf{x}}\mathbf{C}^*_{\mathbf{x}}.\end{aligned} \tag{11}$$

In certain applications it is desirable that the *unbiasedness* condition be satisfied: The expectation of the estimator $\hat{X}(\boldsymbol{p}_k)$ must equal the expectation of the S/TRF $X(\boldsymbol{p})$, i.e.,

$$\overline{\hat{X}(\boldsymbol{p}_k) - X(\boldsymbol{p}_k)} = 0; \tag{12}$$

then, the MMSE concept can be modified to incorporate this condition by means of the

Unbiased MMSE concept: *As in Table 1 with the addition of the unbiasedness condition (12).*

In this case, the variance of the estimation error (4) should be minimized with respect to the weights ξ_i subject to the constraint (12). Constrained minimization leads to the set of equations that includes a Lagrange multiplier μ

$$\mathbf{c_x}\, \xi^* - \mu\, \mathbf{m_x} = \mathbf{c_x^*} \tag{13}$$

where the covariance matrix and vector now involve the centered covariances

$$\mathbf{c_x} = \overline{\delta X\, \delta X^T}\,, \tag{14}$$

with elements $c_{ij} = c_x(\boldsymbol{p}_i;\boldsymbol{p}_j),\ i, j = 1,\ldots,N'$; and

$$\mathbf{c_x^*} = \overline{\delta X(\boldsymbol{p}_k)\, \delta X} = [c_x(\boldsymbol{p}_1;\boldsymbol{p}_k)\ \ldots\ c_x(\boldsymbol{p}_{N'};\boldsymbol{p}_k)]^T, \tag{15}$$

and $\mathbf{m_x}$ is the vector of S/TRF means at the data points

$$\mathbf{m_x} = \overline{X} = [\overline{X(\boldsymbol{p}_1)}\ldots\overline{X(\boldsymbol{p}_{N'})}]^T. \tag{16}$$

The mean value of the S/TRF at the estimation point is given by the equation

$$\overline{X(\boldsymbol{p}_k)} = m_x(\boldsymbol{p}_k) = \boldsymbol{\xi}^T\, \mathbf{m_x}, \tag{17}$$

The weights are obtained from the solution of the system of Eqs. (13) which leads to

$$\xi^* = \mathbf{c_x^{-1}}(\mathbf{c_x^*} + \mu\ \mathbf{m_x}). \tag{18}$$

In view of Eq. (18) the estimation error variance is expressed as

$$\begin{aligned}\sigma_x^2(\boldsymbol{p}_k) &= c_x(\boldsymbol{p}_k;\boldsymbol{p}_k) + \boldsymbol{\xi}^{*T}\, \mathbf{c_x}\, \xi^* - 2\boldsymbol{\xi}^{*T}\, \mathbf{c_x^*} \\ &= c_x(\boldsymbol{p}_k;\boldsymbol{p}_k) - (\mathbf{c_x^*} + \mu\ \mathbf{m_x})^T\, \mathbf{c_x^{-1}}(\mathbf{c_x^*} - \mu\ \mathbf{m_x})\end{aligned} \tag{19}$$

Both the biased and unbiased linear MMSE estimators above require only the mean and covariance functions. While this is a desirable feature in practice, it does not always lead to the best MMSE possible. In general, the mean and covariance do not provide a complete characterization of the underlying probability law. This observation leads to the following:

Proposition 1: *In the general case of an S/TRF $X(\boldsymbol{p})$ with an arbitrary probability law, the linear estimators above are sub-optimal with respect to the MMSE criterion (in the sense that there may exist nonlinear estimators that perform better).*

Multivariate Gaussian S/TRF are an interesting special case: since they are completely determined on the basis of the mean and covariance functions, the following result is valid:

Proposition 2: *If $X(\boldsymbol{p})$ is a multivariate Gaussian S/TRF, the linear estimators above are optimal among all MMSE estimators.*

3.2 Recursive Formulations

In certain cases, the MMSE mapping techniques can be modified if a physical or biological model of the environmental health process is available. Such a situation is examined below.

EXAMPLE 1: Assume that the environmental health process is governed by the model

$$X(s_k, t_{\ell+1}) = \mathcal{P}_s[X(s_k, t_\ell)] + Z(s_k, t_\ell), \tag{20}$$

where $\mathcal{P}_s[\cdot]$ is a linear operator and $Z(s_k, t_\ell)$ is an S/TRF (e.g., white noise) with known statistics; $Z(s_k, t_\ell)$ is statistically independent of $X(s_i, t_j)$ for $j \leq \ell$. Then, the prediction at $(s_k, t_{\ell+1})$ based on data at (s_i, t_j), $i \neq k$, $j \leq \ell$, is given by (Christakos and Bogaert, 1996)

$$\hat{X}(s_k, t_{\ell+1}) = \mathcal{P}_s[\hat{X}(s_k, t_\ell)]. \tag{21}$$

Eq. (21) offers a prediction of the future value of the process X at time $t_{\ell+1}$ in terms of the estimation at the present time t_ℓ. Similarly the estimation error variance is given by

$$\sigma_x^2(s_k, t_{\ell+1}) = \mathcal{P}_s^2[\sigma_x^2(s_k, t_\ell)] + \sigma_z^2, \tag{22}$$

where σ_z^2 is the variance of Z. □

3.3 More Properties

An important goal of environmental health science is to make predictions. The MMSE mapping equations above offer a means to accomplish this goal.

These equations satisfy certain intuitive properties. From a positivist's perspective, Eq. (1) is a model that establishes a relation between observable quantities. The following properties are obvious.

Property 1: MMSE estimators use data that represent the values of a process at various time instants and spatial locations.

An interesting consequence of this property is that in certain practical applications temporal data may compensate for the lack of sufficient spatial data.

Property 2: In the case of unbiased MMSE, the estimates provided by Eq. (1) are on the average equal to the actual values of the process.

Property 3: The systems of estimation equations (8) and (13) account for the spatial locations and time instants of the data. Consequently, the corresponding estimation error variances (11) and (19) depend on the space-time data configurations.

Some other interesting properties are not so obvious. Assume that an estimate is generated at a measured data point p_1, i.e., $p_k \equiv p_1$. Then, the first column (and first row) in Eq. (8) coincides with the covariance vector $\mathbf{C}_\mathbf{x}^*$. Thus the unique solution of Eqs. (8) is $\xi = [1\,0 \ldots 0]^T$ and $\hat{X}(p_1) = X(p_1)$. It follows from Eq. (11) that the estimation error is zero, since $\sigma_x^2(p_1) = \overline{X(p_1)^2} + \xi^T \mathbf{C}_\mathbf{x}\, \xi - 2\xi^T \mathbf{C}_\mathbf{x}^* = 0$. The same result can be obtained for the unbiased estimation system, using Eqs. (13) and (19). This demonstrates the following

Property 4: The MMSE mapping estimators are exact interpolators.

Multiplying the covariances by a constant factor does not change the solution for the weights, and, thus the estimator (1) remains unchanged. However, the estimation error variance for both the biased and unbiased estimators will be multiplied by the same factor.

Property 5: While the MMSE estimates remain unchanged if the covariance is multiplied by a constant factor, the estimation errors are multiplied by the same factor.

Property 6: The MMSE estimation error variances (11) and (19) do not *explicitly* depend on the data values.

Conceptually, the main differences between composite space-time MMSE mapping and purely spatial MMSE mapping are related to the random field parameters (covariances, metrics, etc.) used as inputs in the mapping system, as well as the important spatiotemporal interaction effects incorporated by composite space-time estimators. Numerically, the solution of the space-time mapping system usually requires a larger computational effort than its purely spatial counterpart (due to the increase in the number of parameters, and the larger number of data points that need to be processed).

The performance of the MMSE techniques depends on the availability of the statistical information (means, covariances, etc.) involved in the mapping equations. In practice, inference of the statistics from the data requires the consideration of specific S/TRF models. Below we consider the cases of: (a) homogeneous and stationary model --§4; (b) S/TRF- ν / μ model --§5; and (c) regression S/TRF --§6.

4. HOMOGENEOUS-STATIONARY S/TRF: SPACE-TIME KRIGING FORMS

Assume that the S/TRF $X(\boldsymbol{p})$ is homogeneous and stationary. The mean and the covariance functions can be calculated on the basis of the available data and information using the methods discussed in Chapter III.

In the homogeneous-stationary case all the means are constant, and the covariance functions (both centered and non-centered) are homogeneous and stationary, i.e.,

$$m_x(\boldsymbol{p}_i) = m_x, \tag{1}$$

$$C_x(\boldsymbol{p}_i;\boldsymbol{p}_j) = C_x(\boldsymbol{p}_i - \boldsymbol{p}_j) = C_x(\boldsymbol{r}_{ij}, \tau_{ij}), \tag{2}$$

$$c_x(\boldsymbol{p}_i;\boldsymbol{p}_j) = c_x(\boldsymbol{p}_i - \boldsymbol{p}_j) = c_x(\boldsymbol{r}_{ij}, \tau_{ij}), \tag{3}$$

where $i = 1,\ldots,N'$ and $j = 1,\ldots,N'$ or $j = k$; the space and time lags are given by $\boldsymbol{r}_{ij} = \boldsymbol{s}_i - \boldsymbol{s}_j$ and $\tau_{ij} = t_i - t_j$.

4.1 Biased MMSE Mapping

Eqs. (3.8)-(3.11) can be used without any changes, and taking into account Eqs. (1)-(3) above.

Biased MMSE estimators are useful in certain practical applications, e.g., when the mean of an environmental health process is not known or if there is prior information that requires the implementation of biased estimators. Using the relationship between the covariance and the semi-variogram $\gamma_x(\boldsymbol{r},\tau)$

$$C_x(\boldsymbol{r},\tau) = C_x(\boldsymbol{0},0) - \gamma_x(\boldsymbol{r},\tau), \tag{4}$$

where the zero lag covariance $C_x(\boldsymbol{0},0) = \overline{X^2}$ is a constant in space-time, the MMSE estimator can also be expressed in terms of the semi-variogram.

4.2 Unbiased MMSE Mapping

Eqs. (3.13)-(3.19) remain valid. Now the corresponding means are constant and the centered covariances are homogeneous-stationary as in Eqs. (1)-(3) above. In view of Eq. (1) the unbiasedness condition (3.12) leads to the following constraint

$$\sum_{i=1}^{N'} \xi_i = 1. \tag{5}$$

In light of Eq. (5) the system of Eqs. (3.13) is transformed into

$$\mathbf{c_x}\, \xi^* = \mathbf{c_x^*} \tag{6}$$

where ξ^* represents the *augmented vector of weights* which also includes the Lagrange muliplier, i.e.,

$$\xi^* = [\xi_1 \ldots \xi_{N'}\ \mu]^T, \tag{7}$$

the *augmented data covariance matrix* is given by

$$\mathbf{c_x} = \begin{bmatrix} c_x(\boldsymbol{0};0) & c_x(\boldsymbol{r}_{12};\tau_{12}) & \cdots & \cdots & c_x(\boldsymbol{r}_{1N'};\tau_{1N'}) & 1 \\ c_x(\boldsymbol{r}_{21};\tau_{21}) & c_x(\boldsymbol{0};0) & \cdots & \cdots & c_x(\boldsymbol{r}_{2N'};\tau_{2N'}) & 1 \\ \vdots & \vdots & \vdots & \vdots & \vdots & \vdots \\ \vdots & \vdots & \vdots & \vdots & \vdots & \vdots \\ c_x(\boldsymbol{r}_{N'1};\tau_{N'1}) & c_x(\boldsymbol{r}_{N'2};\tau_{N'2}) & \cdots & \cdots & c_x(\boldsymbol{0};0) & 1 \\ 1 & 1 & \cdots & \cdots & 1 & 0 \end{bmatrix}, \tag{8}$$

and the data/estimation point *covariance vector* is given by

$$\mathbf{c}_{\mathbf{x}}^{*} = [c_x(\boldsymbol{r}_{1k}, \tau_{1k}) \ldots c_x(\boldsymbol{r}_{N'k}, \tau_{N'k}) \ldots 1]^T . \quad (9)$$

The estimation error reduces to

$$\sigma_x^2(p_k) = c_x(\boldsymbol{0},0) - \xi^{*T} \mathbf{c}_{\mathbf{x}} \xi^{*} = c_x(\boldsymbol{0},0) - \mathbf{c}_{\mathbf{x}}^{*T} \xi^{*} = c_x(\boldsymbol{0},0) - \mathbf{c}_{\mathbf{x}}^{*T} \mathbf{c}_{\mathbf{x}}^{-1} \mathbf{c}_{\mathbf{x}}^{*} . \quad (10)$$

Taking into account the relationship $c_x(\boldsymbol{r},\tau) = c_x(\boldsymbol{0},0) - \gamma_x(\boldsymbol{r},\tau)$ between the covariance and the semi-variogram, we find that the equations of the estimation system can be expressed in terms of the semi-variogram as follows

$$\boldsymbol{\gamma}_{\mathbf{x}} \, \xi^{*} = \boldsymbol{\gamma}_{\mathbf{x}}^{*}, \quad (11)$$

where the semi-variogram matrix of the data points is

$$\boldsymbol{\gamma}_{\mathbf{x}} = \begin{bmatrix} \gamma_x(\boldsymbol{0};0) & \gamma_x(\boldsymbol{r}_{12};\tau_{12}) & \cdots & \cdots & \gamma_x(\boldsymbol{r}_{1N'};\tau_{1N'}) & 1 \\ \gamma_x(\boldsymbol{r}_{21};\tau_{21}) & \gamma_x(\boldsymbol{0};0) & \cdots & \cdots & \gamma_x(\boldsymbol{r}_{2N'};\tau_{2N'}) & 1 \\ \vdots & \vdots & \vdots & \vdots & \vdots & \vdots \\ \vdots & \vdots & \vdots & \vdots & \vdots & \vdots \\ \gamma_x(\boldsymbol{r}_{N'1};\tau_{N'1}) & \gamma_x(\boldsymbol{r}_{N'2};\tau_{N'2}) & \cdots & \cdots & \gamma_x(\boldsymbol{0};0) & 1 \\ 1 & 1 & \cdots & \cdots & 1 & 0 \end{bmatrix}, \quad (12)$$

the weight vector is given by

$$\xi^{*} = [\xi_1, \ldots \xi_{N'}, -\mu]^T , \quad (13)$$

and the semi-variogram vector is

$$\boldsymbol{\gamma}_{\mathbf{x}}^{*} = [\gamma_x(\boldsymbol{r}_{1k}, \tau_{1k}) \ldots \gamma_x(\boldsymbol{r}_{N'k}, \tau_{N'k}) \ldots 1]^T . \quad (14)$$

In light of the above, the mapping error in Eq. (19) is now expressed as

$$\sigma_x^2(p_k) = \boldsymbol{\gamma}_{\mathbf{x}}^{*T} \boldsymbol{\gamma}_{\mathbf{x}}^{-1} \boldsymbol{\gamma}_{\mathbf{x}}^{*} . \quad (15)$$

The formulation the unbiased MMSE estimator by means of Eqs. (11)-(15) does not require knowledge of the mean m_x. This method can be called *ordinary space-time kriging* (*OS/TK*), for it is an extension in the space-time domain of the purely spatial ordinary kriging technique, which is popular in geostatistics (e.g., Journel, 1989).

EXAMPLE 1: Consider that hard data are available at the space-time points $\boldsymbol{p}_1 = (s_1, t_1)$ and $\boldsymbol{p}_2 = (s_2, t_2)$. We seek the ordinary kriging estimate at the point $\boldsymbol{p}_k = (s_k, t_k)$. We assume that he spatial lag r and the time lag τ between $\boldsymbol{p}_1$ and $\boldsymbol{p}_k$ are the same as the lags $\boldsymbol{p}_2$ and $\boldsymbol{p}_k$, i.e., $r_{1k} = r_{2k} = r$, $\tau_{1k} = \tau_{2k} = \tau$; in addition, the lags between the data points are $r_{12} = r_{21} = r'$ and $\tau_{12} = \tau_{21} = \tau'$. The semi-variogram is denoted by $\gamma_x(r_{ij}, \tau_{ij})$, $r_{ij} = |s_i - s_j|$ and $\tau_{ij} = |t_i - t_j|$ for $i, j = 1, 2, k$ between the three points are spatially isotropic and temporally stationary. The estimation system (13) becomes

$$\left.\begin{array}{l} \xi_1 \gamma_x(0,0) + \xi_2 \gamma_x(r', \tau') - \mu = \gamma_x(r, \tau) \\ \xi_1 \gamma_x(r', \tau') + \xi_2 \gamma_x(0,0) - \mu = \gamma_x(r, t) \\ \xi_1 + \xi_2 = 1 \end{array}\right\}. \tag{16}$$

It is easily seen by inspection that due to symmetry the solution of the system (16) is $\xi_1 = \xi_1 = \frac{1}{2}$. Hence, from Eq. (3.7) we obtain the following estimate

$$\hat{\chi}_k = \tfrac{1}{2}(\chi_1 + \chi_2), \tag{17}$$

which is the OS/TK estimate. □

If the mean m_x is known, the mapping equations of *simple space-time kriging* (*SS/TK*) can be similarly obtained. In particular, the estimator is given by

$$\hat{X}(\boldsymbol{p}_k) = \boldsymbol{\xi}_O^T \boldsymbol{X}_O, \tag{18}$$

where the weight vector is

$$\boldsymbol{\xi}_O = [\xi_0 \; \xi_1 \ldots \xi_{N'}]^T, \tag{19}$$

and

$$X_0 = [1 \quad X(p_1)...X(p_{N'})]^T. \tag{20}$$

The unbiasedness condition leads to the following constraint on the weights

$$\xi_0 = m_x (1 - \sum_{i=1}^{N'} \xi_i), \tag{21}$$

which, in general different than zero. The estimation system is expressed as

$$\mathbf{c}'_{\mathbf{x}} \, \xi = \mathbf{c}'^{*}_{\mathbf{x}} \tag{22}$$

where $\mathbf{c}'_{\mathbf{x}}$ is given by Eq. (3.14) above and $\mathbf{c}'^{*}_{\mathbf{x}}$ by Eq. (3.15). Finally, the estimation error variance is given by

$$\sigma_x^2(p_k) = c_x(0,0) - \mathbf{c}'^{*T}_{\mathbf{x}} \mathbf{c}'^{-1}_{\mathbf{x}} \mathbf{c}'^{*}_{\mathbf{x}}. \tag{23}$$

The estimate, given by Eq. (18), can now be expressed as

$$\hat{X}(p_k) = m_x (1 - \sum_{i=1}^{N'} \xi_i) + \sum_{i=1}^{N'} \xi_i X(p_i). \tag{24}$$

The above shows that, unlike the previous MMSE estimators, the SS/TK depends explicitly on the mean m_x. However, if $\xi_0 = 0$ the first term in Eq. (24) vanishes, and the SS/TK estimator becomes identical to the OS/TK discussed earlier.

EXAMPLE 2: Consider the same situation as in Example 1 above, with the addition that the mean m_x is also known. The estimation system of Eqs. (22) becomes

$$\left. \begin{aligned} \xi_1 \sigma_x^2 + \xi_2 c_x(r', \tau') = c_x(r, \tau) \\ \xi_1 c_x(r', \tau') + \xi_2 \sigma_x^2 = c_x(r, \tau) \end{aligned} \right\}, \tag{25}$$

where $\sigma_x^2 = c_x(0,0)$. The solution of the above is given by

$$\xi_1 = \xi_2 = c_x(r, \tau) / [\sigma_x^2 + c_x(r', \tau')]. \tag{26}$$

Then, following Eq. (22), the 0-th weight is given by

$$\xi_0 = m_x\{1 - 2c_x(r,\tau)/[\sigma_x^2 + c_x(r',\tau')]\}. \tag{27}$$

Hence, we obtain the SS/TK estimate from Eq. (24) as follows

$$\hat{\chi}_k = m_x[1 - \frac{2c_x(r,\tau)}{\sigma_x^2 + c_x(r',\tau')}] + \frac{c_x(r,\tau)}{\sigma_x^2 + c_x(r',\tau')}(\chi_1 + \chi_2), \tag{28}$$

which, as expected, depends on the mean of the process. □

REMARK 1: In principle SS/TK is more accurate than OS/TK, in the sense that it yields a smaller estimation variance.

5. NONHOMOGENEOUS-NONSTATIONARY S/TRF: THE CASE OF S/TRF- ν/μ

This case of MMSE mapping is concerned with a more realistic situation where the natural or health process of interest $X(\boldsymbol{p})$ shows a heterogeneous spatiotemporal variation. This variation is modelled as an S/TRF- ν/μ (Chapter IV).

Following the fact that the $X(\boldsymbol{p})$, may include polynomial space-time trends represented by $p_{\nu/\mu}(\boldsymbol{s},t)$, the unbiasedness condition (3.12) constrains the weights ξ of the estimator (3.1). More specifically, the weights are subject to the constraint that the linear combination of the polynomial trends evaluated at the data points be equal to the value of the polynomial at the estimation point. The space-time polynomials are a linear superposition of monomials $g_\alpha(\boldsymbol{p})$,

$$g_\alpha(\boldsymbol{p}) = (\prod_{i=1}^n s_i^{\rho_i})t^\zeta, \tag{1}$$

where $\sum_{i=1}^n \rho_i = |\boldsymbol{\rho}| \leq \nu$, $\zeta \leq \mu$, $\alpha = 1,\ldots,\eta$, and $\eta \equiv N_n(\nu/\mu)$ the number of monomials (Chapter IV). The constraint (3.12) can thus be expressed in terms of space-time monomials as follows

$$\sum_{i=1}^{N'} \xi_i\, g_\alpha(\boldsymbol{p}_i) = g_\alpha(\boldsymbol{p}_k). \tag{2}$$

The optimal weights ξ_i in Eq. (3.1) are obtained by constrained minimization of the error variance, i.e., by solving the equations

$$\partial\sigma_x^2(\boldsymbol{p}_k)/\partial\xi_i = 0 \tag{3}$$

for all i subject to the constraints in Eq. (2). The condition $\partial^2\sigma_x^2(\boldsymbol{p}_k)/\partial\xi_i^2 > 0$ must also be satisfied to ensure that the solution of Eqs. (3) is minimum of the loss function. This procedure leads to the following system of equations for the weights ξ_i

$$\mathrm{K}\,\xi^* = \theta\,, \tag{4}$$

where ξ^* is a vector involving the ξ_i and the η Lagrange multipliers μ_α, used to enforce the constraints. The elements of the matrix K involve (i) the values of the generalized covariance for all pairs of data points and (ii) the values of the space-time at these points. The vector θ involve (i) the generalized covariance values between the data points and the estimation point, and (ii) the space-time monomials at the estimation point. The structure of the matrices and vectors involved in the above are presented in detail below; the space time lags are defined as follows: $\boldsymbol{r}_{ij} = \boldsymbol{s}_i - \boldsymbol{s}_j$, $\boldsymbol{r}_{ik} = \boldsymbol{s}_i - \boldsymbol{s}_k$, $\tau_{ij} = t_i - t_j$, $\tau_{ik} = t_i - t_k$.

The *kriging matrix* $\boldsymbol{K}$ involves the data points only, and it has the block structure

$$\boldsymbol{K} = \left[\begin{array}{c|c} \boldsymbol{K}_1 & \boldsymbol{G} \\ \hline \boldsymbol{G}^T & \boldsymbol{0} \end{array}\right], \tag{5}$$

where the block $\boldsymbol{K}_1$ is a *generalized covariance matrix* with dimensions $N' \times N'$.

$$\boldsymbol{K}_1 = \begin{bmatrix} k_x(\boldsymbol{0};0) & k_x(\boldsymbol{r}_{12};\tau_{12}) & \cdots & \cdots & k_x(\boldsymbol{r}_{1N'};\tau_{1N'}) \\ k_x(\boldsymbol{r}_{21};\tau_{21}) & k_x(\boldsymbol{0};0) & \cdots & \cdots & k_x(\boldsymbol{r}_{2N'};\tau_{2N'}) \\ \vdots & \vdots & & & \\ \vdots & \vdots & & & \\ k_x(\boldsymbol{r}_{N'1};\tau_{N'1}) & k_x(\boldsymbol{r}_{N'2};\tau_{N'2}) & \cdots & \cdots & k_x(\boldsymbol{0};0) \end{bmatrix}, \tag{6}$$

and the block $\boldsymbol{G}$ is an orthogonal *matrix of* ν/μ *monomials with* dimensions $N' \times \eta$

$$\boldsymbol{K}_2 = \begin{bmatrix} g_1(\boldsymbol{p}_1) & g_2(\boldsymbol{p}_1) & \cdots & g_\eta(\boldsymbol{p}_1) \\ g_1(\boldsymbol{p}_2) & g_2(\boldsymbol{p}_2) & \vdots & g_\eta(\boldsymbol{p}_2) \\ \vdots & \vdots & \cdots & \vdots \\ g_1(\boldsymbol{p}_{N'}) & g_2(\boldsymbol{p}_{N'}) & \cdots & g_\eta(\boldsymbol{p}_{N'}) \end{bmatrix}. \tag{7}$$

The weight *vector* $\boldsymbol{\xi}^*$ involves both the estimation weights and the Lagrange multipliers

$$\xi^* = \left[\frac{\xi}{\mu}\right], \tag{8}$$

where $\boldsymbol{\xi}$ is given by Eq. (3.5), i.e., $\boldsymbol{\xi} = [\xi_1 ... \xi_{N'}]^T$ and

$$\boldsymbol{\mu} = [\mu_1 ... \mu_\eta]^T. \tag{9}$$

The *estimation point vector* involves both the generalized covariance and the trend monomials

$$\theta = \left[\frac{\theta_1}{\theta_2}\right], \tag{10}$$

where

$$\boldsymbol{\theta}_1 = [k_x(\boldsymbol{r}_{1k}, \tau_{1k}) ... k_x(\boldsymbol{r}_{2k}, \tau_{2k}) \; .... k_x(\boldsymbol{r}_{N'k}, \tau_{N'k})]^T, \tag{11}$$

and

$$\boldsymbol{\theta}_2 = [g_1(\boldsymbol{p}_k) \; g_2(\boldsymbol{p}_k) ... \; g_\eta(\boldsymbol{p}_k)]^T. \tag{12}$$

The kriging matrix K has dimensions $(N' + \eta)^2$. The dimensions of the vectors θ_1 and θ_2 are $N' \times 1$ and $\eta \times 1$, respectively.

The solution of the linear estimation system determines both the weights and the Lagrange multipliers. In light of the analysis in §3, the weights determine the optimal estimate at $\boldsymbol{p}_k$ is given by Eq. (3.4). Minimization of the estimation error variance leads to the following expression for the estimation error variance at each point $\boldsymbol{p}_k$ in space-time

$$\sigma_x^2(\boldsymbol{p}_k) = k_x(\boldsymbol{0}, 0) - \boldsymbol{\xi}^{*T} \boldsymbol{\theta}. \tag{13}$$

The following simple example illustrates the estimation process described above.

EXAMPLE 1: Consider an S/TRF-1/1. Hard data are available at four points $\boldsymbol{p}_i = (s_i, t_i)$, i=1, 2, 3 and 4 (Fig. 1). We seek an estimate at $\boldsymbol{p}_k = (s_k, t_k)$. For simplicity, let the coordinates of the estimation point be $s_k = t_k = 0$; then, $(s_1, t_1) = (-r, \tau)$, $(s_2, t_2) = (r, \tau)$, $(s_3, t_3) = (r, -\tau)$ and $(s_4, t_4) = (-r, -\tau)$. The estimation system (4) is written as

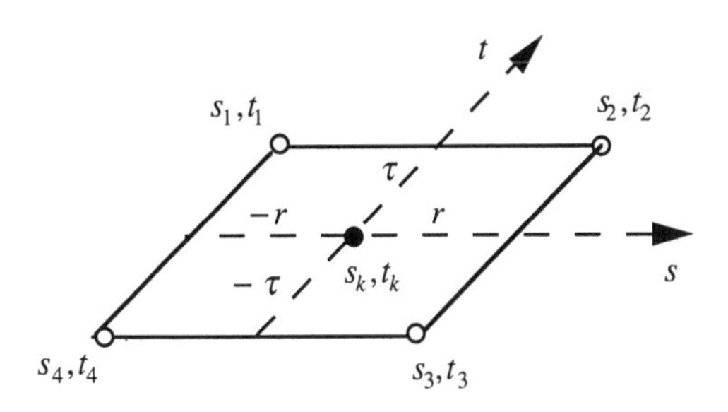

FIGURE 1: The data configuration for Example 1.

$$\left.\begin{array}{l}
\xi_1\, k_x(r_{11},\tau_{11})+\xi_2\, k_x(r_{21},\tau_{21})+\xi_3\, k_x(r_{31},\tau_{31})+\xi_4\, k_x(r_{41},\tau_{41})+\mu_0+\mu_1 s_1 \\
\quad +\mu_2 t_1+\mu_3 s_1 t_1=k_x(r_{k1},\tau_{k1}) \\
\xi_1\, k_x(r_{12},\tau_{12})+\xi_2\, k_x(r_{22},\tau_{22})+\xi_3\, k_x(r_{32},\tau_{32})+\xi_4\, k_x(r_{42},\tau_{42})+\mu_0+\mu_1 s_2 \\
\quad +\mu_2 t_2+\mu_3 s_2 t_2=k_x(r_{k2},\tau_{k2}) \\
\xi_1\, k_x(r_{13},\tau_{13})+\xi_2\, k_x(r_{23},\tau_{23})+\xi_3\, k_x(r_{33},\tau_{33})+\xi_4\, k_x(r_{43},\tau_{43})+\mu_0+\mu_1 s_3 \\
\quad +\mu_2 t_3+\mu_3 s_3 t_3=k_x(r_{k3},\tau_{k3}) \\
\xi_1\, k_x(r_{14},\tau_{14})+\xi_2\, k_x(r_{24},\tau_{24})+\xi_3\, k_x(r_{34},\tau_{34})+\xi_4\, k_x(r_{44},\tau_{44})+\mu_0+\mu_1 s_4 \\
\quad +\mu_2 t_4+\mu_3 s_4 t_4=k_x(r_{k4},\tau_{k4}) \\
\xi_1+\xi_2+\xi_3=1 \\
\xi_1 s_1+\xi_2 s_2+\xi_3 s_3+\xi_4 s_4=s_k \\
\xi_1 t_1+\xi_2 t_2+\xi_3 t_3+\xi_4 t_4=t_k \\
\xi_1 s_1 t_1+\xi_2 s_2 t_2+\xi_3 s_3 t_3+\xi_4 s_4 t_4=s_k t_k
\end{array}\right\}, \qquad (14)$$

Replacing the point coordinates and lags in Eqs. (13) the system simplifies as follows

$$\left.\begin{array}{l}
\xi_1 k_x(0,0)+\xi_2 k_x(2r,0)+\xi_3 k_x(2r,2\tau)+\xi_4 k_x(0,2\tau)+\mu_0-\mu_1 r+\mu_2\tau-\mu_3 r\tau=k_x(-r,\tau) \\
\xi_1 k_x(2r,0)+\xi_2 k_x(0,0)+\xi_3 k_x(0,2\tau)+\xi_4 k_x(2r,2\tau)+\mu_0+\mu_1 r+\mu_2\tau+\mu_3 r\tau_2=k_x(r,\tau) \\
\xi_1 k_x(2r,2\tau)+\xi_2 k_x(0,2\tau)+\xi_3 k_x(0,0)+\xi_4 k_x(2r,0)+\mu_0+\mu_1 r-\mu_2\tau-\mu_3 r\tau=k_x(r,-\tau) \\
\xi_1 k_x(0,2\tau)+\xi_2 k_x(2r,2\tau)+\xi_3 k_x(2r,0)+\xi_4 k_x(0,0)+\mu_0-\mu_1 r-\mu_2\tau+\mu_3 r\tau=k_x(-r,-\tau) \\
\xi_1+\xi_2+\xi_3=1;\; \xi_1+\xi_4=\xi_2+\xi_3;\; \xi_1+\xi_2=\xi_3+\xi_4;\; \xi_1+\xi_3=\xi_2+\xi_4
\end{array}\right\}$$

The solution of Eq. (14) leads to $\xi_i = 1/4$ ($i = 1, 2, 3$ and 4). The estimate is then given by

$$\hat{\chi}_k = \tfrac{1}{4}(\chi_1+\chi_2+\chi_3+\chi_4). \qquad (15)$$

Note that the weights and, therefore, the estimate $\hat{\chi}_k$ as well, are independent of the space-time generalized covariance $k_x(r,\tau)$. □

EXAMPLE 2: Consider the more realistic situation of an S/TRF-1/1 on an $R^2 \times T$ grid. Assume that hard data are available at the space-time points $\boldsymbol{p}_i = (\boldsymbol{s}_i, t_i)$, $i = 1, \ldots, 12$ (Fig. 2). We seek an estimate of the S/TRF at $\boldsymbol{p}_k = (\boldsymbol{s}_k, t_k)$. The local neighborhood of the estimation point is shown in Fig 2. Assume that the S/TRF has a generalized covariance

$$k_x(\boldsymbol{r},\tau) = -\sigma_x^2\,\lambda^4\,\xi^4\,\{\exp[-|\tau|/\lambda] + \theta(\tau)[2\,\tau/\lambda + \tfrac{1}{3}(\,\tau/\lambda)^3\,]\}$$
$$\{4(r/\xi)^{-1}(1-\exp[-r/\xi]) - \exp[-r/\xi] + r/\xi\}. \tag{16}$$

There are six constraints for the estimation weights which are expressed as follows

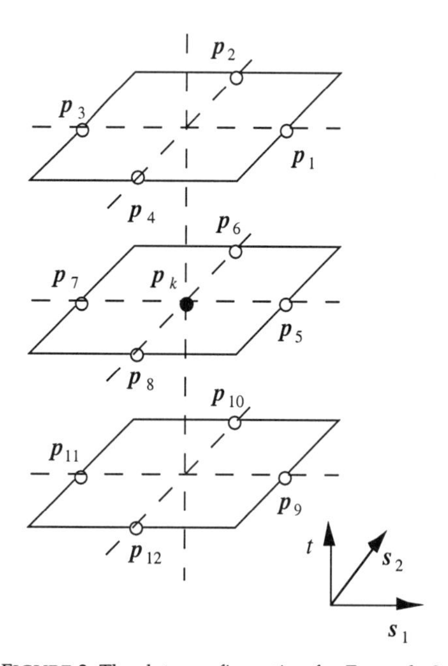

FIGURE 2: The data configuration for Example 2.

$$\left.\begin{array}{l}\sum_{i=1}^{12}\xi_i = 1,\ \sum_{i=1}^{12}\xi_i\,s_{1,i} = s_{1,k},\ \sum_{i=1}^{12}\xi_i\,s_{2,i} = s_{2,k}\\ \sum_{i=1}^{12}\xi_i\,t_i = t_k,\ \sum_{i=1}^{12}\xi_i\,s_{1,i}\,t_i = s_{1,k}\,t_k\\ \sum_{i=1}^{12}\xi_i\,s_{2,i}\,t_i = s_{2,k}\,t_k\end{array}\right\} \tag{17}$$

where $s_i = (s_{1,i}, s_{2,i})$. The kriging matrix $\boldsymbol{K}$ is as shown in Eq. (4). The generalized covariance block $\boldsymbol{K}_1$ is a 12×12 square matrix the elements of which are given by the values $k_x(\boldsymbol{r}_{ij}, \tau_{ij})$ of the generalized covariance. The block $\boldsymbol{G}$ of ν/μ monomials is a 12x6 matrix that involves the values of the filtered monomials at the data points, and it is given explicitly by

$$\boldsymbol{K}_2 = \begin{bmatrix} 1 & s_{1,1} & s_{2,1} & t_1 & s_{1,1}\,t_1 & s_{2,1}\,t_1 \\ 1 & s_{1,2} & s_{2,2} & t_2 & s_{1,2}\,t_2 & s_{2,2}\,t_2 \\ \vdots & \vdots & \vdots & \vdots & \vdots & \vdots \\ \vdots & \vdots & \vdots & \vdots & \vdots & \vdots \\ 1 & s_{1,12} & s_{2,12} & t_{12} & s_{1,12}\,t_{12} & s_{2,12}\,t_{12} \end{bmatrix}. \tag{18}$$

Finally, the estimation point vector θ is given by

$$\boldsymbol{\theta}^T = [k_x(\boldsymbol{r}_{1,k}, \tau_{1,k}), \ldots, k_x(\boldsymbol{r}_{12,k}, \tau_{12,k}), 1, s_{1,k}, s_{2,k}, t_k, s_{1,k}t_k, s_{2,k}t_k]. \tag{19}$$

We assume that the parameters σ, λ and ξ of the generalized covariance are all equal to one in appropriate units, and we solve the linear estimation system, Eqs. (4), using the procedure *linsolve* from the linear algebra library of *Maple*. The following results are obtained for the weights

$$\left.\begin{array}{l} \xi_1 = \xi_2 = \xi_3 = \xi_4 \cong -0.018 \\ \xi_5 = \xi_6 = \xi_7 = \xi_8 \cong 0.287 \\ \xi_9 = \xi_{10} = \xi_{11} = \xi_{12} \cong -0.018 \end{array}\right\}, \tag{20}$$

and for the Lagrange multipliers

$$\left.\begin{array}{l} \mu_1 = -0.109 - 0.067\,t_k \\ \mu_2 = \mu_3 = \mu_5 = \mu_6 = 0 \\ \mu_4 = 0.067 \end{array}\right\}. \tag{21}$$

Certain comments are in order here: First, note that the weights for all the spatial locations on the same time slice are identical. In addition, the Lagrange multipliers for the spatial constraints are zero. Both results are due to the spatial symmetry of the grid and the isotropy of the generalized covariance model. The symmetry is broken in time by the step function in the generalized covariance model. The vanishing of the Lagrange multipliers for the spatial constraints implies that the constraints are automatically satisfied for the specific grid geometry and generalized covariance. The first Lagrange multiplier depends on the estimation time. This dependence can be better understood as follows: The estimation error is defined as the difference between the value of the process and its estimate at $\boldsymbol{p}_k$

$$e(\boldsymbol{p}_k) = X(\boldsymbol{p}_k) - \hat{X}(\boldsymbol{p}_k) = X(\boldsymbol{p}_k) - \sum_{i=1}^{12} \xi_i X(\boldsymbol{p}_i). \tag{22}$$

The estimation error is a homogeneous and stationary S/TRF, because the trends in $X(\boldsymbol{p}_k)$ are filtered out by the weights. Thus, the estimation variance should be a constant independent of $\boldsymbol{p}_k$. Since the estimator is unbiased the estimation variance is given by $\sigma^2(\boldsymbol{p}_k) = \overline{e^2(\boldsymbol{p}_k)}$. In light of Eq. (13) above the estimation variance is expressed as

$$\sigma_x^2(\boldsymbol{p}_k) = k_x(\boldsymbol{0}) - \sum_{i=1}^{12} \xi_i\, k_x(\boldsymbol{p}_k - \boldsymbol{p}_i) - \mu_1 - \mu_2\, s_{1,k}$$
$$-\mu_3\, s_{2,k} - \mu_4 t_k - \mu_5\, s_{1,k} t_k - \mu_6\, s_{2,k} t_k\,. \tag{23}$$

Since $\mu_2 = \mu_3 = \mu_5 = \mu_6 = 0$ and $\mu_1 + \mu_4 t_k = c$ it turns out that $\sigma_x^2(\boldsymbol{p}_k)$ is equal to

$$\sigma_x^2(\boldsymbol{p}_k) = k_x(\boldsymbol{0}) - \sum_{i=1}^{12} \xi_i\, k_x(\boldsymbol{p}_k - \boldsymbol{p}_i) - c \approx 0.017, \tag{24}$$

which is independent of $\boldsymbol{p}_k$. □

6. VECTOR MMSE SPACE-TIME MAPPING

Consider a vector S/TRF-$\boldsymbol{\nu}/\boldsymbol{\mu}$ $X(\boldsymbol{p})$ and assume that observations of $X_\ell(\boldsymbol{p})$ ($\ell = 1, \ldots, M$) are available at points $\boldsymbol{p}_1, \ldots, \boldsymbol{p}_{N_\ell}$. The space-time mapping problem consists in deriving estimates $\hat{X}_\ell$ of X_ℓ ($\ell = 1, \ldots, M$) at unobserved locations $\boldsymbol{p}_k \neq \boldsymbol{p}_1, \ldots, \boldsymbol{p}_{N_\ell}$, which are optimal in the MMSE sense. In particular, we seek estimates $\hat{X}_\ell(\boldsymbol{p}_k)$ that satisfies the following three MMSE conditions:

(a) Linearity, i.e.,

$$\hat{X}_\ell(\boldsymbol{p}_k) = \sum_{i=1}^{M} \boldsymbol{\lambda}_i^T\, \boldsymbol{X}_i, \tag{1}$$

where $\boldsymbol{X}_i = [X_i(p_1),...,X_i(p_{N_i})]^T$ and $\boldsymbol{\lambda}_i = [\lambda_{i,1},...,\lambda_{i,N_i}]^T$ is a vector of weights to be evaluated through the mapping process.

(b) The unbiasedness condition, i.e.,

$$\overline{[\hat{X}_\ell(\boldsymbol{p}_k) - X_\ell(\boldsymbol{p}_k)]} = 0. \tag{2}$$

(c) The minimum mean square error condition, i.e.,

$$\min_{\boldsymbol{\lambda}_\ell} \sigma^2_{X_\ell}(\boldsymbol{p}_k) = \min_{\boldsymbol{\lambda}_\ell} \overline{[\hat{X}_\ell(\boldsymbol{p}_k) - X_\ell(\boldsymbol{p}_k)]^2}. \tag{3}$$

These three conditions constitute a constrained optimization problem, the solution of which leads to the vectors of weights $\boldsymbol{\lambda}_i$ that satisfy the following system of equations (also called *co-kriging* system)

$$\boldsymbol{K}_x \, \boldsymbol{\xi} = \boldsymbol{\Theta}, \tag{4}$$

where $\boldsymbol{\xi}$ is a vector of weights and Lagrange multipliers; $\boldsymbol{K}_x$ and $\boldsymbol{\Theta}$ are, respectively, the co-kriging matrix and the estimation point vector. Explicit expressions for these matrices/vectors are obtained in the following example

EXAMPLE 1: Consider a two-component vector S/TRF consisting of two scalar random fields $X_1(\boldsymbol{p})$ of orders ν_1/μ_1, and $X_2(\boldsymbol{p})$ of orders ν_2/μ_2. The matrices involved in the estimation of the field $X_1(\boldsymbol{p})$ are as follows

$$\boldsymbol{K}_x = \left[\begin{array}{c|c|c|c} \boldsymbol{K}_{11} & \boldsymbol{K}_{12} & \boldsymbol{G}_1 & \boldsymbol{0} \\ \hline \boldsymbol{K}_{21} & \boldsymbol{K}_{22} & \boldsymbol{0} & \boldsymbol{G}_2 \\ \hline \boldsymbol{G}_1^T & \boldsymbol{0} & \boldsymbol{0} & \boldsymbol{0} \\ \hline \boldsymbol{0} & \boldsymbol{G}_2^T & \boldsymbol{0} & \boldsymbol{0} \end{array}\right], \tag{5}$$

where the $\boldsymbol{K}_{ij}$ in Eq. (5) represent the generalized covariance block matrices

$$\boldsymbol{K}_{ij} = \begin{bmatrix} k_{ij}(\boldsymbol{0};0) & k_{ij}(\boldsymbol{r}_{12};\tau_{12}) & \cdots & k_{ij}(\boldsymbol{r}_{1N'};\tau_{1N'}) \\ k_{ij}(\boldsymbol{r}_{21};\tau_{21}) & k_{ij}(\boldsymbol{0};0) & \cdots & k_{ij}(\boldsymbol{r}_{2N'};\tau_{2N'}) \\ & & \vdots & \\ k_{ij}(\boldsymbol{r}_{N'1};\tau_{N'1}) & k_{ij}(\boldsymbol{r}_{N'2};\tau_{N'2}) & & k_{ij}(\boldsymbol{0};0) \end{bmatrix}, \tag{6}$$

with $i,j=1,2$. The $\boldsymbol{G}_i$, $i=1,2$, represent the matrices of ν_i/μ_i monomials

$$\boldsymbol{G}_i = \begin{bmatrix} g_1(\boldsymbol{p}_1) & g_2(\boldsymbol{p}_1) & \cdots & g_{\eta_i}(\boldsymbol{p}_1) \\ g_1(\boldsymbol{p}_2) & g_2(\boldsymbol{p}_2) & \vdots & g_{\eta_i}(\boldsymbol{p}_2) \\ \vdots & \vdots & \cdots & \vdots \\ g_1(\boldsymbol{p}_{N'}) & g_2(\boldsymbol{p}_{N'}) & \cdots & g_{\eta_i}(\boldsymbol{p}_{N'}) \end{bmatrix}; \tag{7}$$

the vector of weights and Lagrange multipliers is

$$\boldsymbol{\xi} = [\lambda_{1,1},\ldots,\lambda_{1,N_\ell}\ \lambda_{2,1},\ldots,\lambda_{2,N_\ell}\ \mu_{1,0},\ldots,\mu_{1,\eta_1}\ \mu_{2,0},\ldots,\mu_{2,\eta_2}]^T, \tag{8}$$

and

$$\boldsymbol{\Theta} = [k_1(\boldsymbol{r}_{1k};\tau_{1K}),\ldots,k_1(\boldsymbol{r}_{N_\ell k};\tau_{N_\ell k}),\ldots,k_{12}(\boldsymbol{r}_{N_\ell k};\tau_{N_\ell k}), g_1(\boldsymbol{p}_k),\ldots,g_{\eta_1}(\boldsymbol{p}_k), 0,\ldots,0]^T \tag{9}$$

is the estimation point vector. □

Finally, the estimation error at $\boldsymbol{p}_k$ is expressed as

$$\sigma^2_{X_\ell}(\boldsymbol{p}_k) = k_\ell(\boldsymbol{0}) - \boldsymbol{\xi}^{\mathrm{T}}\,\boldsymbol{\Theta}. \tag{10}$$

The MMSE system (4) and the estimation error variance (10) involve only the generalized covariances and cross-covariances. Estimates $\hat{X}_\ell(\boldsymbol{p}_k)$ of the processes $X_\ell(\boldsymbol{p})$, $\ell=1,\ldots,M$, are obtained from Eqs. (1) based on the weights $\boldsymbol{\lambda}_i = [\lambda_{i,1},\ldots,\lambda_{i,N_\ell}]^T$, which follow from the solution of the co-kriging system (4). Similarly, the MMSE estimation errors follow from Eqs. (10). Co-kriging incorporates in the estimation of $\hat{X}_\ell(\boldsymbol{p}_k)$ cross-correlated data of other processes $X_{\ell'}(\boldsymbol{p})$ ($\ell \neq \ell'$, $\ell'=1,\ldots,M$). If the estimate $\hat{X}_\ell(\boldsymbol{p}_k)$ is based only on $X_\ell(\boldsymbol{p})$ data the co-kriging system reduces to scalar kriging above.

7. IMPLEMENTATION ISSUES - APPLICATIONS

7.1 Practical Map Generation

The coefficients involved in the Eqs. (5.1)-(5.12) depend on the data point distribution in space-time, the space-time variability, and the space-time coordinates of the estimation point relative to the data points. Both the estimation system of Eqs. (5.4) and the error variance (5.13) do not depend explicitly on the data or the polynomial trends --which are filtered out by the detrending operator. There is an implicit dependence on the data though the generalized covariance model $k_x(\boldsymbol{r}, \tau)$. Hence, the estimation error $\sigma_x^2(\boldsymbol{p}_k)$ can be computed as long as the S/TRF parameters are known. This has significant consequences for space-time sampling. Indeed, assuming that the S/TRF parameters can be estimated (from previous experience with similar situations, etc.), the monitoring network can be optimized before construction of the monitoring stations. Even in cases that these parameters are poorly estimated, a sensitivity analysis can lead to a variety of sampling scenarios for further consideration.

Minimization of $\sigma_x^2(\boldsymbol{p}_k)$ led to the linear estimation system of Eqs. (5.4) that involves the generalized covariance of the S/TRF $X(\boldsymbol{p})$. As we showed in Chapter IV, in practice different domains may have different degrees of space-time heterogeneity, thus requiring S/TRF with varying continuity orders (ν/μ). In this case, the form of the generalized covariance $k_x(\boldsymbol{r}, \tau)$ also varies in different neighborhoods Λ. In view of the incomplete information regarding the S/TRF properties in real case studies, the continuity orders ν and μ can only be justified a posteriori based on the accuracy of the estimates they provide at control points.

Hence, the estimates $\hat{X}(\boldsymbol{p}_k)$ are obtained based on the data within a local neighborhood Λ. The size of Λ is not necessarily uniform across the domain, and it depends on a number of factors including the following: The spatiotemporal distribution of the observations (observation scale); the desired scale for the representation of the variability (mapping scale); the number of data points required for the estimation of the S/TRF parameters. Estimation methods provide useful information only at scales comparable to the observation scale. In practical situations external constraints often help to specify the size of the local neighborhoods.

EXAMPLE 1: Measuring stations that are concentrated around urban areas or certain parts of an ecosystem naturally define local neighborhoods Λ. In other areas, data may be more or less uniformly distributed in space. Let us assume that there are N measurement stations, more or less uniformly distributed over an area D, and that measurements are available in P different instants collected over a time period T. Then it is possible to define local observation scales L_{ob} and T_{ob} in both space and time as $L_{ob} = \alpha\sqrt{D/N}$, where α is a constant factor and $T_{ob} = T/P$. The, there is on the average one measuring station within each block of linear size L_{ob}. If $N_{\min}$ denotes the minimum number of spatial data required for the estimation of the S/TRF parameters, the minimum size of the local neighborhood is given by $L_{\min} = \sqrt{N_{\min}}\, L_{ob}$. The time extent $T_{\min}$ of the neighborhood Λ can be similarly defined as $T_{\min} = P_{\min} T_{ob}$. □

If the data are not uniformly distributed, the local neighborhood Λ can not be estimated by a simple expression as above, but it can still be defined by the minimum number $N_{\min}$ of data required. In this case, the size of the neighborhood varies locally and is determined by the space-time distribution of the available information. The size of the local neighborhoods can be increased beyond the required minimum number $N_{\min}$, in order to include more information. However, the following factors should be considered when increasing the size of the local neighborhood: (a) if distant space-time points are uncorrelated with the estimation point, the information from these points will not improve the accuracy of the estimation; (b) larger local neighborhood sizes lead to considerable increase in computer time due to the higher number of S/TRF parameters.

In order to implement the MMSE mapping approach, a spatiotemporal covariance model needs to be defined and its parameters evaluated within the local neighborhood Λ. As we saw in Chapter IV, generalized covariance models $k_x(\boldsymbol{r}, \tau)$ can be constructed by solving a PDE relating $k_x(\boldsymbol{r}, \tau)$ to the covariance $c_y(\boldsymbol{r}, \tau)$ of a homogeneous and stationary residual. The generalized covariance model involves a number of undetermined coefficients that specify the homogeneous and stationary polynomial terms; i.e., polynomials in r_i ($i = 1,\ldots,n$) and τ. These coefficients are determined from the existing data by means of optimization methods subject to the general permissibility conditions that we discussed in Chapter IV. The choice of the minimization criterion depends on modelling assumptions regarding the distribution of the data and the statistics of the residual random field. Least squares methods do not require any assumptions on the data probability distribution, but they are maximum likelihood estimators only when the errors are normally distributed (Vyas, 1997). One possible optimization approach is suggested in

EXAMPLE 2: The unknown generalized covariance coefficients c_i, $i = 1,\ldots,\eta$, are estimated by means of constrained optimization, which ensures that they represent in an optimal sense the data in the local neighborhood and also satisfy the permissibility conditions. This is accomplished by designating a data point $\boldsymbol{p}_k$ as a *control point* and estimating the value of the process at $\boldsymbol{p}_k$ using the remaining data. The estimation error is given by

$$\delta X(\boldsymbol{p}_k) \equiv \hat{X}(\boldsymbol{p}_k) - X(\boldsymbol{p}_k) = \sum_{i=1}^{N'} \xi_i (1 - \delta_{ik}) X(\boldsymbol{p}_i) - X(\boldsymbol{p}_k), \tag{1}$$

where N' denotes the total number of data in the neighborhood. The error is a space-time increment of order ν/μ as shown in Chapter IV. The variance of the estimation error is a function of the point $\boldsymbol{p}_k$

$$\sigma_{\delta x}^2(\boldsymbol{p}_k) \equiv \overline{\delta X^2(\boldsymbol{p}_k)} = \sum_{i=1}^{N'} \sum_{j=1}^{N'} \xi_i \xi_j k_x(\boldsymbol{r}_{ij}; \tau_{ij}). \tag{2}$$

Note that $\sigma_{\delta x}^2(\boldsymbol{p}_k)$ differs from $\sigma_x^2(\boldsymbol{p}_k)$, since the former is the estimation error at the measured point $\boldsymbol{p}_k$, whereas the latter refers to an unmeasured point. The procedure is repeated for all points where data are available, and for different sets of parameters c_i.

In order to obtain optimal solutions for c_i the *global objective function* F that represents the cumulative error variance is defined as

$$F = \sum_{\boldsymbol{p}_k} [\delta X^2(\boldsymbol{p}_k) - \sigma_{\delta x}^2(\boldsymbol{p}_k)]^2 . \tag{3}$$

Minimization of F with respect to the c_i, i.e., $\partial F/\partial c_i = 0$, where $i = 1, \ldots, \eta$ leads to a linear system of equations in terms of the estimation weights. □

REMARK 1: Map construction is an interactive procedure. The user makes initially certain choices (e.g., regarding the scale and the covariance model). After the numerical estimation procedure is completed, the user receives feedback --in the form of maps-- from the visualization system, based on which the user may change the initial assumptions. In this way the user gains understanding and insight about the space-time patterns of the environmental health process and the S/TRF model used to represent them.

EXAMPLE 3: In a study of springwater ion processes at the Dyle river in Belgium (Christakos and Bogaert, 1996) a rectangular spatiotemporal grid with 31×30 nodes in space and six nodes in time is used. The grid spacing is approximately $1\,km/$node in space and $1\,$year/node in time; the sampling period is six years. The grid can be viewed as the superposition of two-dimensional slices that correspond to snapshots taken at different times. Measurements are available at 68 spatial locations on the grid for all 6 sampling times. The estimation maps for NO_3^- concentration are shown in Fig. 3, and the estimation standard deviation maps shown in Fig. 4. □

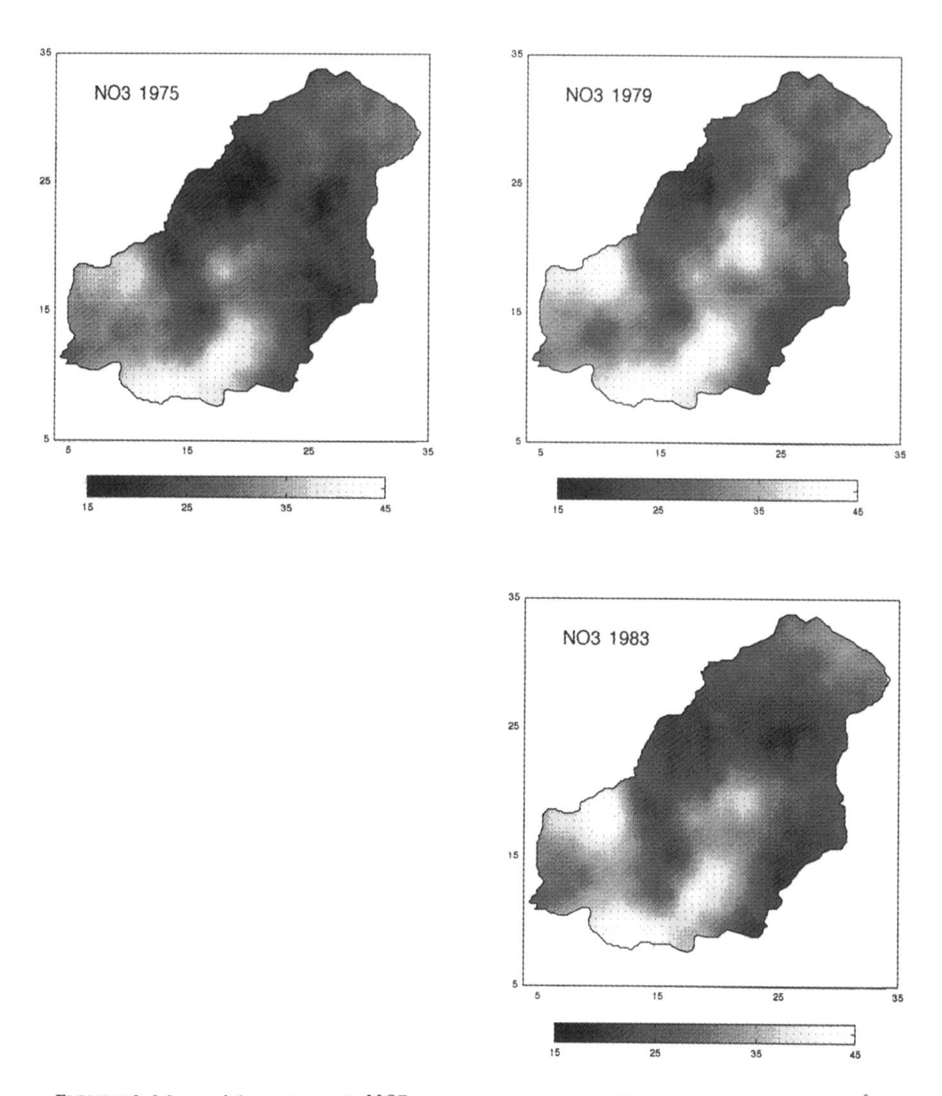

FIGURE 3: Maps of the estimated NO_3^- concentration at year 1975, 1979 and 1983 (in mg/ ℓ).

EXAMPLE 4: Maps of the maximum hourly ozone concentration estimates over Eastern U.S. on July 15 and 16 1995 are shown in Figs. V.1a and 5. An unbiased estimator with a polynomial generalized covariance was used. The amount of data available within each neighborhood Λ influences the quality of estimates (estimation uncertainty increases if the number of data in Λ is reduced). These maps show considerable variation of ozone

TABLE 2: A comparison of spatiotemporal ozone concentration estimation vs. purely spatial estimation at a set of control points

Location	**Dickson, TN**	**Vigo, IN**	**Chatham, GA**	**Escambia, FL**
Identification no.	470430009	181670018	130510021	120330004
Coordinates	36.15N/87.22W	39.29N/87.24W	32.04N/81.03W	30.31N/87.12W
Actual value	0.067 *ppm*	0.071 *ppm*	0.072 *ppm*	0.072 *ppm*
Spatial Observation Scale	200 km	200 km	450 km	500 km
Spatiotemporal Observation Scale	150 km/7days	150km/7days	250 km/13days	250km/13days
Spatial estimate	0.0723 *ppm*	0.0736 *ppm*	0.0681 *ppm*	0.0592 *ppm*
Spatiotemporal estimate	0.0671 *ppm*	0.0717 *ppm*	0.074 *ppm*	0.0719 *ppm*
Spatial est. error	0.0053 *ppm*	0.0026 *ppm*	0.0039 *ppm*	0.0128 *ppm*
Spatiotemporal est. error	0.0001 *ppm*	0.0007 *ppm*	0.002 *ppm*	0.0001 *ppm*

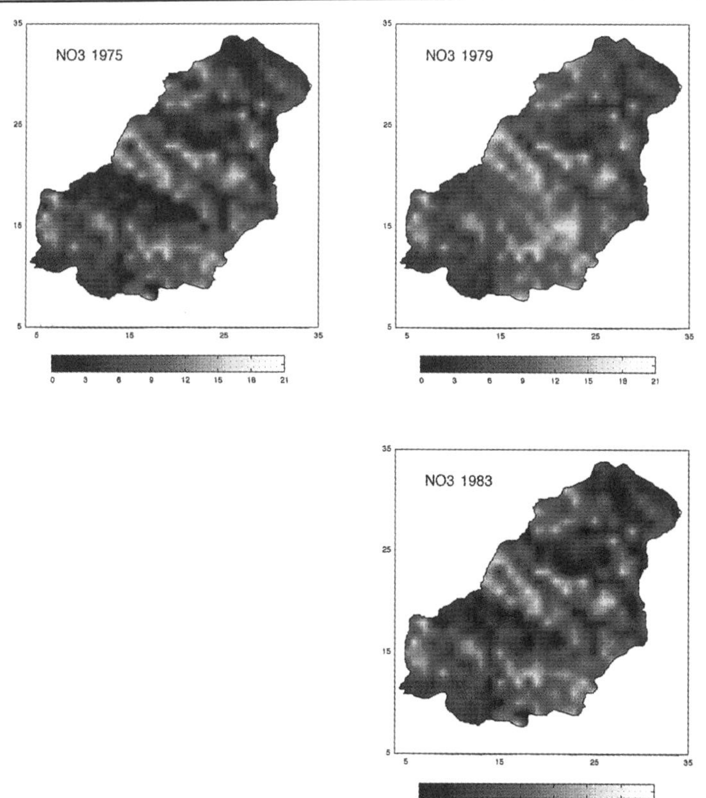

FIGURE 4: Maps of the standard deviation values associated with the NO_3^- estimates of Fig. 3 (in mg/ ℓ). The estimation standard deviation is zero at the data locations.

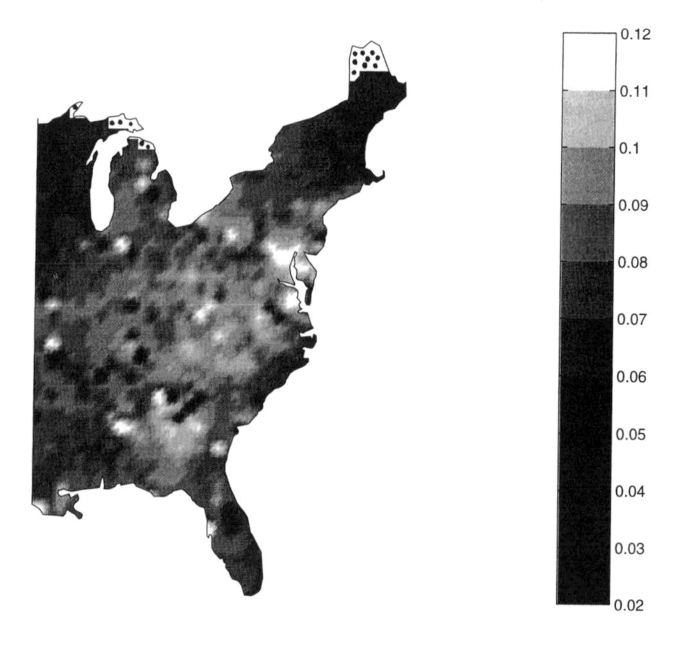

FIGURE 5: Map of maximum hourly ozone concentration (in *ppm*) over Eastern U.S. on July 16th, 1995 (dotted areas indicate regions where estimation was not possible).

concentration across Eastern U.S. (McKee, 1994). The highest ozone concentrations occurred in the mid-Atlantic and in the South (similar results were obtained for the time period 1985-1990 by Eder *et al.*, 1993). Based on the availability of suitable ozone measurements, ozone maps can cover areas much larger than the areas within which conventional air quality models apply. These maps are more sensitive to local variations than trend surface analysis and other global methods (e.g., Bonham-Carter, 1994) and, hence, they do not lead to oversmoothing of the ozone distribution. Accurate estimates of space-time ozone concentrations would assist considerably the development and implementation of regional strategies for attaining ozone standards. In order to evaluate the accuracy of the space-time maps, the ozone concentrations at a set of control points over Eastern U.S. were estimated using the S/TRF method. The results for four representative control points on July 15th, 1995 are shown in Table 2 (page 219 above): The accuracy of the ozone estimates obtained is excellent. The ratio of the spatiotemporal estimation error over the actual ozone concentration value lies in the interval [0.139%, 2.78%]. Unlike previous studies that obtained spatial estimates of hourly ozone at unsampled nodes using data only at the same time period (e.g., Casado *et al.*, 1994), in the present study data in a space-time neighborhood including several different locations and times were used. This led to a composite space-time ozone estimation that is a considerable improvement over previous estimates. In order to perform a numerical comparison between purely spatial and composite space-time maps, the four representative data values of Table 2 were estimated using spatial kriging (Cressie, 1991), as well. The results are shown in Table 2: at all locations the spatiotemporal ozone estimates are considerably more accurate than the purely spatial estimates. The difference in accuracy is especially marked in cases where fewer monitoring stations were available.

Using space-time ozone maps the occurrences of elevated ozone concentrations can be calculated with very good accuracy, and the percentage of residents exposed to ozone levels above the federal standards can be estimated for each region. Sequences of maps can provide continuous visualization of ozone changes in space-time and identify geographic areas of environmental concern. Such maps constitute indispensable tools in ascertaining compliance with ambient air quality standards (e.g., standards specify how many times ozone concentration thresholds can be exceeded within a specific time period), and are important in health impact assessment. Using Eq. (5.13), the estimation error $\sigma_x^2(\boldsymbol{p}_k)$ can also be mapped (Vyas, 1997). Combining this information with the estimates $\hat{X}(\boldsymbol{p}_k)$, several other useful maps can be derived. Assuming that the estimation error $\hat{X}(\boldsymbol{p}_k) - X(\boldsymbol{p}_k)$ is $N(0, \sigma_x)$, confidence interval maps can be defined for the estimates: e.g., the actual concentration values lie in the interval $\hat{X}(\boldsymbol{p}_k) \pm 1.96\,\sigma_x(\boldsymbol{p}_k)$ with 95% confidence. Other useful reliability maps can also be obtained: (i) the space-time probabilities that ozone concentrations do not exceed a certain level; (ii) the probabilities that concentrations stay within certain limits; or (iii) the kth percentiles; i.e., the true values in space-time will not exceed the kth percentile values with probability equal to $100 \times k$. □

In studies relating health to ozone distribution, the air quality sampling network is usually more sparse in space than the available health data sets. Mapping techniques such as the above can be employed to obtain ozone concentration estimates in unmeasured areas.

7.2 Trend Determination

The S/TRF model provides the basis for the effective determination of trends in the ozone data. In particular, space-time polynomials $p_{\nu/\mu}(s,t)$ are constructed that provide the best fit to the ozone data within local neighborhoods. The coefficients of these polynomials are space-time averages of the data, and it is, thus, possible to estimate them by the mapping technique of the previous section (functions other than polynomials can be used as well).

The S/TRF trend determination method differs from a direct polynomial fit to the data in certain important ways: First, it uses the mapping technique to obtain optimal estimates of the ozone values over the whole domain Λ. In this way it takes into consideration important space-time variability information contained in the S/TRF parameters ν, μ and k_x (which the simple least squares fit or other basis function-based techniques do not account for). Then, a fit to both estimated and measured data points is generated which is more accurate than fits based purely on the data points.

EXAMPLE 5: For illustration, a map of S/TRF ozone trend is presented in Fig. 6

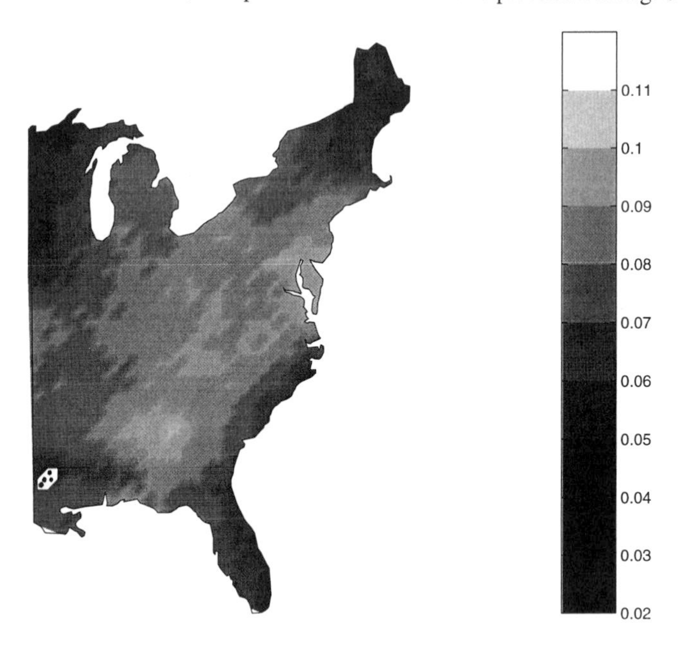

FIGURE 6: Spatiotemporal ozone trend over Eastern U.S. on July 16th, 1995 (in *ppm*; dotted areas indicate regions where estimation was not possible due to numerical problems of ill conditioned matrices).

(Christakos and Vyas, 1998a and b). □

The goals of trend determination are the elimination of artificial fluctuations and the detection of natural trends. However, the elimination of fluctuations introduces smoothing that has some undesired consequences, such as the reduction of the range of values and the variance of the actual data set. Good trend determination techniques should be balanced and not lead to oversmoothing.

EXAMPLE 6: To obtain a numerical comparison between the S/TRF and the commonly used mean filtering ozone trend determination method (Rao *et al.*, 1995), an area is selected in Eastern U.S. with a mean ozone concentration of 0.0711 *ppm*, a range of 0.141 *ppm*, and a standard deviation of 0.019 *ppm* (Christakos and Vyas, 1998a). The ranges and the standard deviations of the ozone trend values produced by the S/TRF and the mean filtering methods are plotted in Figs. 7a and b, respectively, as functions of the spatial range D (km) of the domain Λ ($D = 0$ corresponds to the original unsmoothed data). The temporal extent of Λ is T=5 days. While the S/TRF and the mean filtering methods leave essentially unchanged the mean ozone value (the largest variation from the unsmoothed mean is only

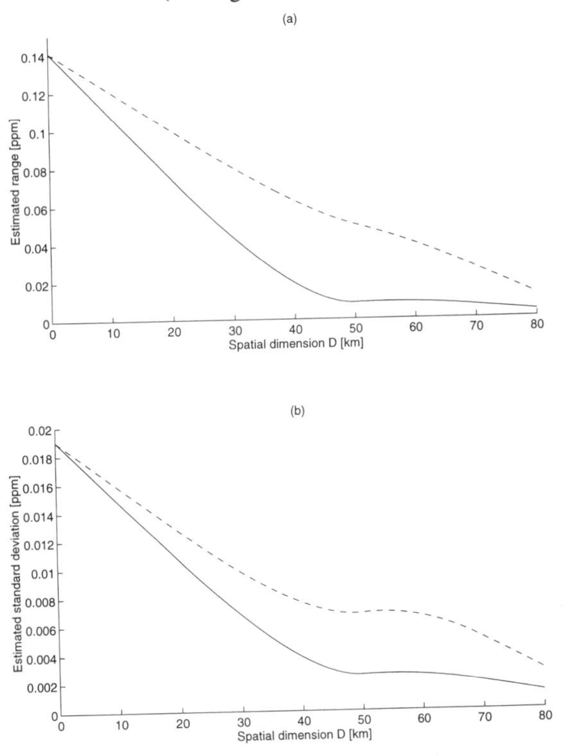

FIGURE 7: Effect of space-time trend determination on (a) the range of smoothed ozone values (*ppm*); and (b) the standard deviation (*ppm* of smoothed ozone data. Two methods were used: mean filtering (continuous line) and the S/TRF-method (dashed line). D (km) is the diameter of the domain Λ.

9.5%), they both have an effect on the range and the standard deviation of the ozone trend values. Mean filtering, however, leads to a much greater reduction of the range and standard deviation of the ozone values than the S/TRF-method. This difference increases as the degree of smoothing increases (larger D). Hence, S/TRF provides a better method of trend determination than mean filtering. □

8. REGRESSION ESTIMATOR

Suppose that we seek an estimate of a process $X(\boldsymbol{p})$ at a location/instant $\boldsymbol{p}_k$ on the basis of data available at N' space-time points ($N' = m \times p$; m is the number of spatial locations and p of time instants). The spatial mean at the time t_j, over a limited area is

$$\mu(s/t_j) = \beta_{s/t_j,1} + \beta_{s/t_j,2}\, alt(\boldsymbol{s}), \tag{1}$$

where $alt(\boldsymbol{s})$ refers to the altitude, and where the coefficients $\beta_{s/t_j,1}$ and $\beta_{s/t_j,2}$ are allowed to vary from one time instant to another.

Similarly, assume that over short periods of time, the temporal mean $\mu(t/\boldsymbol{s}_i)$ at a location $\boldsymbol{s}_i$ can be expressed as

$$\mu(t/\boldsymbol{s}_i) = \beta_{t/s_i,1} + \beta_{t/s_i,2}\, t, \tag{2}$$

where the coefficients $\beta_{t/s_i,1}$ and $\beta_{t/s_i,2}$ may vary from one location to another. As a unique space-time mean $\mu(\boldsymbol{s}_i, t_j)$ is required for estimation, a possibility is to define

$$\mu(\boldsymbol{s}_i, t_j) = \beta_1 + \beta_2\, alt(\boldsymbol{s}_i) + \beta_3\, t_j + \beta_4\, alt(\boldsymbol{s}_i)\, t_j \tag{3}$$

over a limited spatiotemporal neighborhood. Eq. (3) is obtained as the product of the models (1) and (2), and has the same general form as Eqs. (1) and (2) when considered separately along the space and time axes. In vector notations, Eq. (1) can be rewritten

$$\boldsymbol{\mu}_{s/t_j} = \boldsymbol{\Omega}_s\, \boldsymbol{\beta}_{s/t_j}, \tag{4}$$

where $\boldsymbol{\mu}_{s/t_j} = [\mu(s_1/t_j), \ldots, \mu(s_m/t_j)]^T$, $\boldsymbol{\beta}_{s/t_j} = [\beta_{s/t_j,1}, \beta_{s/t_j,2}]^T$, $\Omega_{s,i1} = 1$, and $\Omega_{s,i2} = alt(\boldsymbol{s}_i)$, $i = 1, \ldots, m$. Similarly, Eq. (2) can be expressed as

$$\boldsymbol{\mu}_{t/s_i} = \boldsymbol{\Omega}_t \boldsymbol{\beta}_{t/s_i}, \tag{5}$$

where $\boldsymbol{\mu}_{t/s_i} = [\mu(t_1/s_i),\ldots,\mu(t_p/s_i)]^T$, $\boldsymbol{\beta}_{t/s_i} = [\beta_{t/s_i,1}, \beta_{t/s_i,2}]^T$, $\Omega_{t,i1} = 1$, and $\Omega_{t,i2} = t_i$, $i = 1,\ldots,p$. The mean vector $\boldsymbol{\mu} = \left[\mu(s_1,t_1),\ldots, \mu(s_m,t_1),\ldots, \mu(s_1,t_p),\ldots, \mu(s_m,t_p)\right]^T$ corresponding to Eq. (3) can be expressed as (Bogaert and Christakos, 1997a and 1997b)

$$\boldsymbol{\mu} = (\boldsymbol{\Omega}_t \otimes \boldsymbol{\Omega}_s)\boldsymbol{\beta} \tag{6}$$

where $\otimes$ is the Kronecker delta (Searle, 1971) and $\boldsymbol{\beta} = \left[[\boldsymbol{\beta}_{t/s_i}], [\boldsymbol{\beta}_{s/t_j}]\right]$. Eq. (6) gives a satisfactory definition of the mean within a finite estimation neighborhood. The β values may differ from one neighborhood to another, thus allowing a certain amount of flexibility.

A space-time estimate $\hat{X}(\boldsymbol{p}_k)$ of $X(\boldsymbol{p}_k)$ is an unbiased linear estimate of the form

$$\hat{X}(\boldsymbol{p}_k) = \boldsymbol{\xi}^T \boldsymbol{X}, \tag{7}$$

where $\boldsymbol{X} = [X(\boldsymbol{p}_1),\ldots, X(\boldsymbol{p}_{N'})]^T$ is the space-time vector of measurements and $\boldsymbol{\xi}$ is a vector of weights ξ_i, i.e., $\boldsymbol{\xi} = [\xi_1,\ldots,\xi_{N'}]^T$. The corresponding mapping variance is

$$\sigma_x^2(\boldsymbol{p}_k) = 2\boldsymbol{\xi}^T(\boldsymbol{\gamma}_Y + \boldsymbol{1}_t \otimes \boldsymbol{\gamma}_{M_1} + \boldsymbol{\gamma}_{M_2} \otimes \boldsymbol{1}_s) - \boldsymbol{\xi}^T(\boldsymbol{\Gamma}_Y + \boldsymbol{1}_t\boldsymbol{1}_t^T \otimes \boldsymbol{\Gamma}_{M_1} + \boldsymbol{\Gamma}_{M_2} \otimes \boldsymbol{1}_s\boldsymbol{1}_s^T)\boldsymbol{\xi}, \tag{8}$$

where $\boldsymbol{\Gamma}_Y$ is an $N' \times N'$ square matrix $\Gamma_{Yij} = \gamma_Y(\boldsymbol{r}_{ij}, \tau_{ij})$, $\boldsymbol{\gamma}_Y$ is a vector of size $N' \times 1$, $\boldsymbol{\gamma}_Y = \left[\gamma_Y(\boldsymbol{r}_{01}, \tau_{01}),\ldots, \gamma_Y(\boldsymbol{r}_{0N'}, \tau_{0N'})\right]^T$, $\boldsymbol{\Gamma}_{M_1}$ is an $m \times m$ matrix, $\Gamma_{M_1,ij} = \gamma_{M_1}(\boldsymbol{r}_{ij})$, $\boldsymbol{\Gamma}_{M_2}$ is an $p \times p$ matrix $\Gamma_{M_{2,kl}} = \gamma_{M_2}(\tau_{kl})$, $\boldsymbol{\gamma}_{M_1} = [\gamma_{M_1}(\boldsymbol{r}_{01}),\ldots,\gamma_{M_1}(\boldsymbol{r}_{0m})]^T$ is a $m \times 1$ vector, $\gamma_{M_2} = [\gamma_{M_2}(\tau_{01}),\ldots,\gamma_{M_2}(\tau_{0p})]^T$ is a $p \times 1$ vector, and $\boldsymbol{1}_s$, $\boldsymbol{1}_t$ are unit vectors of size m and p, respectively. Minimization of Eq. (8) with respect to $\boldsymbol{\xi}$ leads to the estimation system

$$\begin{bmatrix} \boldsymbol{\Gamma}_Y + \boldsymbol{1}_t\boldsymbol{1}_t^T \otimes \boldsymbol{\Gamma}_{M_1} + \boldsymbol{\Gamma}_{M_2} \otimes \boldsymbol{1}_s\boldsymbol{1}_s^T & \boldsymbol{\Omega}_t \otimes \boldsymbol{\Omega}_s \\ \boldsymbol{\Omega}_t^T \otimes \boldsymbol{\Omega}_s^T & \boldsymbol{0} \end{bmatrix} \begin{bmatrix} \boldsymbol{\Xi} \\ \boldsymbol{\nu} \end{bmatrix} = \begin{bmatrix} \boldsymbol{\gamma}_Y + \boldsymbol{1}_t \otimes \boldsymbol{\gamma}_{M_1} + \boldsymbol{\gamma}_{M_2} \otimes \boldsymbol{1}_s \\ \boldsymbol{\chi}_0 \end{bmatrix}, \tag{9}$$

where $\chi_0 = [1, alt(s_0), t_0, alt(s_0)\, t_0]^T$ is a 4×1 vector, $\boldsymbol{0}$ is a 4×4 null matrix, and ν is a 4×1 vector of Lagrangians. In deriving Eqs. (8) and (9) we have assumed that the same number of time measurements are available at all stations, although this is not necessary.

9. BME vs. MMSE ESTIMATORS

The best spatiotemporal MMSE estimator of $X(\boldsymbol{p}_k)$ is the conditional mean [Eq. (2.6)]

$$\hat{X}(\boldsymbol{p}_k) = \overline{X(\boldsymbol{p}_k)|\chi_{\text{hard}}} \,. \tag{1}$$

If $X(\boldsymbol{p})$ is a Gaussian S/TRF, the Eq. (1) reduces to a linear estimator, which is optimal among all MMSE estimators. Typically, these MMSE estimators use hard data $\boldsymbol{\chi}_{\text{hard}}$ and involve the space-time mean and covariance functions.

On the other hand, if the general knowledge consists of the mean and covariance functions, the BME posterior pdf is Gaussian. The latter is symmetric function, hence the BME estimate is by definition the conditional mean, i.e., the same as the MMSE estimate.

$$\hat{\chi}_{MMSE} \;=\; \overline{X(\boldsymbol{p}_k)|\boldsymbol{\chi}_{\text{hard}}} \;\overset{\text{Gaussian}}{=}\; \begin{cases} \xi^T \boldsymbol{\chi}_{\text{hard}} \\ \hat{\chi}_{BME} \end{cases}, \tag{2}$$

where ξ is the vector of weights of the data points. Thus, the following is valid.

Proposition 1: *For a Gaussian S/TRF, the MMSE estimator $\hat{\chi}_{MMSE}$ obtained on the basis of hard data and general knowledge of the mean and covariance functions coincide with the BME mode estimate $\hat{\chi}_{BME}$ derived using the same hard data and general knowledge. The Gaussian estimator is linear.*

In the following sections we see that popular linear MMSE estimators, such as the various types of space-time *kriging*, are indeed special cases of the BME estimator.

10. BME vs. KRIGING ESTIMATORS

The BME theory preserves most features of simpler estimators, which are obtained as its limiting cases. Assuming only hard data χ_{hard}, the BME equation (Chapter V) reduces to

$$\partial \Im[\boldsymbol{\chi}_{\text{map}}] / \partial \chi_k \big|_{\chi_k = \hat{\chi}_k} = 0 \,. \tag{1}$$

By considering statistical moments up to second order, the BME analysis can reproduce several of the kriging estimators. A few examples follow

EXAMPLE 1: When only hard data χ_{hard} exist and the general knowledge includes the mean, variance and centered covariance, Eq. (1) leads to

$$\sum_{i=1}^{N',k} c_{ii}^{-1}[\frac{\partial g_{ii}(\chi_i)}{\partial \chi_k}]_{\chi_k=\hat{\chi}_k} + \sum_{i=1}^{N',k}\sum_{j=1,\neq i}^{N',k} c_{ij}^{-1}[\frac{\partial g_{ij}(\chi_i,\chi_j)}{\partial \chi_k}]_{\chi_k=\hat{\chi}_k} = 0; \tag{2}$$

$g_{ii} = (\chi_i - m_i)^2$ and $g_{ij} = (\chi_i - m_i)(\chi_j - m_j)$, $i \neq j$. Eq. (2) yields the BME estimate

$$\hat{\chi}_k = m_k - \sum_{i=1}^{m}(\chi_i - m_i)c_{ik}^{-1}/c_{kk}^{-1}, \tag{3}$$

which coincides with the simple kriging (SK) estimator of §5 above. Consider the same situation as in Example 4.2, where hard data at points $\boldsymbol{p}_1 = (\boldsymbol{s}_1, t_1)$ and $\boldsymbol{p}_2 = (\boldsymbol{s}_2, t_2)$ were available. We seek the BME estimate at $\boldsymbol{p}_k = (\boldsymbol{s}_k, t_k)$. The S/TRF is homogeneous-stationary with constant mean m_x and variance σ_x^2. The space-time lag between $\boldsymbol{p}_1$ and $\boldsymbol{p}_k$ is the same as between $\boldsymbol{p}_2$ and $\boldsymbol{p}_k$ so that $c_{1k} = c_{2k} = c_{k1} = c_{k2} = c_x(r,\tau)$ and $c_{12} = c_{21} = c_x(r',\tau')$. Under these circumstances, the SK estimate is [see, Eq. (4.28)]

$$\hat{\chi}_k = m_x + \frac{c_x(r,\tau)}{\sigma_x^2 + c_x(r',\tau')}[(\chi_1 - m_x) + (\chi_2 - m_x)], \tag{4}$$

On the other hand, the BME Eq. (3) yields

$$\hat{\chi}_k = m_x - (\chi_1 - m_x)c_{1k}^{-1}/c_{kk}^{-1} - (\chi_2 - m_x)c_{2k}^{-1}/c_{kk}^{-1}, \tag{5}$$

where the $c_{1k}^{-1} = c_{2k}^{-1}$ and c_{kk}^{-1} are given by

$$\left.\begin{array}{l} c_{1k}^{-1} = |\boldsymbol{c}|^{-1}(c_{21}c_{k2} - \sigma_x^2 c_{k1}) \\ c_{2k}^{-1} = |\boldsymbol{c}|^{-1}(c_{k1}c_{k2} - \sigma_x^2 c_{k2}) \;\text{ and }\; c_{kk}^{-1} = |\boldsymbol{c}|^{-1}(\sigma_x^4 - c_{12}^2) \end{array}\right\}, \tag{6}$$

with $|\boldsymbol{c}| = c_{1k}(c_{21}c_{k2} - \sigma_x^2 c_{k1}) - c_{2k}(\sigma_x^2 c_{k2} - c_{12}c_{k1}) + \sigma_x^2(\sigma_x^4 - c_{12}^2)$. Since $c_{1k} = c_{2k} = c_{k1} = c_{k2} = c_x$ and $c_{12} = c_{21} = c_x'$, Eq. (6) leads to $c_{1k}^{-1}/c_{kk}^{-1} = c_{2k}^{-1}/c_{kk}^{-1} = -c_x/(\sigma_x^2 + c_x')$. By replacing the latter into Eq. (5) we obtain Eq. (4). This proves that the BME and the SK estimates are in this case identical. □

EXAMPLE 2: Let us revisit the analysis of the porosity data in Example V.4.1. Using the 56 hard porosity data, SK produces the map of Fig. 8 which is considerably less accurate

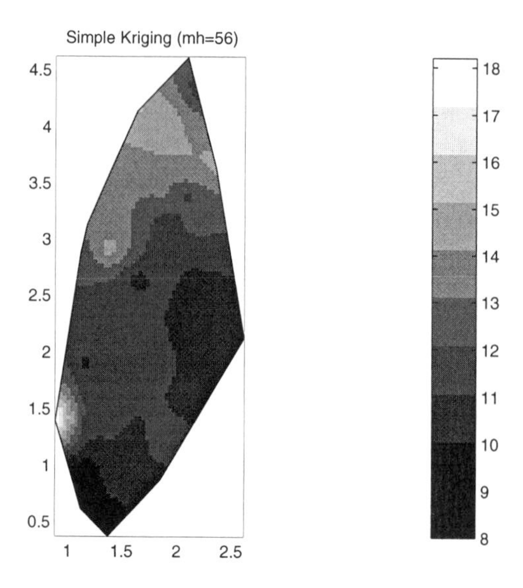

FIGURE 8: The SK porosity map using 56 hard data (in %).

than the corresponding BME map of Fig. V.7. □

EXAMPLE 3: Consider an S/TRF-1/1 as in Example 5.1. Hard data are available at points $\boldsymbol{p}_i = (s_i, t_i)$, i=1, 2, 3 and 4 in space-time (Fig. 5.1). We seek the BME estimate at the point $\boldsymbol{p}_k = (s_k, t_k)$ and compare it with the kriging estimate of Example 5.1. For simplicity let $s_k = t_k = 0$; then, $(s_1, t_1) = (-r, \tau)$, $(s_2, t_2) = (r, \tau)$, $(s_3, t_3) = (r, -\tau)$ and $(s_4, t_4) = (-r, -\tau)$. In order to compare the BME estimate with the kriging estimate of Example 5.1 we must formulate the BME problem in a way that is consistent with the kriging problem of Example 5.1. An S/TI-1/1 consistent with the analysis in Example 5.1 is

$$\psi = Q[\boldsymbol{\chi}_{\text{map}}] = \chi_k - \tfrac{1}{4}\textstyle\sum_{i=1}^{4} \chi_i \,. \tag{7}$$

Then, the g_α functions for the problem are $g_0(\boldsymbol{\chi}_{\text{map}}) = 1$ and

$$g_1(\boldsymbol{\chi}_{\text{map}}) = Q^2[\boldsymbol{\chi}_{\text{map}}] = [\chi_k - \tfrac{1}{4}\textstyle\sum_{i=1}^{4} \chi_i]^2 . \tag{8}$$

The known prior statistic is

$$\overline{g_1(\boldsymbol{x}_{\text{map}})} = \overline{[x_k - \tfrac{1}{4}\sum_{i=1}^{4} x_i]^2} = -\tfrac{1}{2}\mu_1^{-1}. \tag{9}$$

In light of Eq. (8), the BME Eq. (1) becomes (assuming $\mu_1 \neq 0$)

$$\hat{\chi}_k = \tfrac{1}{4}\sum_{i=1}^{4} \chi_i, \tag{10}$$

which is the same as the kriging estimate [see, Eq. (5.15) in Example 5.1]. □

EXAMPLE 4: To compare BME vs. indicator kriging (IK; Deutsch and Journel, 1992), the following experiment was designed: For a fixed set of 13 spatial points (Fig. 9), 500 realizations of a random field $X(\boldsymbol{p})$ were generated. The simulated values follow a multivariate (0, 1) Gaussian law; the covariance function is the Gaussian

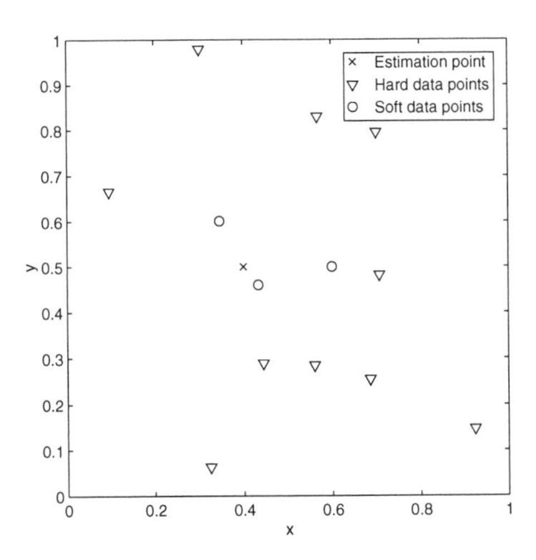

FIGURE 9: Data configuration in R^2. Measurements (hard data) are available at points indicated by ∇; soft data are available at points O. The estimation point is indicated by $\times$.

$$c_x(\boldsymbol{r}) = \exp(-r^2/a^2), \tag{11}$$

with $a = 1$. At three points which are kept fixed for all simulations, only interval (soft) data are provided. We assume that there are thirteen intervals $[\alpha_j, \alpha_{j+1}]$. The intervals are defined by $\alpha_1 = -\infty$, $\alpha_{13} = +\infty$, $\alpha_2 = F_x^{-1}(0.01)$, $\alpha_{12} = F_x^{-1}(0.99)$ and $\alpha_{j+2} = F_x^{-1}(0.1j)$

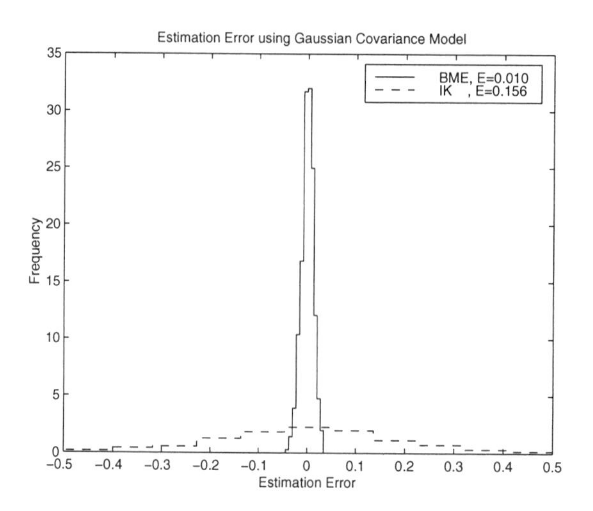

FIGURE 10: Estimation error distributions, BME vs. IK.

for $j = 1,...,9$; $F_x^{-1}(p)$ is the p-quantile of the standard Gaussian law. Each point is randomly assigned an interval $\chi(\boldsymbol{p}_i) \in [\alpha_j, \alpha_{j+1}]$. Estimates are sought at the point $\boldsymbol{p}_k$ -- the same for all simulations. The ten points with known process values constitute the hard data points, while the three locations with interval data are the soft data points. For both BME and IK we assume that all the parameters of the multivariate Gaussian law are known in advance, in order to avoid methodological issues. For the same reason, the same intervals have been chosen for all points, where the limits of the intervals will also be the IK threshold values. BME does not require that the same intervals be used for all points. Since we assume that complete information about the distribution is available, we can use simple IK to compute the probabilities and indicator covariances directly from the bivariate Gaussian law. BME can provide directly an estimate at point $\boldsymbol{p}_k$ (e.g., the mode of the posterior pdf) as the solution of the BME equation. Since neither the mode nor the mean can be reliably determined for the IK posterior distributions (e.g., non-monotonicity and extreme discretization of the cdf), we will use the median of the distributions. The results are shown in Fig. 10. All estimated values are centered with respect to the known values at $\boldsymbol{p}_k$; thus, each plot gives the corresponding estimation error distribution. Is evident from these plots that BME performs considerably better than IK. □

Chapter VII: Stochastic Partial Differential Equation Modelling of Flow and Transport

"I learned many years ago never to waste time trying to convince my colleagues."

A. Einstein

1. INTRODUCTION

Several natural processes are involved in the transport of pollutants from a source through different environmental compartments and media, leading ultimately to exposure of human populations. Environmental modelling is concerned with the study of these processes and the prediction of pollutant levels at specified space-time points in the environment.

The evolution of natural and biological processes is governed by differential equations that describe the motion of concentrations and fluxes in space and time. The coefficients of these equations, which also are functions of space and/or time, represent properties of the media within which the processes take place. If the spatiotemporal variability of the coefficients as well as the initial/boundary conditions are known precisely, the processes are represented by means of deterministic partial differential equations (PDE). However, a complete characterization of the variability is impossible due to measurement errors that lead to uncertainty, and to various constraints that limit the number of sampling points in space-time. Such processes are represented by means of stochastic PDE (SPDE).

Exact solutions of SPDE are not in general available in explicit form. However, explicit solutions can provide useful physical intuition, even in cases that they are valid only for simplified models and not adequate for realistic applications. In complex situations, two approaches are commonly used for the solution of SPDE: while the first focuses on obtaining solutions that are valid for specific *realizations* of the coefficient S/TRF, the second approach focuses on the estimation of *stochastic moments.* Both approaches have advantages and drawbacks, and they should be viewed as complementary tools for determining the behavior of stochastic solutions.

Realization-based approaches focus on numerical solutions of the SPDE. They are the only option if information on the behavior of specific realizations is required, and if the

SPDE can not be solved explicitly by analytical methods. However, numerical methods become computationally intensive if small scale details are to be resolved in large scale systems. In addition, numerical methods is not efficient for moment estimation if the ergodicity assumption does not hold. In such cases the SPDE must be solved repeatedly for many realizations (samples) of the coefficients in order for ensemble averages to be evaluated. Numerical investigations of this type are called *Monte Carlo* after the name of the tiny principality with the famous casino. An advantage of Monte Carlo methods is that they can be easily conditioned to available data.

Moment-based approaches focus on solving the deterministic equations that govern the stochastic moments of the natural processes represented by the SPDEs. The moment equations can usually be solved explicitly only if the correlation functions of the coefficients satisfy certain symmetry requirements (such as homogeneity and isotropy). Moment equations, though, may suffer from the well-known *closure problem*: The equation satisfied by the Nth-order moment involves higher order moments for all N. Hence, an infinite open-ended hierarchy of moment equations is obtained. Closure is usually achieved using truncated perturbation series or non-perturbative approximations.

In the following we will focus our discussion on two of the most advanced methods for studying SPDEs of flow and transport: *Diagrammatic* and *space transformation* methods.

Diagrammatic methods represent natural processes by diagrams or graphs, which stand for complicated analytical expressions, just as Chinese ideograms stand for whole phrases. Space transformation methods reduce the original PDE into an ordinary (ODE) which is easier to solve. Before going into details, we illustrate the main concepts by means of the following examples.

EXAMPLE 1: Consider the SPDE

$$L[X(\mathbf{p})] = f(\mathbf{p}), \tag{1}$$

where $\mathbf{p} = (\mathbf{s}, t)$, $X(\mathbf{p})$ is the S/TRF modelling the natural process of interest, L is a linear space-time differential operator that in general involves S/TRFs, and $f(\mathbf{p})$ is a known source function, likely also a S/TRF. It is possible to decompose L as $L = L_0 + L_R$, where L_0 is a linear differential operator with constant coefficients and L_R is the random part, the coefficients of which are zero-mean S/TRF. Assume that $L_0[X_0(\mathbf{p})] = f(\mathbf{p})$ can be solved for $X_0(\mathbf{p})$, as it is usually the case, and let the deterministic Green's function $G_0(\mathbf{p} - \mathbf{p}') = G_0(\mathbf{s} - \mathbf{s}', t - t')$ satisfy the equation

$$L_0[G_0(s-s',t-t')] = \delta(s-s')\delta(t-t'). \tag{2}$$

In light of Eq. (2), the solution of the initial SPDE (1) can be expressed in a closed form as

$$X(s,t) = X_0(s,t) - \int ds' \int_{-\infty}^{t} dt' \, G_0(s-s',t-t') \, L_R[X(s',t')], \tag{3}$$

where L_R operates at the space-time point $\boldsymbol{p}' = (s',t')$. For notational convenience, the closed-form solution will also be expressed in a unified space-time notation as

$$X(\boldsymbol{p}) = X_0(\boldsymbol{p}) - \int d\boldsymbol{p}' \, G_0(\boldsymbol{p}-\boldsymbol{p}') \, L_R[X(\boldsymbol{p}')]. \tag{4}$$

Eq. (4) is physically meaningful, provided that the functions $L_R[X(\boldsymbol{p}')]$ are sufficiently smooth for the space-time integral to exist. Even if the smoothness criterion is satisfied, the integral equation can not be solved explicitly for $X(\boldsymbol{p})$ due to the spatiotemporal variability of the random operator. This is exactly what happens in the case of flow and transport equations in heterogeneous porous media: the hydraulic conductivity and the velocity correlations have a complicated dependence on the spatial and temporal coordinates. Similarly, in mathematical models of epidemics and insect dispersal (Murray, 1993), spatiotemporal heterogeneity affects the form of the model coefficients (e.g., dispersion coefficient). In such cases explicit model solutions usually are not possible. If the solution $X(\boldsymbol{p})$ does not differ significantly from $X_0(\boldsymbol{p})$ --that is if the fluctuations induced by the random part L_R are weak-- accurate explicit approximations of $X(\boldsymbol{p})$ can be obtained in terms of truncated perturbation expansions. The latter are generated from Eq. (4) by recursively expanding $X(\boldsymbol{p}')$ on the right hand side leading to the *Neumann-Born* perturbation series

$$X(\boldsymbol{p}) = X_0(\boldsymbol{p}) + (-1)\int d\boldsymbol{p}_1 \, G_0(\boldsymbol{p}-\boldsymbol{p}_1) \, L_R[X_0(\boldsymbol{p}_1)] + (-1)^2 \int d\boldsymbol{p}_1 \int d\boldsymbol{p}_2 \, G_0(\boldsymbol{p}-\boldsymbol{p}_1)$$
$$L_R[G_0(\boldsymbol{p}_1-\boldsymbol{p}_2)] \, L_R[X_0(\boldsymbol{p}_2)] + O(L_R^3), \tag{5}$$

where $O(L_R^3)$ denotes terms that are of order three or higher in the random operator L_R. Information about the moments $\overline{X^n}(\boldsymbol{p})$, where the bar denotes the stochastic average over the joint pdf of all S/TRF in L_R, is also useful. For moment calculations, explicit expressions have a definite advantage over numerical simulations for two reasons: First, numerical calculations computationally expensive, because they involve averaging over a large number of realizations (e.g., in the range of several hundred to several thousand for typical groundwater problems). Secondly, the evaluation of stochastic moments involves integrals of smooth correlation functions instead of the fluctuating $L_R[X(\boldsymbol{p}')]$ terms. In the next section we gain more insight into the mathematical details of moment calculations by means of a dispersion equation. □

REMARK 1: A comment is in order here. Perturbation expansions such as Eq. (5) provide approximate solutions for $X(\boldsymbol{p})$ to any desired perturbation order. However, if L_R is a stochastic operator containing S/TRF coefficients with significant variability, the multiple integrals involved in perturbation expansions such as Eq. (5) can not be evaluated explicitly. Thus, it is more efficient to solve numerically the SPDE for specific realizations instead of evaluating perturbation approximations.

The following example offers an intuitive introduction of the diagrammatic approach in a simple subsurface flow situation. This approach is the subject of §§3-6 of this chapter.

EXAMPLE 2: Consider 1-D steady-state flow with constant mean hydraulic conductivity $\overline{K}$. The general form of the flow solution is

$$J(s) = J_0(s) - \int_{-\infty}^{s} du\, G_0(s-u) w(u) J(u), \tag{6}$$

where $J(s)$ is the hydraulic head gradient, $J_0(s)$ is the bare head gradient, $G_0(s)$ is its associated Green's function, and $w(s) = df(s)/ds$ where $f(s)$ is the zero mean fluctuation of the log-conductivity $\ln K(s)$. We begin by representing the hydrologic processes in diagrammatic form. If no change in K occurs between two points 0 and s of the porous medium, the head gradient is represented by a line connecting 0 and s, viz. 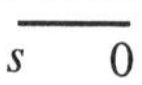 s —— 0. A dot ∘ denotes that a change in K occurs at some point within the medium, i.e., the diagram s ——∘—— 0 (dot at 1) implies that one change in K occurs at point 1. If hydraulic conductivity changes at two points (1 and 2), we draw the diagram s ——∘——∘—— 0 (dots at 1, 2), and so on. According to the superposition principle, the head gradient at point s is the sum of all possible correlation pairings represented by the above diagrams. If we denote this sum by a double line ═══, the hydraulic gradient is expressed in diagrammatic terms as

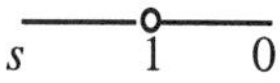

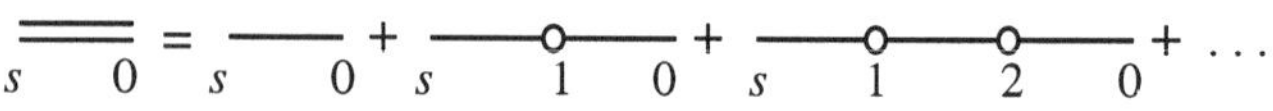

$$\underset{s\qquad 0}{=\!=\!=} \;=\; \underset{s\qquad 0}{\text{———}} + \underset{s\qquad 1\qquad 0}{\text{——}\circ\text{——}} + \underset{s\qquad 1\qquad 2\qquad 0}{\text{——}\circ\text{——}\circ\text{——}} + \ldots \tag{7}$$

Notice that the diagrams to the right of $\overline{}$ in the second and subsequent terms form an expression which is equal to the infinite series representing $=\!=\!=$, so that the following concise diagrammatic equation is obtained

$$\underset{s\qquad 0}{=\!=\!=\!=} \; = \; \underset{s\qquad 0}{-\!\!-\!\!-\!\!-} \; + \; \underset{s\qquad 1\qquad 0}{-\!\!-\!\!-\!\!\circ\!=\!=\!=} \quad . \tag{8}$$

The above pictorial argument is not a rigorous derivation, but it serves as an introduction. The rules by means of which diagrammatic expressions are given mathematical meaning are further explained below. We emphasize that diagrams contain exactly the same information as the underlying physical model. However, they represent information concisely and more efficiently than cumbersome mathematical expressions. Other advantages of the diagrammatic formalism will be mentioned forthwith. □

Space transformation methods for solving SPDEs modelling flow and transport in environmental media and compartments is the second group of advanced methods to be discussed in this chapter (§7). These methods simplify the study of multidimensional problems by reducing them to unidimensional ones, which are easier to solve analytically or computationally.

EXAMPLE 3: Consider an n-dimensional function --neglecting temporarily the time dimension-- defined in terms of its values on a cubic lattice with N nodes per side. Hence, the representation of the function requires N^n numbers. If an ST representation is used instead, values of the 1-D function along N_L transform lines are required. Assuming that $N_d \cong N$ points are used for the discretization of the 1-D function, the ST representation requires $N_L \times N_d$ numbers. Accurate representations can be obtained with a relatively small number, up to a few hundred, of transform lines. This can lead to significant computational gains. □

2. PERTURBATION EXPANSIONS

We investigate perturbation expansions of SPDE in the context of a *reactive-dispersive transport* equation used in biochemical systems (e.g., Haken, 1983). We assume here that the flow field and the diffusion coefficient are uniform, that the backward reaction rate can be neglected, and that the forward reaction rate is a fluctuating S/TRF. Similar equations represent the transport of pollutants in groundwater. In the case of subsurface transport, however, velocity fluctuations and backward reaction due to desorption can not be ignored. With the above assumptions, the transport equation is expressed as

$$\partial C(s,t)/\partial t - D\nabla^2 C(s,t) + A(s,t)\,C(s,t) + V \cdot \nabla C(s,t) = \phi(s,t), \tag{1}$$

where $C(s,t)$ represents the concentration of the chemical, D the diffusion coefficient, $\phi(s,t)$ the source of the chemical, V the flow velocity, and $A(s,t)$ the fluctuating reaction rate (the reaction removes the chemical from the fluid). We assume that $A(s,t)$ is a Gaussian random field $N(\overline{A}, \sigma_A^2)$ with $\overline{A} >> \sigma_A$. The Gaussian hypothesis is an idealization, since negative values of the reaction rate imply release of chemical into the fluid. However, if the coefficient of variation of $A(s,t)$ satisfies the condition $\mu_A << 1$, the fraction of negative $A(s,t)$ values in each realization is small, and its impact on the calculation is negligible. Without loss of generality we assume a point source $\phi(s,t) = \delta(s - s_o)\,\delta(t - t_o)$ at point (s_o, t_o).

In the dispersion Eq. (1) the operator $L = \partial/\partial t - D\nabla^2 + V \cdot \nabla + A(s,t)$ includes a deterministic component $L_0 = \partial/\partial t - D\nabla^2 + V \cdot \nabla + \overline{A}$ and a random component $L_R = \alpha(s,t) = A(s,t) - \overline{A}$. The deterministic equation represents dispersion in a homogeneous medium with a uniform reaction coefficient equal to $\overline{A}$. Hence, the concentration field is identical to the Green's function $G_0(s - s_o, t - t_o)$ of the diffusion-reaction operator L_0, that is $C_0(s,t) = G_0(\boldsymbol{r}, \tau)$, where $\boldsymbol{r} = \boldsymbol{s} - \boldsymbol{s}_o$ and $\tau = t - t_o$ represent the space and time displacements from the origin. The deterministic Green's function $G_0(\boldsymbol{r}, \tau)$ is also known as *free* or *bare Green's function* in the literature; both names indicate that $G_0(\boldsymbol{r}, \tau)$ is not dressed with the fluctuations. The solution for the Green's function in the case of a conservative medium ($\overline{A} = 0$) is Gaussian with variance equal to $\sigma^2 = 2D\tau$, i.e.,

$$G_0^c(\boldsymbol{r}, \tau) = \theta(\tau) \exp[-(\boldsymbol{r} - V\tau)^2 / 4D\tau] / (4\pi D\tau)^{3/2}, \tag{2}$$

shown in Fig. 1. In the case of dispersion in a medium with a uniform, non-zero, reaction rate the Green's function $G_0^c(\boldsymbol{r}, \tau)$ is exponentially reduced with the time lag,

$$G_0(\boldsymbol{r}, \tau) = G_0^c(\boldsymbol{r}, \tau) \exp[-\overline{A}\,\tau]. \tag{3}$$

Green's functions describe propagation between two points in the fluid. It can be shown that they satisfy the following general identity that relates propagation between two points in terms of compound propagations which involve an intermediate point

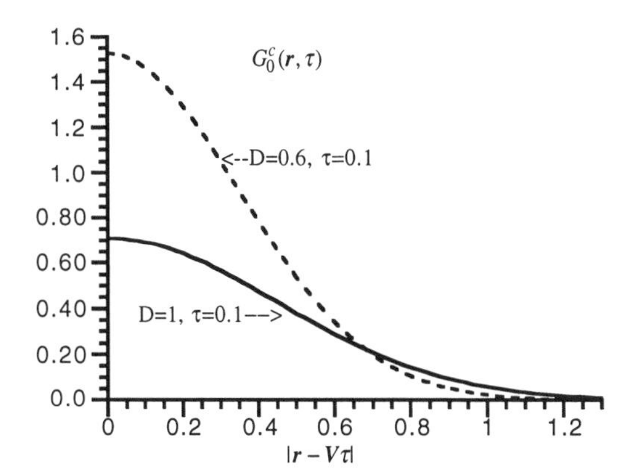

FIGURE 1: Plot of the Green's function $G_0^c(r,\tau)$ vs. the space-time separation for two different values of the diffusion coefficient.

$$G_0(\boldsymbol{s}-\boldsymbol{s}',t-t') = \int d\boldsymbol{s}_1\, G_0(\boldsymbol{s}-\boldsymbol{s}_1,t-t_1)\, G_0(\boldsymbol{s}_1-\boldsymbol{s}',t_1-t'). \tag{4}$$

The physical interpretation of this identity is that propagation from $\boldsymbol{s}'$ to $\boldsymbol{s}$ can be expressed as a superposition of two propagation events from $\boldsymbol{s}'$ to $\boldsymbol{s}_1$ and from $\boldsymbol{s}_1$ to $\boldsymbol{s}$ over all intermediate locations $\boldsymbol{s}_1$. The integral of the Green's function satisfies the sum rule

$$\int d\boldsymbol{s}\, G_0(\boldsymbol{s}-\boldsymbol{s}',t-t') = \exp[-\overline{A}\,(t-t')]. \tag{5}$$

Hence, the Green's function can also be viewed as the probability that a particle moves from $\boldsymbol{s}'$ to $\boldsymbol{s}$ over a time interval equal to $t-t'$. The probability integral is normalized to one for $t=t'$, but its value subsequently decreases due to removal of the chemical by the reaction. In the case of conservative dispersion, $A(\boldsymbol{s},t)=0$, the probability integral is stationary and normalized to one.

As we discussed in the previous section, the concentration S/TRF can be expressed by means of a Neumann-Born perturbation series as follows

$$C(\boldsymbol{s},t) = C_0(\boldsymbol{s},t) + c_1(\boldsymbol{s},t) + c_2(\boldsymbol{s},t) + \ldots, \tag{6}$$

where $c_N(s,t)$ denotes the perturbation correction of order N in the reaction rate $\alpha(s,t)$. The moments of the concentration can be obtained by evaluating the stochastic average of all the perturbation corrections, i.e.,

$$\overline{C(s,t)} = C_0(s,t) + \overline{c_1(s,t)} + \overline{c_2(s,t)} + \ldots \tag{7}$$

Therefore, the Nth-order approximation to $C(s,t)$ is given by the following truncated series

$$C_N(s,t) = C_0(s,t) + \sum_{l=1}^{N} c_l(s,t). \tag{8}$$

In view of Eqs. (7) and (1.5), where $X(\boldsymbol{p})$ is replaced by $C(\boldsymbol{p})$, the 1st-order approximation for Eq. (8) is given by

$$C_1(s,t) = C_0(s,t) - \int d\boldsymbol{s}_1 \int_{t_o}^{t} dt_1\, G_0(s - s_1, t - t_1)\, \alpha(s_1,t_1)\, G_0(s_1 - s_o, t_1 - t_o). \tag{9}$$

The lower time integration limit is equal to t_o because the concentration $C_0(s_1,t_1)$ is zero for $t_1 < t_o$, i.e., before the chemical starts dispersing. The 1st-order perturbation correction vanishes, because the fluctuations of the reaction rate have zero mean.

In 2nd-order, the mean concentration perturbation becomes

$$\overline{c_2(s,t)} = \sigma_A^2 \int_{t_o}^{t} dt_1 \int_{t_o}^{t_1} dt_2 \int d\boldsymbol{s}_1 \int d\boldsymbol{s}_2 G_0(s - s_1, t - t_1)\, G_0(s_1 - s_2, t_1 - t_2)$$
$$C_0(s_2,t_2)\, \rho_A(s_1 - s_2, t_1 - t_2), \tag{10}$$

where $\rho_A(s_1 - s_2)$ is the correlation function of the reaction rate. Since all the functions in the above expression are either positive or zero, the leading-order perturbation correction leads to an increase of the mean chemical concentration above the $C_0(s,t)$.

REMARK 1: This result may seem strange, because the fluctuations in $A(s,t)$ are symmetrically distributed about $\overline{A}$, so that one might expect a zero overall effect. Instead, the mean concentration increases above $C_0(s,t)$, which implies a reduced effective reaction rate. This reduction can be accounted, neglecting the effect of fluctuation correlations, by the convexity of the exponential function $\exp(-\overline{A}\,\tau)$.

EXAMPLE 1: Consider a homogeneous medium with binary probability for the reaction coefficient $\alpha(s,t)$, i.e., $p(\alpha_1) = p(\alpha_2) = 1/2$. The concentration in each realization α_i is

$$\mathrm{C_i}(s,t) = \mathrm{e}^{-\alpha_i(t-t_o)}\, G_0^c(s - s_o, t - t_o). \tag{11}$$

The mean concentration is the average over the two realizations given by

$$\overline{\mathrm{C}(s,t)} = \tfrac{1}{2} G_0^c(s - s_o, t - t_o)[\mathrm{e}^{-\alpha_1(t-t_o)} + \mathrm{e}^{-\alpha_2(t-t_o)}]. \tag{12}$$

The 0th-order approximation is the concentration in a uniform medium with reaction rate $\overline{A}$ which is given by

$$\mathrm{C_0}(s,t) = \exp[-\overline{A}(t - t_o)]\, G_0^c(s - s_o, t - t_o), \tag{13}$$

where $\overline{A} = (\alpha_1 + \alpha_2)/2$. The convexity of the exponential function leads to the inequality

$$\tfrac{1}{2}[\mathrm{e}^{-\alpha_1(t-t_o)} + \mathrm{e}^{-\alpha_2(t-t_o)}] \geq e^{-(\alpha_1+\alpha_2)(t-t_o)/2}. \tag{14}$$

The above can easily be verified using Fig. 2. In view of the convexity inequality, it

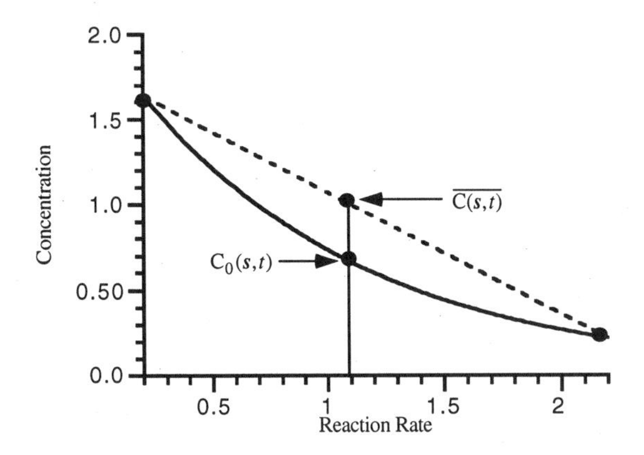

FIGURE 2: For a convex function the value at the midpoint of an interval $[a_1, a_2]$ is less than the average of the function evaluated at the endpoints.

follows that $\overline{\mathrm{C}(s,t)} \geq \mathrm{C_0}(s,t)$. □

The Nth-order perturbation correction is given by the multiple integral

$$\overline{c_N(\boldsymbol{s},t)} = (-1)^N \int d\boldsymbol{p}_1 \ldots \int d\boldsymbol{p}_N \ldots G_0(\boldsymbol{p}-\boldsymbol{p}_1)\ldots G_0(\boldsymbol{p}_N-\boldsymbol{p}_o)\overline{\alpha(\boldsymbol{p}_1)\ldots\alpha(\boldsymbol{p}_N)}. \quad (15)$$

In order to make progress with explicit calculations, the N-point correlation function $\overline{\alpha(\boldsymbol{p}_1)\ldots\alpha(\boldsymbol{p}_N)}$ must be evaluated. For a zero-mean Gaussian S/TRF the N-point function can always be expressed by means of products of two-point functions as follows

$$\overline{\alpha(\boldsymbol{p}_1)\ldots\alpha(\boldsymbol{p}_N)} = \delta_{N,2l}\,\sigma_A^{2l} \sum_P \rho_A(\boldsymbol{p}_{1(1)}-\boldsymbol{p}_{1(2)})\ldots\rho_A(\boldsymbol{p}_{l(1)}-\boldsymbol{p}_{l(2)}), \quad (16)$$

where the sum in the above expression is over all possible ways of pairing N coordinates, and $\boldsymbol{p}_{i(j)}$, $j=1,2$ denote the space-time coordinates of the i-th pair, where $i=1,\ldots,l$. The number of different pairings N_{2l} is given by

$$N_{2l} = (2l)!/l!2^l. \quad (17)$$

By direct evaluation it follows that the number of pairs for each perturbation order is $N_0=1$, $N_2=1$, $N_4=3$, $N_6=15$, and so on. The number of pairs increases fast with the order of the perturbation. This trend complicates the explicit representation and evaluation of high-order perturbation corrections.

REMARK 2: Most studies of moment equations in environmental health science are limited to approximations of the fluctuation effects using low order perturbation analysis. In the case of subsurface hydrology, leading order approximations are standard (e.g., Gelhar, 1993). Calculations of 2nd-order effects have also been investigated; the calculations are straightforward but lengthy and complicated (Dagan, 1992; Deng and Cushman, 1995). The *diagrammatic* method (Christakos *et al.*, 1993a, b, c and d; 1995; Oliver and Christakos, 1996; Hristopulos and Christakos, 1997), on the other hand, provides concise representation of cumbersome mathematical expressions and the power of non-perturbative approximations.

3. THE DIAGRAMMATIC APPROACH

The diagrammatic method was introduced by Feynman for summing perturbation series of electron interactions in quantum electrodynamics (Feynman, 1962). The Feynman diagrams were later applied to problems of classical statistical mechanics such as fluid turbulence (see refs. in McComb, 1990). The first attempt to use the diagrammatic theory in porous media flow is due to King (1987). In stochastic hydrology the diagrammatic method was introduced for flow moment calculations by Christakos *et al.* (1993a, b and c; 1995). It has also been used successfully in investigations of effective hydraulic conductivity (Hristopulos and Christakos, 1997), in calculations of transient fluid mixing (Zhang, 1995), and in a renormalization-group analysis of macrodispersivity (Jaekel and Vereecken, 1997).

The basic idea in the diagrammatic method is that geometric objects (diagrams) can be used in order to represent mathematical expressions and to calculate with them. Mathematics can be viewed as a language that consists of an alphabet of symbols, and a set of operation rules that define how simple symbols are combined. Diagrams can also be used for the same purpose, i.e., as tools for representing mathematical expressions. They provide a compact notation, and in addition they help to interpret the physical meaning of different terms and their relative importance in calculations.

3.1 The Diagrammatic Language - Physical Space

The power of the diagrammatic approach is most evident in moment calculations, in which it helps to deal effectively with the proliferation of higher-order perturbation terms. We begin by considering the Nth-order perturbation correction of the mean concentration for the problem of reactive dispersion, see Eqs. (2.15) and (2.16) above

$$\overline{c_N(\boldsymbol{p})} = \delta_{N,2l}\, \sigma_A^N \int d\boldsymbol{p}_1 \ldots \int d\boldsymbol{p}_N\; G_0(\boldsymbol{p}-\boldsymbol{p}_1)\ldots G_0(\boldsymbol{p}_N-\boldsymbol{p}_o)$$
$$\sum\nolimits_P \rho_A(\boldsymbol{p}_{1(1)}-\boldsymbol{p}_{1(2)})\ldots\rho_A(\boldsymbol{p}_{l(1)}-\boldsymbol{p}_{l(2)}) \tag{1}$$

(note that only the even terms are considered). The right-hand side of Eq. (1) involves the following symbols:

1. An initial point $\boldsymbol{p}_o$ and a final point $\boldsymbol{p}$.
2. The integration points $\boldsymbol{p}_1,\ldots,\boldsymbol{p}_n$.
3. The Green's functions $G_0(\boldsymbol{p}_i-\boldsymbol{p}_j)$.
4. The two-point correlation functions $\rho_A(\boldsymbol{p}_{i(1)}-\boldsymbol{p}_{i(2)})$.
5. The fluctuation variance σ_A^2.

In the diagrammatic approach these symbols are represented by means of geometric objects (diagrams or graphs). There is a geometric correspondence between diagrams and symbols in the mathematical expression. The main factor is the number of points involved in a mathematical symbol: Single points $\boldsymbol{p}_i$, e.g., should be represented by vertices, two-point functions by lines, four-point functions by shapes with four vertices, e.g., squares, and so on. The geometrical considerations still allow significant freedom in assigning diagrams to mathematical symbols. Vertices can be depicted as points or circles, lines can be thin or thick, straight or curved, continuous or broken; circles can be open (white) or solid (black). There is no fundamental reason for selecting one diagram over another with similar geometry, (e.g., the same number of vertices) which explains why different authors may use different diagrams for the same purpose.

Mathematical calculations require rules for evaluating diagrammatic expressions. These rules should lead to the mathematical expressions from the diagrams; hence, they are specific to the particular SPDE. For the reactive dispersion equation, the following set of diagrammatic rules can be used:

REPRESENTATION RULES:

1. The endpoints of the diagram denote the external points $\boldsymbol{p}_o$ and $\boldsymbol{p}$.
2. Open circles denote the internal points $\boldsymbol{p}_1,\ldots,\boldsymbol{p}_n$ (vertices). Each vertex is associated with the fluctuation standard deviation σ_A.
3. Straight lines (G-lines) between points denote the Green's functions.
4. Triangular arcs with two vertices denote the two-point correlation functions.
5. Thick lines ($\overline{G}$-lines) denote the mean Green's function and the mean concentration.
6. Thick lines with a superimposed number (N) denote the Nth-order approximation.

We supplement the representation rules that define the diagrammatic alphabet with a set of operation rules that permit manipulation and evaluation of diagrams.

OPERATION RULES:

1. Simple diagrams can be combined in order to construct composite graphs.
2. Multiplication of all the individual diagrams of a graph is implied.
3. Space-time integration is implied at all vertices $\boldsymbol{p}_1,\ldots,\boldsymbol{p}_n$.

At this point we have all the necessary components for writing diagrammatic representations of the stochastic moment equations. The truncated perturbation series for $\overline{C_4(\boldsymbol{p})}$ is given in diagrammatic form in Fig. 3. Diagrammatic rules are useful for translating between diagrammatic graphs and mathematical expressions. For example, the

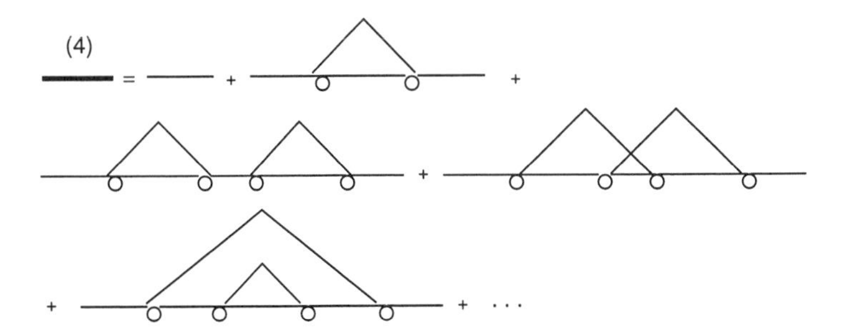

FIGURE 3: Diagrammatic representation of $\overline{C_4(\boldsymbol{p})} = \overline{C_4(\boldsymbol{s},t)}$. The first term to the right of the equality sign is the 0th-order approximation, the second term is the leading order perturbation correction, and the last three terms represent the 2nd-order (in variance) correction.

3rd graph on the right of the equality sign in Fig. 3 represents the first term in the perturbation correction $\overline{c_4(\boldsymbol{s},t)}$. By means of the representation rules the diagrams involved in this graph correspond to mathematical functions as shown in Fig. 4. In

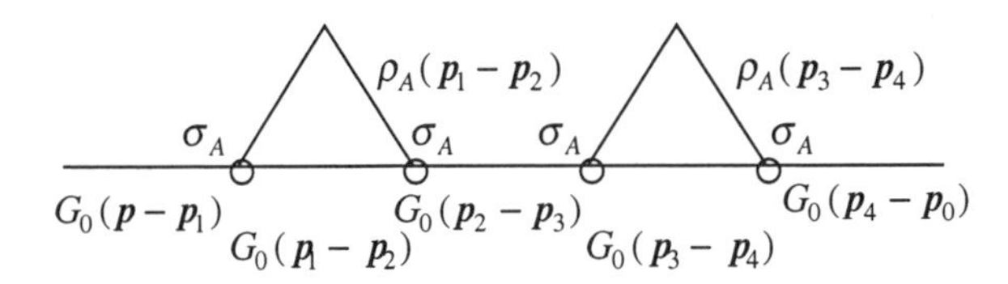

FIGURE 4: The first of the three diagrammatic terms in $\overline{c_4(\boldsymbol{s},t)}$.

addition, by means of the operation rules the following mathematical expression is obtained for the diagram

$$\overline{c_{4(1/3)}(\boldsymbol{p})} = \sigma_A^4 \int d\boldsymbol{p}_1 \ldots \int d\boldsymbol{p}_4 \, G_0(\boldsymbol{p} - \boldsymbol{p}_1) G_0(\boldsymbol{p}_1 - \boldsymbol{p}_2) G_0(\boldsymbol{p}_2 - \boldsymbol{p}_3) G_0(\boldsymbol{p}_3 - \boldsymbol{p}_4)$$
$$G_0(\boldsymbol{p}_4 - \boldsymbol{p}_o) \rho_A(\boldsymbol{p}_1 - \boldsymbol{p}_2) \rho_A(\boldsymbol{p}_3 - \boldsymbol{p}_4), \quad (2)$$

where $\overline{c_{4(1/3)}(\boldsymbol{p})}$ denotes the first of the three diagrams that compose $\overline{c_4(\boldsymbol{p})}$.

3.2 Diagrams in Frequency Space

Stochastic moment equations involve multiple integrals in space and time over the vertices $\boldsymbol{p}_i$. In the case of a stochastically homogeneous and stationary medium the correlation functions depend purely on the space-time separations. In addition, in infinite media the

Green's functions also depend purely on space-time lags. In the case of finite media they include corrections that decay with increasing distance from the boundaries. When the approximation of an infinite, stochastically homogeneous and stationary medium is valid, the moments are expressed in terms of convolution integrals which can be simplified using integral transforms. The standard representations involve the Fourier transform (FT) in space and time. When the time dependence involves a step function at the origin, the Laplace transform (LT) is more appropriate.

EXAMPLE 1: The dispersion equation in a uniform medium leads to the following expression for the Fourier-Laplace transform (F-LT) of the Green's function

$$\tilde{G}_0(\boldsymbol{\kappa},\omega) = (\omega + D\kappa^2 + \bar{A} + i\,\boldsymbol{\kappa}\cdot V)^{-1}, \tag{3}$$

where ω denotes the temporal frequency, $\boldsymbol{\kappa}$ the spatial frequency (momentum) and $\kappa = |\boldsymbol{\kappa}|$. It is a good exercise to verify that the above expression yields the right answer in real (configuration) space. We can write the above as $\tilde{G}_0(\boldsymbol{\kappa},\omega) = [\omega + g(\boldsymbol{\kappa})]^{-1}$, where $g(\boldsymbol{\kappa}) = D\kappa^2 + \bar{A} + i\,\boldsymbol{\kappa}\cdot V$. By inverting the LT, the following expression is obtained for the spatial FT

$$\tilde{G}_0(\boldsymbol{\kappa},\tau) = \exp[-g(\boldsymbol{\kappa})\,\tau]. \tag{4}$$

The inverse FT is given by means of $G_0(\boldsymbol{r},\tau) = \int_{\boldsymbol{\kappa}} \tilde{G}_0(\boldsymbol{\kappa},\tau)\exp(i\,\boldsymbol{\kappa}\cdot\boldsymbol{r})$, Since the integrand is separable we can write in n spatial dimensions,

$$G_0(\boldsymbol{r},\tau) = e^{-\bar{A}\,\tau} \prod_{i=1}^{n} \int_{\kappa_i} \exp[i\,\kappa_i\,(r_i - V_i\,\tau) - D\,\tau\,\kappa_i^2]. \tag{5}$$

Hence, the calculation is reduced to evaluating an 1-D integral. By completing the square in the exponent the 1-D integral is expressed as (e.g., Byron and Fuller, pp. 46-47)

$$\int_{\kappa_i} \exp[i\,\kappa_i\,(r_i - V_i\,\tau) - D\,\tau\,\kappa_i^2] = I(\tau D)\exp[-(r_i - V_i\,\tau)^2 / 4\,\tau D], \tag{6}$$

where $I(\tau D) = \int_{\kappa_i} \exp[-\tau D(\kappa_i - i\,\kappa_{0i})^2]$ and $\kappa_{0i} = (r_i - V_i\,\tau)^2 / 2\,\tau D$. The $I(\tau D)$ is evaluated by using the complex plane contour shown in Fig. 5 and the residue theorem. The range R is assumed to tend to infinity. The integral $I(\tau D)$ is along the contour segment Γ_3. Since there are no enclosed poles the contour integral is zero. Integrals along

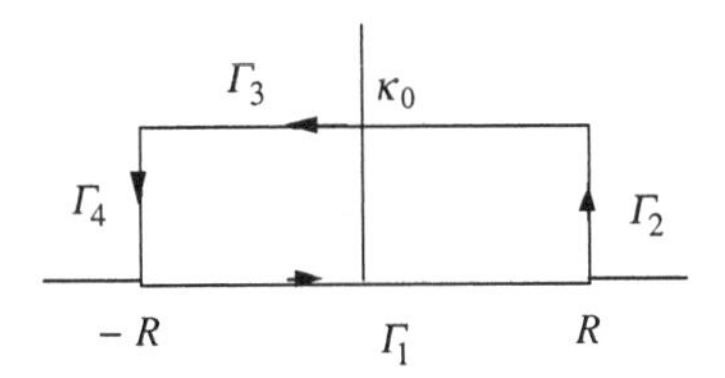

FIGURE 5: The integration contour in the complex plane.

the segments Γ_2 and Γ_4 vanish as $R \to \infty$, provided that $\tau > 0$. Hence, the $I(\tau D)$ is equal to the integral along the real line Γ_1, $I(\tau D) = \sqrt{\pi / D\tau}$. This leads to Eq. (2.2) for Green's function. □

The diagrams that represent a particular term in frequency domain look identical to the diagrams that correspond to the same term in the space-time domain. However, the diagrammatic rules are slightly modified in the frequency domain. The representation and operation rules for calculating with diagrams in frequency space are as follows:

REPRESENTATION RULES:

1. Open circles mark the vertices; each vertex is associated with a fluctuation σ_A.
2. G-lines denote the F-LT of Green's functions evaluated at $\boldsymbol{w} = (\boldsymbol{\kappa}, \omega)$, where ω is the temporal frequency and $\boldsymbol{\kappa}$ the spatial frequency (momentum).
3. Triangular arcs with vertices denote the F-LT of two-point correlation functions.
5. $\overline{G}$ -lines denote the F-LT of the mean Green's function $\overline{\tilde{G}(\boldsymbol{\kappa}, \omega)}$ or the concentration $\overline{\tilde{C}(\boldsymbol{\kappa}, \omega)}$.
6. Solid lines with a superimposed (N) denote the Nth-order approximation $\overline{\tilde{C}_N(\boldsymbol{\kappa}, \omega)}$.

OPERATION RULES:

1. Simple diagrams can be combined to form composite graphs.
2. Multiplication of all the individual diagrams of a graph is implied.
3. The incoming and outgoing lines of a graph have identical spatial and temporal frequencies $(\boldsymbol{\kappa}, \omega)$. These are the external frequencies of the graph.
4. Both spatial and temporal frequencies are conserved at the vertices of the graph. Frequency conservation implies that $\sum_{i \in \text{in}} \boldsymbol{\kappa}_i = \sum_{j \in \text{out}} \boldsymbol{\kappa}_j$, and $\sum_{i \in \text{in}} \omega_i = \sum_{j \in \text{out}} \omega_j$.
5. Integration is assumed over the internal frequencies of the graph.

EXAMPLE 2: In Figs. 6a-b the diagrams that correspond to the perturbation corrections

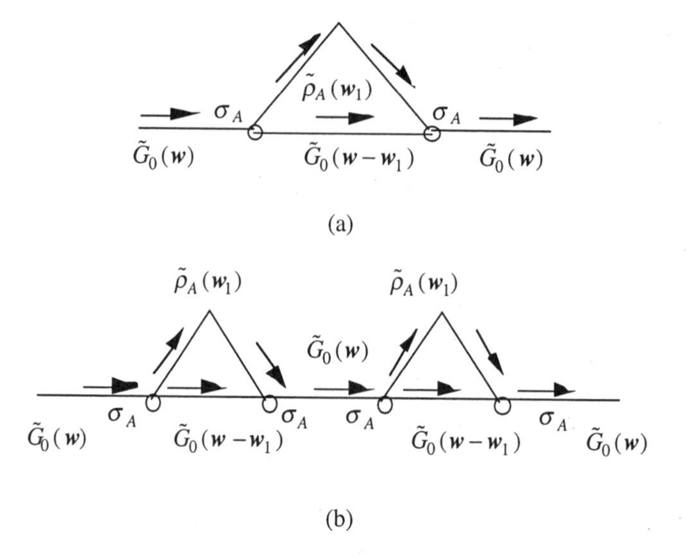

FIGURE 6: The diagram that represents (a) the term $\overline{\tilde{c}_2(w)}$ and (b) the term $\overline{\tilde{c}_{4(1/3)}(w)}$ in frequency space, $w = (\kappa, \omega)$.

$\overline{\tilde{c}_2(\kappa, \omega)}$ and $\overline{\tilde{c}_{4(1/3)}(\kappa, \omega)}$ are shown. □

REMARK 1: Certain comments are in order regarding the diagrams in Figs. 6a-b. Arrows are used as an accounting device for imposing the frequency conservation at the vertices. The direction of the arrows determines whether a frequency should be counted as incoming or outgoing. Notice that both spatial and temporal frequencies are conserved at the vertices. For example, check the leftmost vertex: There is an incident Green's function with momentum κ, and two outgoing functions: a Green's function with momentum $\kappa - \kappa_1$ and a correlation function with momentum κ_1 so that the momenta of the outgoing functions add up to the momentum of the incident Green's function.

In view of the above rules the 2nd-order perturbation correction $\overline{c_2(s,t)}$ is given by

$$\overline{\tilde{c}_2(\kappa, \omega)} = \sigma_A^2 \, \tilde{G}_0^2(\kappa, \omega) \int_{\kappa_1} \int_{\omega_1} \tilde{G}_0(\kappa - \kappa_1, \omega - \omega_1) \tilde{\rho}_A(\kappa_1, \omega_1). \tag{7}$$

Similarly, the F-LT of the first term of the 4th-order perturbation correction is given by

$$\overline{\tilde{c}_{4(1/3)}(\boldsymbol{\kappa},\omega)} = \sigma_A^4\, \tilde{G}_0^3(\boldsymbol{\kappa},\omega)[\int_{\boldsymbol{\kappa}_1}\int_{\omega_1} \tilde{G}_0(\boldsymbol{\kappa}-\boldsymbol{\kappa}_1,\omega-\omega_1)\tilde{\rho}_A(\boldsymbol{\kappa}_1,\omega_1)]^2. \tag{8}$$

It is obvious from the above that the diagrammatic representation provides a powerful means for visualizing the correlations between fluctuations at different space-time points. The diagrammatic approach does not provide, however, explicit evaluation of the integrals involved in the diagrammatic expressions. This should be accomplished by standard analytical or numerical techniques.

4. SELF-CONSISTENT EQUATIONS

4.1 The Basic Concepts

We have shown above how diagrams can be used for representing terms in a perturbation expansion. These diagrammatic representations can lead to insightful reorganizations of the perturbation series that emphasize important features and provide approximations which involve all perturbation orders. Such approximations are called *partial summations*, because they estimate the series by means of a summable sub-series. Partial summations involve an infinite number of terms of all perturbation orders. To illustrate the difference between low-order perturbation approximations and partial summations we discuss an example.

EXAMPLE 1: Consider the infinite series $S=\sum_{i=1}^{\infty} c_i$. It is always possible to express a term c_i as $c_i = c_i^{(1)} + c_i^{(2)}$. Hence, we can write S as $S = S^{(1)} + S^{(2)}$, where $S^{(k)} = \sum_{i=1}^{\infty} c_i^{(k)}$, $k=1,2$. The infinite series $S^{(1)}$ and $S^{(2)}$ represent partial sums of the series S. If the partial sum $S^{(1)}$ is dominant, i.e., $S^{(1)} >> S^{(2)}$, then $S^{(1)}$ is an accurate approximation of S, i.e., $S \cong S^{(1)}$. The infinite series $S^{(1)}$ contains perturbation terms up to all orders, and it thus remains an accurate approximation even when the perturbation parameter is not small. On the other hand, low order approximations estimate the infinite series S by truncating after a small number of terms, $S \cong S_M = \sum_{i=1}^{M} c_i$. □

Low order perturbation approximations are accurate only if the perturbation series converges fast. For environmental health processes this condition is usually met in the case of weak heterogeneity. On the other hand, partial summations can lead to more informative results for strong heterogeneity. The question that must be addressed is how the partial sum is selected. In many cases there are clear physical reasons that guide this selection. In other cases the selection may not be obvious and different partial sums should be investigated (Christakos *et al.*, 1995; Hristopulos and Christakos, 1997). In order to gain more insight, consider the following example.

EXAMPLE 2: Assume that $S = \sum_{i=1}^{\infty} \overline{X^i}$ where the random variable X is $X = \overline{X} + x$. It is assumed that the pdf of the variable X is unknown, and that the fluctuations are weak, i.e., $\overline{x^n} << (\overline{X})^n$ for all n. Then, the infinite series S can be approximated by means of the partial sum $S^{(1)} = \sum_{i=1}^{\infty} \overline{X}^i = \overline{X}/(1-\overline{X})$. Let us consider a slightly different approach for summing the series $S^{(1)}$. Since $S^{(1)} = \overline{X} + \overline{X}^2 + \ldots$, it can be expressed in closed form as the solution of the algebraic equation $S^{(1)} = \overline{X} + \overline{X}\, S^{(1)}$. This expression is a *self-consistent* equation in the series $S^{(1)}$. □

The concept of self-consistent equations is very useful.

Self-consistency: *It implies that a function is related to itself via an algebraic or integral relation, the solution of which determines the function.*

In the diagrammatic analysis of SPDE, partial summations are obtained as solutions of integral self-consistent equations. At this point we return to the transport equation, and we examine the Green's function that satisfies

$$\partial G(s-s',t-t')/\partial t - D\nabla^2 G(s-s',t-t') + V \cdot \nabla G(s-s',t-t') = \delta(s-s')\delta(t-t'). \quad (1)$$

In the case of a unit point source the equation for the concentration field is identical to Eq. (1). The average Green's function is given in diagrammatic form in Fig. 7. The equivalent

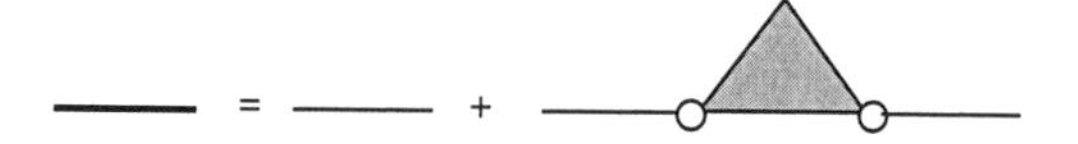

FIGURE 7: The diagrammatic equation for the Green's function

mathematical expression for the equation in Fig. 7 is

$$\overline{G(p-p')} = G_0(p-p') + \int dp_1 \int dp_2\, G_0(p-p_1)\Sigma(p_1-p_2)G_0(p_2-p'). \quad (2)$$

We emphasize that this equation is an exact expression for the Green's function. The function $\Sigma(s_1 - s_2, t_1 - t_2)$ contains the fluctuation effects and is called the *self-energy*. The self-energy is represented by the dotted triangle in Fig. 7. For applications in subsurface hydrology the term *porous media description operator* (PMDO) has also been used (e.g., Christakos *et al.*, 1993a, b, c and d; 1995), in order to emphasize that the function $\Sigma(s_1 - s_2, t_1 - t_2)$ describes the correlations within the porous medium. In general, the term self-energy is used to denote any part of a diagram that is connected to the rest of the diagram by means of two lines. The self-energy represents an infinite series of two-point correlations given by the diagrams in Fig. 8.

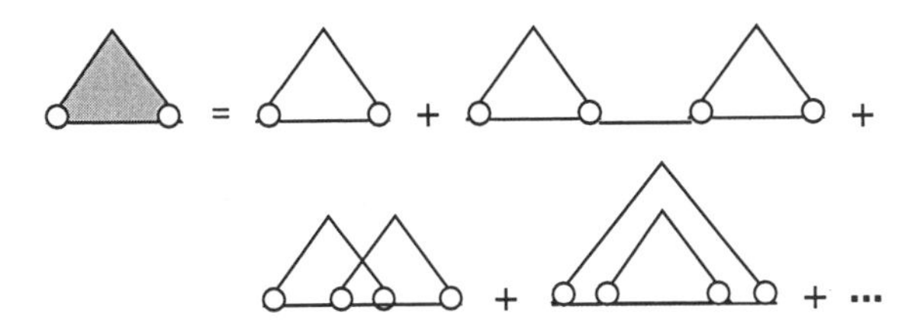

FIGURE 8: The diagrams that represent the self-energy up to 4th-order.

The self-energy includes two different types of graphs: *reducible* graphs, which can be separated by cutting a single G-line; and *irreducible* graphs which can not be separated by cutting a single G-line. For example, while the second diagram on the right side of the equality in Fig. 8 is reducible, the three remaining diagrams are irreducible. The *proper self-energy* is defined as the sum of all irreducible self-energy diagrams and is denoted by $\Sigma^*(p_1 - p_2)$. We use a shaded triangle with two vertices to denote the proper self-energy. The self-energy $\Sigma(p_1 - p_2)$ can be expressed as an infinite series of proper self-energy terms by means of the diagrammatic equation shown in Fig. 9. The equivalent mathematical expression for the equation in Fig. 9 is

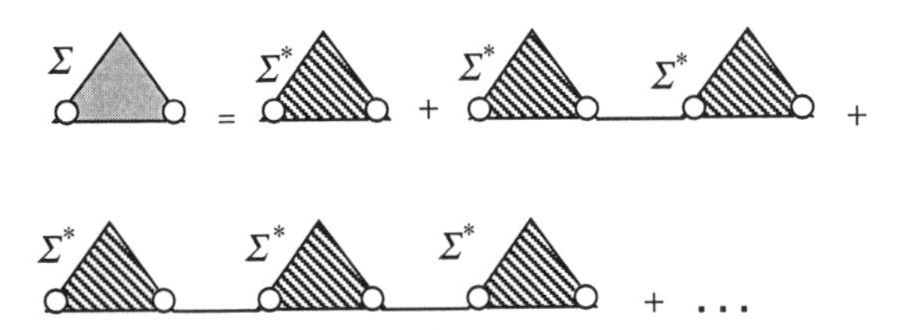

FIGURE 9: The self-energy is expressed in terms of the proper self-energy diagrams.

$$\Sigma(p-p') = \Sigma^*(p-p') + \int dp_1 \int dp_2\, \Sigma^*(p-p_1) G_0(p_1-p_2) \Sigma^*(p_2-p') + \int dp_1 \int dp_2 \int dp_3 \int dp_4\, \Sigma^*(p-p_1) G_0(p_1-p_2) \Sigma^*(p_2-p_3) G_0(p_3-p_4) \Sigma^*(p_4-p') + \ldots \tag{3}$$

Correspondingly, an infinite series is obtained for Green's function in terms of the proper self-energy as follows

$$\overline{G(p-p')} = G_0(p-p') + \int dp_1 \int dp_2\, G_0(p-p_1) \Sigma^*(p_1-p_2) G_0(p_2-p') + \int dp_1 \int dp_2 \int dp_3 \int dp_4\, G_0(p-p_1) \Sigma^*(p_1-p_2) G_0(p_2-p_3) \Sigma^*(p_3-p_4) G_0(p_4-p') + \ldots \tag{4}$$

The diagrammatic expression for the above infinite series is shown in Fig. 10. By inspection of the diagrammatic series it follows that the Green's function expansion is obtained by adding iteratively G-lines and proper self-energy diagrams. Thus, the equation in the second line of Fig. 10 is obtained. Note that the second line is a closed-form expression that resembles strongly the self-consistent equation in Example 2 above.

The self-consistent diagrammatic equation in the second line of Fig. 10 is given by the following mathematical expression

$$\overline{G(p-p')} = G_0(p-p') + \int dp_1 \int dp_2\, G_0(p-p_1) \Sigma^*(p_1-p_2) \overline{G(p_2-p')}. \tag{5}$$

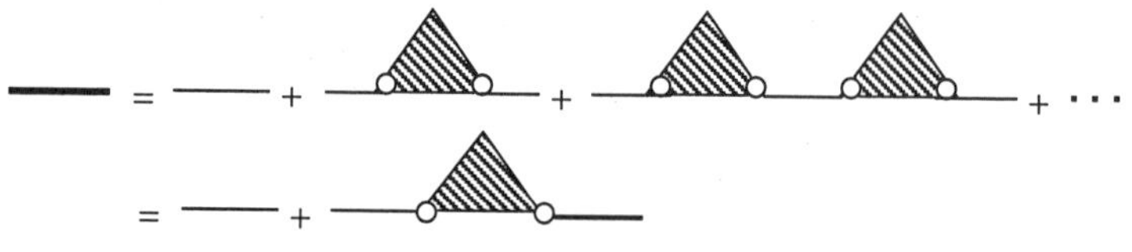

FIGURE 10: The proper self-energy series for the average Green's function.

This is known as the *Dyson equation*, and is an exact integral expression for the average Green's function. The Dyson equation is an example of the intuitive power of diagrams that can lead to insightful, non-perturbative approximations of SPDE solutions. By means of the F-LT, Dyson equation leads to the following algebraic equation in frequency space

$$\overline{\tilde{G}(\boldsymbol{\kappa},\omega)} = \tilde{G}_0(\boldsymbol{\kappa},\omega) + \tilde{G}_0(\boldsymbol{\kappa},\omega) \Sigma^*(\boldsymbol{\kappa},\omega) \overline{\tilde{G}(\boldsymbol{\kappa},\omega)}. \tag{6}$$

If both the Green's function and the self-energy are scalar, then

$$\overline{\tilde{G}(\boldsymbol{\kappa},\omega)} = \frac{1}{[\tilde{G}_0(\boldsymbol{\kappa},\omega)]^{-1} - \Sigma^*(\boldsymbol{\kappa},\omega)}, \tag{7}$$

which is the solution of the Dyson equation.

4.2 The Asymptotic Limit

It is often adequate to know the asymptotic behavior of the Green's function, i.e., for distances and times large compared to the characteristic scales of the natural process. This is determined from the behavior of the Green's function F-LT at small spatial and temporal frequency. Hence, provided that $\Sigma^*(\boldsymbol{\kappa},\omega)$ is analytic around zero,

$$\overline{\tilde{G}(\boldsymbol{\kappa},\omega)} \cong \frac{1}{[\tilde{G}_0(\boldsymbol{\kappa},\omega)]^{-1} - \Sigma^*(0) - \boldsymbol{\sigma}_1^* \cdot \boldsymbol{\kappa} - \boldsymbol{\sigma}_2^* : \boldsymbol{\kappa}\boldsymbol{\kappa}}, \tag{8}$$

where the constants $\Sigma^*(0)$, $\boldsymbol{\sigma}_1^*$ and $\boldsymbol{\sigma}_2^*$ are given by $\Sigma^*(0) = \Sigma^*(\boldsymbol{w}=\boldsymbol{0})$, $\boldsymbol{\sigma}_1^* = \nabla_{\boldsymbol{\kappa}} \Sigma^*(\boldsymbol{w})\big|_{\boldsymbol{w}=\boldsymbol{0}}$ and $\boldsymbol{\sigma}_2^* = \frac{1}{2}\nabla_{\boldsymbol{\kappa}}\nabla_{\boldsymbol{\kappa}} \Sigma^*(\boldsymbol{w})\big|_{\boldsymbol{w}=\boldsymbol{0}}$; and the product $\boldsymbol{\sigma}_2^* : \boldsymbol{\kappa}\boldsymbol{\kappa}$ denotes the double sum $\sum_{i,j=1}^{n} \sigma_{2,ij}^* \kappa_i \kappa_j$. The self-energy insertions renormalize the effective values of the reaction rate, the velocity and the diffusion coefficient. By inspection of Eq. (8) it follows that $\overline{\tilde{G}}$ has the same form as $\tilde{G}_0$ with $\overline{A}$, $\boldsymbol{V}$ and D replaced by effective counterparts.

EXAMPLE 3: We examine the asymptotic behavior of the reaction-dispersion equation in light of the above analysis. Assume that the reaction rate fluctuations are exponentially correlated in space and have a vanishing correlation length in time, i.e., $\rho_A(\boldsymbol{r},\tau) = \exp(-r/\xi)\,\delta(\tau)$. The F-LT of the average Green's function at large times and distances is

$$\overline{\tilde{G}(\boldsymbol{\kappa},\omega)} \cong [\omega + \overline{A} - \Sigma^*(0) + i\,\boldsymbol{\kappa}\cdot(\boldsymbol{V} + i\boldsymbol{\sigma}_1^*) + (D\boldsymbol{I} - \boldsymbol{\sigma}_2^*) : \boldsymbol{\kappa}\boldsymbol{\kappa}]^{-1}. \tag{9}$$

The asymptotic behavior is determined by the behavior of the proper self-energy for zero frequency. The average Green's function can be expressed as a renormalized version of the deterministic Green's function

$$\overline{\tilde{G}(\boldsymbol{\kappa}, w)} \cong \frac{1}{\omega + A_{eff} + i\,\boldsymbol{\kappa}\cdot \boldsymbol{V}_{eff} + \boldsymbol{D}_{eff} : \boldsymbol{\kappa\kappa}}, \tag{10}$$

where $A_{eff} = \overline{A} - \Sigma^*(0)$, $\boldsymbol{V}_{eff} = \boldsymbol{V} + i\boldsymbol{\sigma}_1^*$ and $\boldsymbol{D}_{eff} = D\boldsymbol{I} - \boldsymbol{\sigma}_2^*$. Next, we investigate a low order approximation of the proper self-energy based on the following diagram

$$\Sigma^* \cong \Sigma^{*(1)} =$$  $$= \sigma_A^2 \int_{\boldsymbol{\kappa}_1} \tilde{G}_0(\boldsymbol{\kappa} - \boldsymbol{\kappa}_1, \omega)\tilde{\rho}_A(\boldsymbol{\kappa}_1). \tag{11}$$

This approximation is justified at least if $\sigma_A < \overline{A}$. Explicit evaluation of the integral above for $\boldsymbol{\kappa} = 0$ and $\omega = 0$ leads to

$$\Sigma^{*(1)}(0) = \frac{\sigma_A^2}{\overline{A}} \frac{1 + \tau_r/\tau_d}{(\tau_r/\tau_a)^2 - (1 - \tau_r/\tau_d)^2}, \tag{12}$$

where the τ_i $(i = r, d, \mathrm{a})$ represent characteristic times for reaction, diffusion and advection given by $\tau_r = 1/\overline{A}$, $\tau_d = \xi^2/D$ and $\tau_\mathrm{a} = \xi/V$, respectively. If $\Sigma^{*(1)}(0)$ is positive the effective reaction rate is less than $\overline{A}$. Let us consider in more detail the physical mechanisms leading to these results. As a first approximation we neglect diffusion so that $\tau_\mathrm{d} = \infty$ and the zero frequency self-energy becomes

$$\Sigma^{*(1)}(0) = \sigma_\alpha^2 / \overline{A}\,[(\tau_r/\tau_\mathrm{a})^2 - 1], \tag{13}$$

which is positive if $\tau_r/\tau_\mathrm{a} > 1$, i.e., if the characteristic reaction time exceeds the characteristic advection time. In this case advection occurs faster than reaction, and leads to a reduced effective reaction rate. On the other hand, if τ_r/τ_a is below the threshold the reaction rate is increased. If diffusion is not negligible, the threshold of τ_r/τ_a for $\Sigma^{*(1)}(0)$ to be positive is lowered, since diffusion competes with advection. □

5. DIAGRAMMATIC METHODS IN SUBSURFACE HYDROLOGY

In stochastic hydrology the diagrammatic method was introduced for flow moment calculations by Christakos *et al.* (1993a, b and c; 1995). These publications contain detailed calculations of hydraulic head moments (e.g., mean and covariance) by means of various selective summations, comparisons of diagrammatic approximations with exact 1-D results, as well as an intuitive introduction to diagrammatic techniques. Here we will not elaborate on the diagrammatic representations and approximations of covariance functions; the corresponding rules for covariances are similar to the ones presented above for the calculation of Green's functions, and they are given in detail in Christakos *et al.* (1995) and Oliver and Christakos (1996). Application of the diagrammatic technique to the estimation of effective hydraulic conductivity of stochastically homogeneous aquifers (Hristopulos and Christakos, 1997) leads to an increased set of diagrams. This effect is due to the coupling of two random fields, hydraulic conductivity and hydraulic gradient, that results in the generation of cross-moments. Using diagrams the terms that have the greater impact in the effective hydraulic conductivity calculation can be determined. Diagrams are grouped according to their topological properties. Thus, it is possible to systematically enumerate the diagrams in the diagrammatic expansion that correspond to the same value.

The topological grouping of diagrams is useful, because it allows assessing the relative importance of different groups as the perturbation order increases. Diagrams are also used for determining the effective properties of heterogeneous media by means of a process known as *renormalization group* analysis (*RNG*; Wilson and Kogut, 1974). In very general terms, RNG methods provide the means for analyzing the behavior of physical systems under coarse graining. There are various approaches for implementing RNG. The most straightforward approach focuses on the study of coarse graining transformations in configuration space. This approach is best suited for numerical implementation. For heterogeneous media it has been used to calculate effective fluid permeability (King, 1989). A different approach is *frequency-space renormalization* which operates on the FTs of the fluctuations. This approach is more suitable for analytical calculations; it has been used in calculations of transient fluid mixing (Zhang, 1995), and of macrodispersivity (Jaekel and Vereecken, 1997). In the section below we present a brief introduction to RNG analysis.

6. RENORMALIZATION GROUP ANALYSIS - CHANGE OF SCALE

The RNG approach provides a systematic framework for studying change of scale transformations and the effects of coarse graining behavior on physical systems. Turbulent transport is a typical process in which fluctuations occur on many scales, and the RNG methods were first applied to this problem by Yakhot and Orszag (1986). Applications of RNG to various other transport problems in random media have been studied (e.g., Aronovitz and Nelson, 1984; Fisher, 1984; Bouchaud and Georges, 1990; Avellaneda and Majda 1992a and b). Transport of environmental pollutants in the subsurface occurs at low Reynolds numbers unlike transport in fully turbulent flows. However, estimation of the fluctuation effects at different physical scales is a problem that subsurface solute transport shares with turbulence, and the RNG methods are useful for investigating the macrodispersivity (Dean *et al.*, 1994; Jaekel and Vereecken, 1997). A thorough presentation of RNG schemes and applications is beyond the scope of this book. The interested reader should consult the references given herein for more details.

6.1 The Basic Concepts

Heterogeneous systems are characterized by spatial and temporal fluctuations at many scales. Exact knowledge of the small scale variability is not required for the macroscopic characterization of the system, but the large scale behavior is influenced by the statistics of the small scale fluctuations. The procedure by means of which the system properties are evaluated at increasing scales is called *coarse graining*. In the coarse graining procedure the small scale information that is relevant for the large scale behavior is identified, and estimates of the parameters that characterize large scale behavior (e.g., effective parameters or scaling exponents) are obtained. The RNG provides a physically motivated coarse graining approach. There are various RNG schemes, but they are all based on the same underlying philosophy. One usually distinguishes between *configuration space* RNG methods (which operate in space-time) and *Fourier space* RNG methods (which operate on the FTs).

In order to understand intuitively how RNG works in configuration space let us consider a spatial random field $X(\boldsymbol{s})$ on a n-D hypercube with spacing between nodes equal to α and total of N_0 nodes. The set of the lattice random field values is denoted by $\{X\}_0$. One possible coarse graining operation is to generate a hypercube that contains N_0/b^n nodes, where $b>1$ is an integer, with lattice spacing $\ell=\alpha b$. An averaging operation assigns to each node in the coarse grained lattice a value equal to a local average of the $\{X\}_0$. This operation, called variable scaling, generates a coarse grained variable. Variable scaling must preserve the probability measure in the coarse grained system. Finally, a length scaling operation restores the spacing of the coarse grained lattice to the initial spacing α. The combination of length scaling and variable scaling operations constitutes a single renormalization step.

The RNG operation can be described as $\{X\}_1 = R_b\{X\}_0$, where $\{X\}_1$ denotes the set of renormalized variables and R_b the renormalization transformation. The RNG operation is applied recursively, and the asymptotic limit $b \rightarrow \infty$ is evaluated at the end of the calculation. The transformation R_b satisfies the semigroup property $R_{2b}\{X\}_n = R_b^2\{X\}_n$, where $R_b^2 = R_b R_b$. The operations involved in an RNG step lower the resolution of the system by averaging over the small scale fluctuations and changing the overall scale. Assume that after a large number of RNG steps the system tends to $X_* = \lim_{n\to\infty} R_b^n\{X\}_0$, such that $X_* = R_b\{X_*\}$. Then, the X_* is called a *fixed point* of the RNG. This implies that at the asymptotic limit, corresponding to very large length scales, the system and its renormalized copy are identical. Since the unit of length is rescaled by a factor b at every step, the system is in fact self-similar at the RNG fixed point.

EXAMPLE 1: Consider a random walk in R^n, and let the increments $\boldsymbol{r}$ be governed by the Gaussian probability density

$$f(\boldsymbol{r}) = (2\pi\sigma_0^2)^{-n/2} \exp[-(\boldsymbol{r}-\boldsymbol{r}_0)^2/2\sigma_0^2]. \tag{1}$$

The variable scaling step of the RNG for the random walk process involves constructing the pdf $f(\boldsymbol{r}')$, where $\boldsymbol{r}'$ is the coarse grained increment $\boldsymbol{r}' = \beta(N)\sum_{i=1}^{N}\boldsymbol{r}_i$, and $\beta(N)$ is a normalization function. The coarse grained pdf is given by

$$f(\boldsymbol{r}') = \int d\boldsymbol{r}_1 ... d\boldsymbol{r}_N \; f(\boldsymbol{r}_1)...f(\boldsymbol{r}_N)\; \delta[\boldsymbol{r}' - \beta(N)\sum_{i=1}^{N}\boldsymbol{r}_i]. \tag{2}$$

By means of the FT representation of the delta function the increments are integrated over, and we obtain

$$f(\boldsymbol{r}') = [2\pi N\sigma_0^2\beta^2(N)]^{-n/2} \exp\{-\frac{[\boldsymbol{r}' - N\beta(N)\boldsymbol{r}_0]^2}{2N\beta^2(N)\sigma_0^2}\}. \tag{3}$$

The above pdf is of the same form as $f(\boldsymbol{r})$ with rescaled parameters $\sigma' = \sqrt{N}\beta(N)\sigma_0$ and $\boldsymbol{r}_0' = N\beta(N)\boldsymbol{r}_0$. The second step, length scaling, involves choosing an appropriate expression for the normalization function. If we choose to preserve the mean square of the increment's deviation from the mean, then $\beta(N) = 1/\sqrt{N}$, and the rescaled pdf is given by

$$f(\boldsymbol{r}') = (2\pi\sigma_0^2)^{-n/2} \exp[-(\boldsymbol{r}' - \sqrt{N}\boldsymbol{r}_0)^2/2\sigma_0^2]. \tag{4}$$

The renormalized pdf is identical in form to the original, and the full RNG equations are $R[\sigma_0] = \sigma_0$ and $R[\boldsymbol{r}_0] = \sqrt{N}\boldsymbol{r}_0$. The variance is called a *marginal variable*, because it is invariant under the RNG procedure, whereas the drift is a *relevant variable,* since it scales as $\sqrt{N}$. In the case of *zero drift*, i.e., $\boldsymbol{r}_0 = 0$, the original and the renormalized pdf are identical. Thus, it is impossible to distinguish, based on statistical properties, a random walk from its renormalized copy. This is what is meant by self similarity in a stochastic context. □

REMARK 1: The RNG transform is related with limit theorems in probability. Consider independent identically distributed variables $\boldsymbol{r}_i$, $i = 1,\ldots,N$ with zero mean and variance σ_0^2. The central limit theorem (CLT) states that at the asymptotic limit the pdf of the coarse grained variable $\boldsymbol{r}' = \sum_{i=1}^{N} \boldsymbol{r}_i \big/ \sqrt{N}$ is normal with variance σ_0^2. If a normalizing factor N^{μ} with $\mu \neq 1/2$ is used, the variance of the coarse grained variable becomes either zero or infinite at the asymptotic limit. The connection between the RNG approach and the CLT is obvious, since they both address the issue of coarse graining transformations. However, the RNG approach is more general, because, unlike the CLT, it is not constrained by the assumption of statistical independence. Finally, the existence of fixed points of the RNG transform hints to the connection between the RNG and the theory of stable processes in probability (Gnedenko and Kolmogorov, 1954). □

For explicit calculations with moment equations, the frequency space RNG is preferred. In this approach an upper spatial frequency cutoff Λ is introduced. The fluctuations are analyzed into a low frequency component for $0 < \kappa < \Lambda / b$ and a high frequency component $\Lambda / b < \kappa < \Lambda$. The high-frequency component is eliminated from the moment equations, usually by means of approximations that become valid at the limit $\kappa \to 0$. This operation, which is equivalent to *variable scaling,* leads to coarse grained equations satisfied by the low frequency component. At the next step, the upper cutoff is restored by means of the frequency rescaling operation $\boldsymbol{\kappa}' = b\,\boldsymbol{\kappa}$ --this is the equivalent of *length scaling* in configuration space. These steps lead to moment equations with renormalized coefficients that are functions of the scaling variable b. The final step involves evaluating the change in the renormalized coefficients due to an infinitesimal change of the scaling variable which leads to differential equations. Solution of the differential equations at the limit of infinite b (or, equivalently, of zero Λ) yields the asymptotic values of the effective parameters. This approach is close in spirit to the RNG used in theories of critical phenomena for the calculations of scaling exponents.

A different approach, which helps to illustrate the mathematical implementation of the physical concepts, has been used in the analysis of transport in a potential flow field (Dean *et al.*, 1994). In the following several pages we will consider the dispersion of a contaminant in a random velocity field given by $\boldsymbol{v}(\boldsymbol{s}) = -\sigma \nabla \phi(\boldsymbol{s})$, where $\phi(\boldsymbol{s})$ denotes the velocity potential. As we discussed above the contaminant concentration due to a point unit source is given by the Green's function. The steady-state equation for the Green's function is given by

$$D\nabla^2 G(s-s') + \sigma \nabla\phi(s)\cdot \nabla G(s-s') = -\delta(s-s'). \tag{5}$$

It is assumed that the velocity potential satisfies $\overline{\phi(s)} = 0$ and $\overline{\phi(s)\,\phi(s')} = \rho(s-s')$. The Green's function in a homogeneous medium of uniform diffusivity D satisfies the Poisson equation

$$D\nabla^2 G_0(s-s') = -\delta(s-s'), \tag{6}$$

which in Fourier space leads to the simple solution

$$\tilde{G}_0(\boldsymbol{\kappa}) = 1/D\kappa^2. \tag{7}$$

If the correlation function $\rho(\boldsymbol{r})$ is short ranged, it can be shown that the asymptotic behavior of the average Green's function is $O(\kappa^{-2})$, and that the fixed point of the RNG transformation is the Gaussian distribution, which corresponds to a generalized central limit theorem for correlated random fields.

$$\lim_{\kappa\to 0} \overline{\tilde{G}}(\boldsymbol{\kappa}) = 1/D_{eff}\,\kappa^2. \tag{8}$$

In view of Dyson's self-consistent equation, the average Green's function can be expressed as follows

$$\overline{\tilde{G}}(\boldsymbol{\kappa}) = 1/[D\kappa^2 - \Sigma^*(\boldsymbol{\kappa})], \tag{9}$$

where $\Sigma^*(\boldsymbol{\kappa})$ is the self-energy of the one-particle-irreducible diagrams. It then follows from Eqs. (8) and (9) that

$$D_{eff} = D - d\Sigma^*(\kappa)/d\kappa^2\Big|_{\kappa=0}. \tag{10}$$

Hence, in order to calculate the static behavior of the Green's function one needs to estimate the leading term in a Taylor expansion of the self-energy. First, we define the diagrammatic rules for this problem.

DIAGRAMMATIC RULES:

(i) Green's functions are denoted by solid straight lines.
(ii) Correlation functions are denoted by broken lines, and they carry a factor $\tilde{\rho}(\boldsymbol{\kappa}_i)$.
(iii) Spatial frequencies are conserved at each vertex.
(iv) Each vertex of the form shown in Fig. 11a carries a factor $\sigma(\boldsymbol{\kappa}-\boldsymbol{\kappa}_1)\cdot\boldsymbol{\kappa}_1$, and each

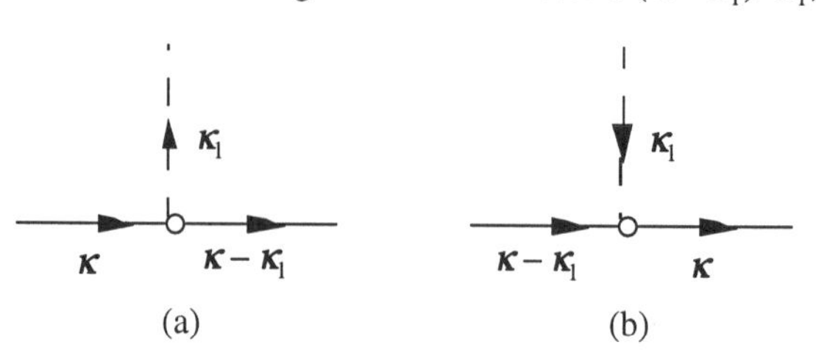

FIGURE 11: Diagrammatic representation of a vertex

vertex in Fig. 11b carries a factor $-\sigma\boldsymbol{\kappa}_1\cdot\boldsymbol{\kappa}$.
(v) Integration is implied over the internal frequencies.

The diagrammatic representation of the proper self-energy is then given in Fig. 12

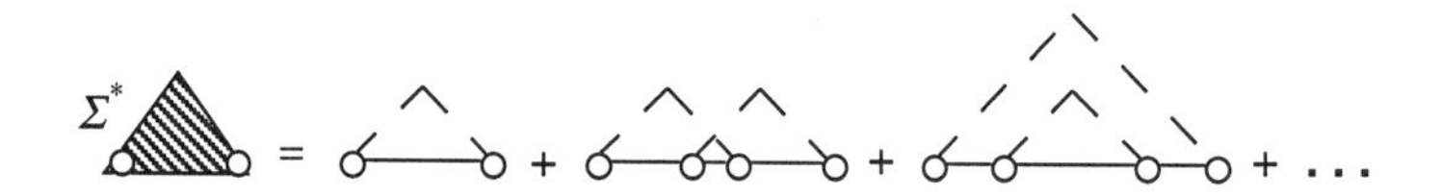

FIGURE 12: Diagrammatic representation of the proper self-energy

6.2 Diagrammatic Approximations for the Self-energy

In order to put the RNG analysis in context we first investigate certain diagrammatic approximations based on the ideas presented in §4 above. The leading-order approximation for the proper self-energy is given by the integral

$$\Sigma^{*(1)}(\boldsymbol{\kappa})=-\frac{\sigma^2}{D}\int_{\boldsymbol{\kappa}_1}\tilde{\rho}(\boldsymbol{\kappa}_1)\,\boldsymbol{\kappa}_1\cdot(\boldsymbol{\kappa}-\boldsymbol{\kappa}_1)\,\boldsymbol{\kappa}\cdot\boldsymbol{\kappa}_1\big/(\boldsymbol{\kappa}-\boldsymbol{\kappa}_1)^2\,. \tag{11}$$

The leading term in the Taylor expansion of the integral, which is required for the evaluation of the effective diffusivity in Eq. (10), is

$$\Sigma^{*(1)}(\kappa) = -D^{-1}\sigma^2 \int_{\boldsymbol{\kappa}_1} \tilde{\rho}(\boldsymbol{\kappa}_1)(\boldsymbol{\kappa}_1 \cdot \boldsymbol{\kappa})^2 / \boldsymbol{\kappa}_1^{\,2} + O(\kappa^4). \tag{12}$$

Hence, the leading correction in the effective diffusivity is given by

$$D_{eff}^{(1)} = D - \sigma^2 / n\, D \tag{13}$$

(note that $\rho(0) = 1$). This approximation coincides with the leading order perturbation. As expected, it becomes inaccurate as σ increases, and it is unstable if $\sigma^2 > n\,D$.

Dyson's equation provides a self-consistent relation for the Green's function that involves the proper self-energy. However, in practice one uses approximate schemes for the self-energy such as the leading order approximation above. The proper self-energy expressions involve the Green's function G_0. It is also possible to replace G_0 with the self-consistent mean Green's function $\overline{G}$ (this leads to the PMDO-2 operator in Christakos *et al.*, 1993a, b and c; 1995). Then, Dyson's equation is written as

$$\overline{\tilde{G}}(\boldsymbol{\kappa}) \cong [D\,\kappa^2 - \Sigma_{s-c}^{*(1)}(\boldsymbol{\kappa})]^{-1}, \tag{14}$$

and the self-energy is given by the following nonlinear integral equation

$$\Sigma_{s-c}^{*(1)}(\boldsymbol{\kappa}) = -\sigma^2 \int_{\boldsymbol{\kappa}_1} \tilde{\rho}(\boldsymbol{\kappa}_1)\tilde{\rho}(\boldsymbol{\kappa}_1)[\boldsymbol{\kappa}_1 \cdot (\boldsymbol{\kappa} - \boldsymbol{\kappa}_1)\,\boldsymbol{\kappa} \cdot \boldsymbol{\kappa}_1] / [D(\boldsymbol{\kappa} - \boldsymbol{\kappa}_1)^2 - \Sigma_{s-c}^{*(1)}(\boldsymbol{\kappa} - \boldsymbol{\kappa}_1)]. \tag{15}$$

This equation can be solved numerically (Dean *et al.*, 1994) for $\sigma \leq 1$, but the solution does not converge for $\sigma > 1$.

6.3 Frequency Space RNG Treatment of Transport

The RNG transform in momentum space is implemented by using an upper frequency cutoff Λ and analyzing the FT velocity field $\tilde{\phi}(\boldsymbol{\kappa})$ into low and high frequency components $\tilde{\phi}(\boldsymbol{\kappa}) = \tilde{\phi}_\Lambda^<(\boldsymbol{\kappa}) + \tilde{\phi}_\Lambda^>(\boldsymbol{\kappa})$, so that $\tilde{\phi}_\Lambda^<(\boldsymbol{\kappa}) = \tilde{\phi}(\boldsymbol{\kappa})$ for $\kappa \leq \Lambda$, $\tilde{\phi}_\Lambda^<(\boldsymbol{\kappa}) = 0$ for $\kappa > \Lambda$ and $\tilde{\phi}_\Lambda^>(\boldsymbol{\kappa}) = \tilde{\phi}(\boldsymbol{\kappa})$ for $\kappa > \Lambda$, $\tilde{\phi}_\Lambda^>(\boldsymbol{\kappa}) = 0$ for $\kappa \leq \Lambda$. The fast fluctuations (i.e., the Fourier components for frequencies exceeding the cutoff) are first eliminated. This leads to a renormalized Greens' function $\tilde{G}_\Lambda(\boldsymbol{\kappa})$, that satisfies the Dyson equation with a renormalized proper self-energy term (Jaekel and Vereecken, 1997). The integrals in the renormalized proper self-energy are evaluated for internal wavevectors with magnitudes less than an arbitrary cutoff Λ.

The RNG transformation leads to an equation for $\tilde{G}_{\Lambda}$ that is similar to the initial Green's function equation, but also includes new terms involving higher powers and derivatives of $\tilde{\phi}_{\Lambda}^{<}$. If the correlation function $\rho(\boldsymbol{r})$ is short-ranged, the asymptotic behavior is Fickian diffusion. In this case, an $O(\kappa^2)$ approximation of the proper self-energy using the leading order renormalized diagram is a permissible approach, which leads to accurate approximations for the effective diffusivity. Since the $\tilde{G}_{\Lambda}(\boldsymbol{\kappa})$ is expressed in terms of a renormalized proper self-energy $\Sigma_{\Lambda}^{*}(\boldsymbol{\kappa})$, the bare diffusivity D is replaced by a renormalized diffusivity $D(\Lambda)$, and the fluctuation variance σ by a renormalized variance $\sigma(\Lambda)$.

Next, the change in the effective diffusivity due to the elimination of an infinitesimal frequency shell is calculated. This leads to a differential equation for the renormalized diffusivity $D(\Lambda)$ in terms of the cutoff Λ. The differential equation is solved, and the effective diffusivity is obtained at the limit as the cutoff tends to zero, $D_{eff} = D(\Lambda = 0)$. The physical meaning of the zero cutoff limit is that repetitive coarse graining eliminates all the fluctuations when the size of the coarse grained block approaches the size of the physical domain (i.e., the support). The main steps in the RNG procedure are as follows:

(i) The fluctuation random field is analyzed into low frequency and high frequency components with respect to an arbitrary cutoff Λ

$$\left.\begin{aligned} \phi_{\Lambda}^{>}(s) &= \int_{|\boldsymbol{\kappa}|>\Lambda} d\boldsymbol{\kappa}\, e^{i\boldsymbol{\kappa}\cdot s}\tilde{\phi}(\boldsymbol{\kappa}) \\ \phi_{\Lambda}^{<}(s) &= \int_{|\boldsymbol{\kappa}|<\Lambda} d\boldsymbol{\kappa}\, e^{i\boldsymbol{\kappa}\cdot s}\tilde{\phi}(\boldsymbol{\kappa}) \end{aligned}\right\}. \tag{16}$$

(ii) The stochastic average over the fluctuations that exceed the cutoff is evaluated. If nonlinear terms in $\phi_{\Lambda}^{<}$ are ignored, the resulting equation for the partially averaged Green's function G_{Λ} is similar to the equation for G_0 ,

$$D(\Lambda)\nabla^2 G_{\Lambda}(s,s') + \sigma(\Lambda)\nabla\phi_{\Lambda}^{<}(s)\cdot\nabla G_{\Lambda}(s,s') = -\delta(s-s'). \tag{17}$$

At the limit of infinite cutoff, $\phi_{\Lambda}^{<}(s) \to \phi(s)$, the initial equation for the Green's function G_0 is recovered; this sets the initial condition to $D(\Lambda = \infty) = D$.

(iii) The mean Green's function $\bar{\tilde{G}}_{\Lambda}(\boldsymbol{\kappa})$ satisfies the renormalized Dyson equation

$$\bar{\tilde{G}}_{\Lambda}(\boldsymbol{\kappa}) = [\tilde{G}_0^{-1}(\boldsymbol{\kappa}) - \Sigma_{\Lambda}^{*}(\boldsymbol{\kappa})]^{-1}, \tag{18}$$

where the renormalized proper self-energy $\Sigma_{\Lambda}^{*}(\boldsymbol{\kappa})$ involves the same diagrams as $\Sigma^{*}(\boldsymbol{\kappa})$, but the integrals are truncated by the cutoff. The renormalized diffusivity is given by

$$D(\Lambda) = D - d\Sigma_{\Lambda}^{*}(\boldsymbol{\kappa})/d\kappa^2\Big|_{\kappa=0}. \tag{19}$$

(iv) The $\Sigma_{\Lambda}^{*}(\boldsymbol{\kappa})$ is evaluated explicitly to leading order in σ^2. By eliminating an infinitesimal frequency shell, one obtains a differential equation which expresses the variation of $D(\Lambda)$ as a function of the cutoff Λ. Therefore the RNG equation is

$$\frac{dD(\Lambda)}{d\Lambda} = -\frac{\partial}{\partial\Lambda}\frac{\partial^2\Sigma_{\Lambda}^{*}(\kappa)}{\partial\kappa^2}\bigg|_{\kappa=0}, \tag{20}$$

Two approximation schemes for the proper self-energy are investigated in the following.

6.4 Perturbative RNG Schemes

The simplest RNG scheme is based on the *one-loop* approximation which involves the $\Sigma_{\Lambda}^{*(1)}(\boldsymbol{\kappa})$ shown in Fig. 13. The thick lines indicate that (i) the renormalized mean Green's

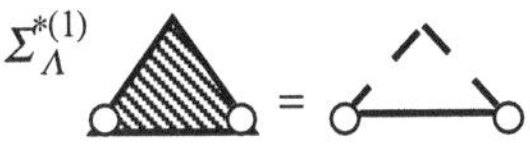

$\Sigma_{\Lambda}^{*(1)}$ [diagram]

FIGURE 13: One-loop approximation of the renormalized proper self-energy

function $\bar{\tilde{G}}_{\Lambda}(\boldsymbol{\kappa})$ is used, and (ii) that the integrals over the internal frequencies of the graph are cutoff by Λ. This leads to the following equation for the renormalization contribution from a narrow frequency shell $\Lambda - \delta\Lambda < |\boldsymbol{\kappa}_1| < \Lambda$

$$\Sigma_{\Lambda}^{*(1)}(\boldsymbol{\kappa}) - \Sigma_{\Lambda-\delta\Lambda}^{*(1)}(\boldsymbol{\kappa}) = \frac{\sigma^2}{D(\Lambda)}\int_{\boldsymbol{\kappa}_1 \in shell} \tilde{\rho}(\boldsymbol{\kappa}_1)\frac{\boldsymbol{\kappa}_1\cdot(\boldsymbol{\kappa}-\boldsymbol{\kappa}_1)\,\boldsymbol{\kappa}\cdot\boldsymbol{\kappa}_1}{(\boldsymbol{\kappa}-\boldsymbol{\kappa}_1)^2}. \tag{21}$$

If both sides of Eq. (21) are divided by $\delta\Lambda$ and the limit $\delta\Lambda \to 0$ is taken, a differential RNG equation is obtained for the effective diffusivity as follows

$$\frac{dD^2(\Lambda)}{d\Lambda} = \frac{2\sigma^2 S_n \Lambda^{n-1}\tilde{\rho}(\Lambda)}{(2\pi)^n n}. \tag{22}$$

The solution of Eq. (22) is given by

$$D_{eff} = D(1 - 2\sigma^2/n D^2)^{1/2}. \tag{23}$$

Eq. (23) gives to leading order $D_{eff}^{(1)} = D(1 - \sigma^2/n D^2)$, consistently with perturbation theory, but it fails when $\sigma^2 \geq n D^2/2$. The instability is due to the fact that in the one-loop approximation the variance is independent of the cutoff frequency. This condition is not realistic, since it is known that coarse graining changes the fluctuation variance. Thus, an improved RNG scheme should incorporate renormalization of the variance.

The renormalization of the variance is accomplished by means of a procedure called *vertex renormalization*. The bare (unrenormalized) vertex of Fig. 11a carries the factor $\sigma_{ij}^{(0)}\, \kappa_{1i}(\kappa_j - \kappa_{1j})$, where $\sigma_{ij}^{(0)} = \sigma\,\delta_{ij}$ is a diagonal tensor. The renormalized vertex is denoted by $\sigma_{ij}(\boldsymbol{\kappa}, \boldsymbol{\kappa}_1; \Lambda)$. The asymptotic behavior is determined by the low frequencies and, thus, the vertex can be approximated by $\sigma_{ij}(\Lambda) = \sigma_{ij}(\boldsymbol{0},\boldsymbol{0};\Lambda)$. The change $\delta\sigma_{ij}(\Lambda) = \sigma_{ij}(\Lambda) - \sigma_{ij}(\Lambda - \delta\Lambda)$ in the vertex function for an infinitesimal increment of the cutoff is given by the diagram in Fig. 14, where the integral of the internal wavevector $\boldsymbol{\kappa}_2$ is

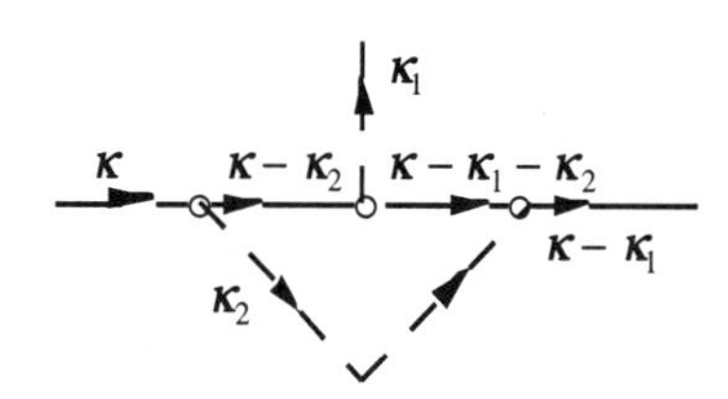

FIGURE 14: Diagram for vertex renormalization

over the shell $\Lambda - \delta\Lambda < |\boldsymbol{\kappa}_2| < \Lambda$. Evaluation of the integral leads to the following RNG equation for the rate of change of $\sigma(\Lambda)$ with the cutoff

$$\frac{d\sigma(\Lambda)}{d\Lambda} = \frac{S_n\, \Lambda^{n-1}}{(2\pi)^n\, n} \frac{\sigma^3(\Lambda)\tilde{\rho}(\Lambda)}{D^2(\Lambda)}, \tag{24}$$

subject to the initial condition $\sigma = \sigma(\infty)$. Eq. (24) complements the RNG equation for the renormalized diffusivity given by

$$\frac{dD^2(\Lambda)}{d\Lambda} = \frac{2 S_n}{(2\pi)^n\, n} \sigma^2(\Lambda)\tilde{\rho}(\Lambda)\Lambda^{n-1}. \tag{25}$$

Note that Eq. (25) is almost identical with Eq. (22), the only difference being that Eq. (25) involves the renormalized vertex $\sigma(\Lambda)$ instead of the bare σ. The system of Eqs. (24) and (25) is further expressed as

$$\frac{dD(\Lambda)}{d\Lambda} = \frac{D(\Lambda)}{\sigma(\Lambda)}\frac{d\sigma(\Lambda)}{d\Lambda}, \tag{26}$$

which implies that $\sigma(\Lambda)D(\infty) = D(\Lambda)\sigma(\infty)$, and

$$\frac{dD(\Lambda)}{d\Lambda} = \frac{\sigma^2}{D^2}\frac{S_n}{(2\pi)^n n} D(\Lambda)\tilde{\rho}(\Lambda)\Lambda^{n-1}. \tag{27}$$

The effective diffusivity is obtained by solving Eq. (27) using the inverse FT integral $\int_{\boldsymbol{\kappa}} \tilde{\rho}(\boldsymbol{\kappa}) = \rho(0)$ which leads, for isotropic correlations, to the expression

$$D_{eff} = D(\Lambda = 0) = De^{-\sigma^2/nD^2}. \tag{28}$$

The above result is independent of the specific form of the correlation function. The effective diffusivity obtained by Eq. (28) is scalar, meaning that diffusion is asymptotically isotropic. This is due to (a) lack of a velocity trend, and (b) isotropy of the fluctuations.

REMARK 2: It is important to emphasize that Eq. (28) is well-defined even for large heterogeneity. Numerical simulations (Dean *et al.*, 1994) show that the RNG estimate with vertex renormalization agrees well with numerical estimates at least for $\sigma/D \leq 2$. Surprisingly, for this specific transport problem leading order perturbation was found to estimate the effective diffusivity more accurately than the self-consistent diagrammatic approximations and the one-loop RNG without vertex renormalization.

6.5 RNG Analysis of Subsurface Pollution Transport

In the case of realistic groundwater pollutant transport the velocity field is incompressible, i.e., $\nabla \cdot \mathbf{v}(s) = 0$. In addition, the pollutants are advected in a velocity field with a finite mean, and the velocity correlation functions are commonly anisotropic due to the presence of preferential flow directions. These physical differences modify considerably the mathematical details of the RNG analysis of effective dispersivity, even though the same concepts are used. The effective dispersivity in this case is represented by a tensor characterized by a longitudinal component in the direction of the mean flow and a transverse component in the orthogonal directions. The effective dispersivity at the field scale is called *macrodispersivity*. A standard explicit expression for the macrodispersivity (Gelhar and Axness, 1983) is equivalent to the solution of the Dyson equation using a one-loop approximation for the proper self-energy. This solution is known to severely underestimate the transverse macrodispersivity observed in field experiments.

A recent calculation of the macrodispersivity (Jaekel and Vereecken, 1997) using the one-loop RNG scheme verified that the self-consistent approximation is accurate for the longitudinal component. However, the RNG estimate of the transverse component is significantly larger than the self-consistent approximation. An additional difference between the two theories is that the RNG transverse component depends on the correlation lengths of the hydraulic log-conductivity, while the self-consistent transverse component is independent. The results of the RNG analysis agree well with macrodispersivity estimates obtained from the Borden site experiments. The one-loop RNG scheme is believed to be accurate, since the vertex renormalization is negligible due to the incompressibility condition. Nonetheless, incompressible flow lacks the mathematical simplicity of potential flow, and even the one-loop RNG equations must be solved by numerical means.

In the above discussions we have implicitly assumed that the diffusion is asymptotically Fickian, so that far from the source the mean square displacement of the pollutant is a linear function of time, i.e., $\overline{r^2} \propto t$. This is reminiscent of the linear dependence of the variance in the CLT. In fact, there is a strong correspondence between diffusion and limit theorems: if the discrete displacement increments are viewed as random variables, the total displacement represents the sum of the increments, and it is governed asymptotically by a limit theorem. The asymptotic diffusion law is a central issue in the study of groundwater transport. Some of the first theories (Dagan, 1984) assumed that for large times the transport is always Fickian. However, in certain cases anomalous diffusion with $\overline{r^2} \propto t^{\beta}$, where $\beta \neq 1$ (Aronovitz and Nelson, 1984) is known to occur. In fact, Matheron and de Marsily (1980) used physical arguments to show that anomalous diffusion with $\beta = 3/2$ is expected in a perfectly stratified medium. In the case of anomalous diffusion, approximate RNG schemes are not adequate for the prediction of the scaling exponent α (Avellaneda and Majda, 1992a). The crucial difference between Fickian and anomalous diffusion is the velocity correlation function: Fickian diffusion is obtained asymptotically for short range velocity correlations which are integrable. This requires that the correlation function decay as $r^{-\nu}$, where $\nu > n$ at large distances. Then, the characteristic mixing time is given by $\tau_m \propto \xi / |\overline{V}|$, where $|\overline{V}|$ is the mean flow velocity, and ξ is the maximum correlation length. The asymptotic limit is obtained for $t >> \tau_m$, while in the preasymptotic regime deviations from Fickian diffusion are observed. The asymptotic limit also requires that the size of the pollutant plume be much larger than ξ.

Long range velocity correlation functions are not integrable (do not have correlation lengths) and this leads to anomalous diffusion. The asymptotic limit is never attained in anomalous diffusion. Hence the process in non-ergodic, and the ensemble (stochastic) average of the pollutant displacement is not necessarily a good estimator of the sample average. Nonetheless, the stochastic average remains a useful qualitative measure. Long range correlations change the qualitative features of the diffusion process: While short range correlations permit representing the asymptotic behavior by Fickian diffusion with a renormalized diffusivity D_{eff}, long range correlations do not permit the renormalization of the diffusivity tensor, and they lead to a change in the asymptotic scaling law. In addition, anomalous diffusion changes the shape of the concentration profile, which is Gaussian for Fickian diffusion and non-Gaussian in the opposite case (Koch and Brady, 1989).

REMARK 3: In subsurface hydrology, most of the theoretical work on groundwater solute transport has focused on determining the effective diffussivity for Fickian diffusion scenarios (Cushman *et al.*, 1994; Dagan, 1984, 1988; Gelhar and Axness, 1984; Neuman *et al.*, 1987). However, the assumption that asymptotic conditions prevail in the field has been criticized as unrealistic, and non-ergodic transport theories have recently received attention (Kitanidis, 1988; Zhang and Neuman, 1995). Nonetheless, even these theories consider the preasymptotic regime of short range correlated fluctuations, and not the more difficult case of anomalous diffusion.

7. THE SPACE TRANSFORMATION TECHNIQUE

Numerical investigations of environmental systems are often hampered by the heterogeneity of natural processes. A large number of nodes is required for an accurate representation of the spatiotemporal variability. The solution of the resulting equations is a computationally intensive operation. Hence, it is often convenient to reduce the spatial dimensionality of the problem. This can be achieved with the help of space transformation operators (ST; Christakos, 1984b; 1986; 1992; Christakos and Hristopoulos, 1994; 1997).

7.1 Basic Definitions

STs are integral operators just like the FTs and LTs. In the case of the latter the variability in space and time is mapped into the frequency domain. In the case of space transformations mapping involves the projections of the n-dimensional function onto planes orthogonal to transform lines determined by direction vectors in R^n. The projections involve the integrals of the function over hyperplanes (planes if $n=3$) defined by means of the equation $\Pi_{\boldsymbol{\theta}}(s)=\sigma-s\cdot\boldsymbol{\theta}=0$, where $\boldsymbol{\theta}$ denotes the direction vector and σ the ordinate of their projection onto the transform line. The projections along a single transform line can be viewed as functions of a single variable (i.e., the projection length). The n-dimensional function can be reconstructed exactly from the 1-D functions in all directions. However, in many cases the function or a random field can be accurately reconstructed by means of projections along a finite set of lines. This effectively reduces the dimensionality of the reconstructed function (Christakos and Panagopoulos, 1992).

In certain cases, PDEs can be reduced to ODEs by means of STs. The PDE solution can then be reconstructed from the ODE solutions along a finite set of transform lines. Next, we give a more mathematical definition of the ST.

Definition 1: *Let R^n denote the n-dimensional Euclidean space, and $\boldsymbol{\theta}=(\theta_1,\ldots,\theta_n)$ denote a direction vector on the surface of the generalized unit hypersphere in n dimensions denoted by $S^n\subset R^n$. If $X_n(s,t)$ denotes an S/TRF on $R^n\times T$, the ST is defined by means of the Radon operator*

$$T_n^1:\ X_n(s,t)\rightarrow\int ds\,X_n(s,t)\,\delta(\sigma-s\cdot\boldsymbol{\theta}),\tag{1}$$

where $\sigma\in R^1$ and δ is the delta function.

The Radon operator is also known as the plane wave integral (e.g., Helgason, 1980; John, 1955). The integration in Eq. (1) is carried out over the entire space unless specific limits are indicated. The $X_n(s,t)$ are suitable functions (ordinary or generalized), so that the integral in Eq. (1) exists. In order for this to be true the functions $X_n(s,t)$ must decay fast at infinity, or have a compact support. The ST T_n^1 represents the projection of the n-dimensional function on hyperplanes that are perpendicular to a specific direction. Let S_n denote the surface area of the n-dimensional unit sphere. The following operator is useful for evaluating the inverse ST

$$\Omega=\frac{(-1)^{m-1}}{2(2\pi)^{2m-1}}\begin{cases}-\dfrac{S_{2m+1}}{2\pi}\dfrac{\partial^{2m}}{\partial\sigma^{2m}}[\cdot], & \text{if } n=2m+1\\[2ex] S_{2m}\,\mathcal{H}\{\dfrac{\partial^{2m-1}}{\partial\sigma^{2m-1}}[\cdot]\}, & \text{if } n=2m\end{cases},\tag{2}$$

where $\mathcal{H}$ denotes the Hilbert operator.

Definition 2: *Consider the S/TRF $X_n(s,t)$ in $R^n \times T$. The ST operators T_n^1 and Ψ_n^1, that reduce the $X_n(s,t)$ to an S/TRF in $R^1 \times T$ are defined as*

$$T_n^1[X_n](\sigma,\boldsymbol{\theta},t) = \hat{X}_{1,\boldsymbol{\theta}}(\sigma,t), \tag{3}$$

and

$$\Psi_n^1[X_n](\sigma,\boldsymbol{\theta},t) = \Omega T_n^1[X_n](\sigma,\boldsymbol{\theta},t) = X_{1,\boldsymbol{\theta}}(\sigma,t). \tag{4}$$

The STs T_n^1 and Ψ_n^1 are completely defined if the integrals expressions in Eqs. (3) and (4) are known for all $\boldsymbol{\theta}$.

7.2 Some Properties of STs

It follows from Eqs. (2) and (3) that the T_n^1 consists of an infinite set of integrals that involve delta function kernels. The delta functions define the transform plane. We present without proof certain properties for the T_n^1. Proofs can be obtained directly from the definition of T_n^1 using the properties of generalized functions (an excellent review of the theory of generalized functions may be found in Gel'fand and Shilov, 1964).

The ST operators (3) and (4) can be inverted as follows

$$\Psi_1^n[X_{1,\boldsymbol{\theta}}](s,t) = S_n^{-1}\int_{S_n} d\boldsymbol{\theta}\, X_{1,\boldsymbol{\theta}}(s\cdot\boldsymbol{\theta},t) = X_n(s,t), \tag{5}$$

and

$$T_1^n[\hat{X}_{1,\theta}](s,t) = \Psi_1^n \Omega[\hat{X}_{1,\theta}](s,t) = X_n(s,t). \tag{6}$$

In the spectral domain, the ST turn out to have simple algebraic forms. Indeed, for the T_n^1 we have

$$T_1^n[\tilde{X}_n](\kappa,\boldsymbol{\theta},t) = \tilde{X}_n(\boldsymbol{\kappa},t) = \tilde{\hat{X}}_{1,\boldsymbol{\theta}}(\kappa,\boldsymbol{\theta},t) \tag{7}$$

along the spatial frequency $\boldsymbol{\kappa} = \kappa \boldsymbol{\theta}$, where $\tilde{X}$ denotes the FT of X; and

$$T_l^n[\tilde{\hat{X}}_{l,\theta}](\boldsymbol{\kappa},t) = \tilde{\hat{X}}_{l,\theta}(\kappa,t) = \tilde{X}_n(\boldsymbol{\kappa},t). \tag{8}$$

Similarly, for the Ψ_n^l and Ψ_l^n it can be shown that,

$$\Psi_n^l[\tilde{X}_n](\kappa,t) = S_n(\kappa/2\pi)^{n-1}\tilde{X}_n(\boldsymbol{\kappa},t) = \tilde{X}_{l,\theta}(\kappa,t), \tag{9}$$

and

$$\Psi_l^n[\tilde{X}_{l,\theta}](\boldsymbol{\kappa},t) = S_n^{-1}(2\pi/\kappa)^{n-1}\tilde{X}_{l,\theta}(\kappa,t) = \tilde{X}_n(\boldsymbol{\kappa},t), \tag{10}$$

respectively.

In the case of a nonrandom isotropic function $f_n(s,t)$ with infinite support, e.g., certain correlation functions-- the T_n^l is also isotropic, namely the function $\hat{f}_{1,\boldsymbol{\theta}}(\sigma,t)$ is independent of the orientation of the vector $\boldsymbol{\theta}$.

EXAMPLE 1: Consider the isotropic Gaussian $f_3(s) = \exp(-s^2/\ell^2 - t^2/\tau^2)$. The T_n^l transformation is given by

$$T_3^1[f_3(s)] = \pi\ell^2\exp(-\sigma^2/\ell^2 - t^2/\tau^2). \tag{11}$$

The isotropy of the ST breaks down in the case of a finite support that lacks spherical symmetry (e.g., rectangular support). The ST depends on the orientation of the transform line and the shape of the support. Note that finite supports can be used as spatial filters that permit to define the ST even for non-integrable functions. The tradeoff is that compact supports introduce boundary effects. □

The T_n^l satisfies the following scaling property under dilations of the spatial vector $\boldsymbol{s}$

$$T_n^1[X_n(\lambda s,t)] = \lambda^{1-n}\hat{X}_{1,\theta}(\lambda\sigma,t). \tag{12}$$

The shifting property represents the change caused by a translation of the position vector $\boldsymbol{s}$ in space

$$T_n^1[X_n(s - a, t)] = \hat{X}_{1,\theta}(\lambda \sigma - \theta \cdot a, t); \tag{13}$$

The ST includes redundant information, a fact that is reflected in the property

$$\hat{X}_{1,\theta}(\sigma, t) = \hat{X}_{1,-\theta}(-\sigma, t); \tag{14}$$

this property expresses the fact that the mapping from the configuration space $R^n \times T$ onto the space $S^n \times R \times T$ leads to a double covering of the physical space if both the projection length and the transform line direction vectors are unrestricted. The redundancy is lifted by restricting the ST to positive projection lengths $\sigma > 0$, or to direction vectors contained within one hemisphere of the unit hypersphere in S^n.

The STs of random field gradients are particularly useful in the study of SPDEs that govern three dimensional groundwater flow and solute transport. The STs of partial derivatives can be expressed in terms of the direction cosines of the transform line and the derivative of the ST with respect to the projection length. This is a very useful property, because it allows transforming PDEs into ODEs. In particular,

$$T_3^1[\partial \mathrm{X}_n / \partial \mathrm{s}_i](\sigma, \theta, t) = \theta_i \, \partial \hat{\mathrm{X}}_{1,\theta}(\sigma, t) / \partial \sigma, \tag{15}$$

and

$$\Psi_3^1[\partial \mathrm{X}_3 / \partial \mathrm{s}_i](\sigma, \theta, t) = \Omega \mathrm{T}_3^1[\partial \mathrm{X}_3(s, t) / \partial \mathrm{s}_i](\sigma, \theta, t) = \theta_i \, \partial \mathrm{X}_{1,\theta}(\sigma, t) / \partial \sigma. \tag{16}$$

Physical processes usually occur in three spatial dimensions. The inverse ST T_1^3 of the function $\hat{\mathrm{X}}_{1,\theta}(\sigma, t)$ consists --according to Eq. (6) above-- of two steps: (i) application of the Ω operator leads to $\mathrm{X}_{1,\theta}(\sigma, t)$, and (ii) the Ψ_1^3 operator reconstructs the S/TRF $X_3(s)$. It follows from Eq. (2) that the Ω operator in 3-D is given by

$$X_{1,\theta}(\sigma, t) = \Omega[\hat{X}_{1,\theta}(\sigma, t)] = -\frac{1}{2\pi} \frac{\partial^2 \hat{X}_{1,\theta}(\sigma, t)}{\partial \sigma^2}. \tag{17}$$

EXAMPLE 2: The unit vectors of the transform lines in 3-D can be represented in terms of the polar angle χ and the azimuthal angle ϕ as follows

$$\boldsymbol{\theta} = (\theta_1, \theta_2, \theta_3) = (\sin\chi\cos\phi, \sin\chi\sin\phi, \cos\chi). \tag{18}$$

As we discussed above a single covering is obtained by considering unit vectors $\boldsymbol{\theta}$ in one hemisphere while leaving the projection length σ unconstrained. In terms of the polar and azimuthal angles,

$$X_3(s,t) = \Psi_1^3[X_{1,\boldsymbol{\theta}}(\sigma,t)](s) = \frac{1}{4\pi}\int_0^{\pi} d\phi \int_0^{\pi} d\chi \sin\chi\,[X_{1,\boldsymbol{\theta}}(\sigma,t) + X_{1,\boldsymbol{\theta}}(-\sigma,t)]_{\sigma=s\cdot\boldsymbol{\theta}}, \tag{19}$$

which is the image of the operator Ψ_1^3 in three dimensions. □

7.3 Numerical Implementations of ST

Numerical implementations of STs involve discretized representations, which permit evaluation of the transformations on a numerical grid. In practice, derivatives are estimated by means of finite differences, the inverse is evaluated from a finite set of projections, and integrals are estimated by means of discrete sums. Numerical evaluation of the integrals in the T_3^1 requires values of the function at off-grid points. In the case of a discrete function these can be obtained by means of interpolation schemes. However, oversampling of S/TRF realizations by interpolation may lead to artificial smoothing and variance reduction. In the case of S/TRF Monte Carlo integration methods could provide a more efficient integration scheme. Important numerical parameters for the calculation of integrals are: (i) The number of transform lines, (ii) the discretization of the 3-D grid (i.e., the number of nodes), and (iii) the discretization of the transform line (i.e., the number of points per line). For a discussion of these numerical issues and accuracy tests of T_3^1 calculations see (Hristopulos *et al.*, 1998).

All applications of the STs involve at some point an inversion operation which reconstructs the 3-D spatial structure of the S/TRF from its projections on all possible directions in S^n (Christakos, 1984a; 1987a and b). There are various methods for implementing the inversion in both the spectral and configuration domains (Jain, 1989). Spectral methods are more appropriate in the case of infinite supports, where they have the potential for fast numerical codes. On the other hand, for problems that involve S/TRF realizations in bounded domains inversion methods in configuration space are more useful. We present below a numerical algorithm for calculation of the inverse in configuration space.

The 2nd-order derivative involved in the Ω operator is usually estimated by means of

$$\frac{\partial^2 \hat{X}_{1,\boldsymbol{\theta}}(\sigma,t)}{\partial \sigma^2} \cong \frac{1}{\alpha^2}[\hat{X}_{1,\boldsymbol{\theta}}(\sigma+\alpha,t)+\hat{X}_{1,\boldsymbol{\theta}}(\sigma-\alpha,t)-2\hat{X}_{1,\boldsymbol{\theta}}(\sigma-\alpha,t)], \tag{20}$$

where α represents the lattice spacing. Various methods exist for generating a finite set of transform lines. A commonly used discretization scheme leads to a uniform distribution of the projections over the surface of the unit sphere. The direction vectors $\boldsymbol{\theta}_{jk}$ are determined by means of the angles (χ_j, ϕ_{jl}) given by

$$\chi_j = (j - \frac{1}{2})\frac{\pi}{N_p},\ j = 1,2,\ldots,N_p \tag{21}$$

and

$$\phi_{jl} = 2\pi l / K_j,\ l = 1,2,\ldots,K_j. \tag{22}$$

The number K_j of azimuthal angles is chosen so that a uniform distribution of projections is obtained. An approximately uniform distribution is obtained by means of

$$K_{i+1}/K_i = \lfloor \sin\chi_i / \sin\chi_{i+1} \rfloor,\quad i = 1,\ldots,N_p - 1 \tag{23}$$

where $\lfloor\ \rfloor$ denotes the integer part, and K_1 is an arbitrary number. The total number of transform lines generated by this method is equal to $N_L = \sum_{j=0}^{N_p} K_j$. Finally, numerical inversion of the inverse ST involves the approximation of the integral in Eq. (19) by means of the following summation over the N_L lines

$$X_3(\boldsymbol{s}) = (N_L)^{-1} \sum_{i=1}^{N_L} \hat{X}_{1,\boldsymbol{\theta}_i}(\boldsymbol{s}\cdot\boldsymbol{\theta}_i). \tag{24}$$

A different scheme that is based on randomly oriented lines has been used in turning bands simulations (Tompson *et al.*, 1989). However, the projections of randomly generated lines tend to accumulate near the poles, thus leading to an uneven coverage of the surface of the unit sphere.

REMARK 1: Some issues related to numerical ST calculations should be mentioned: Reconstruction from a finite number of projections poses the issue of non-uniqueness of the reconstructed function (Louis, 1986). Fortunately, at least for smooth functions, arbitrarily good approximations can be constructed by increasing the number of ST lines. It is also possible to improve the solution by imposing conditions based on optimality criteria or *a priori* information. In addition, the inverse problem is technically ill-posed, thus leading to solutions that may be unstable to small perturbations of the projections (e.g., Tikhonov and Goncharsky, 1987). This ill-posedness is due to the numerical approximation of the 2nd-order derivative in T_3^1. STs of random field realizations are more prone to exhibit ill-posedness due to the random nature of fluctuations. The stability of ST solutions for random fields can be improved by using frequency filters that eliminate fast fluctuations (Jain, 1989).

In certain cases (e.g., ST solutions of 2nd-order differential equations; Christakos and Hristopulos, 1994; 1997), the solution is obtained in terms of an expression that does not involve the differential operator Ω, thus overcoming the problem of ill-posedness. Details on the numerical implementation of the inverse ST in problems of groundwater flow are given in (Hristopulos *et al.*, 1998).

7.4 Applications of STs in SPDE Solving

The ST method can be used to simplify the solution of SPDEs by reducing dimensionality. In the case of purely spatial PDEs, the ST approach leads to ODEs along the transform lines. The ODEs involve derivatives with respect to the projection length σ along transform lines and can be solved more efficiently than the initial PDE, often leading to explicit expressions. The PDE solution can be obtained by inverting the ODE solutions along the transform lines. This approach works best for linear PDEs with constant coefficients (Christakos, 1992). PDEs with variable coefficients (e.g., S/TRF) are considerably more complicated (Christakos and Hristopulos, 1994; 1997). We discuss some of the issues below. In the case of space-time PDEs the ST approach leads to a PDE with a single space derivative. We illustrate the ST approach for solving PDEs by means of a test problem.

EXAMPLE 3: Assume that the concentration $c(\boldsymbol{s})$ obeys the following model PDE in two spatial dimensions

$$\boldsymbol{\vartheta} \cdot \nabla c(\boldsymbol{s}) + \phi(\boldsymbol{s}) c(\boldsymbol{s}) = 0, \tag{25}$$

subject to the boundary condition $c(s=0)=1$, where $\boldsymbol{\vartheta}$ is a unit vector. This equation specifies that the concentration gradient in the direction $\boldsymbol{\vartheta}$ is proportional to the local concentration with coefficient $\phi(\boldsymbol{s})$. Let us assume that $\boldsymbol{\vartheta}=(1/\sqrt{2},1/\sqrt{2})$ and $\phi(\boldsymbol{s})=\alpha(s_1+s_2)$, so that the $\phi(\boldsymbol{s})$ is constant along the lines defined by (s_1+s_2) =constant. The equation can then be solved exactly by means of the well-known method of *separation of variables (SoV)*, which leads to

$$c(\boldsymbol{s})=\exp[-\alpha s^2/\sqrt{2}+\beta(s_1-s_2)], \tag{26}$$

where β is an arbitrary constant. If $\beta=0$ the isotropic solution $c(\boldsymbol{s})=\exp(-\alpha s^2/\sqrt{2})$ is recovered; the T_3^1 of the latter is given by $\hat{c}_{1,\boldsymbol{\theta}}(\sigma)=\sqrt{\pi}\exp(-\sigma^2)$. By application of the T_3^1 the Eq. (25) is transformed into the integro-differential equation

$$(\boldsymbol{\vartheta}\cdot\boldsymbol{\theta})\partial\hat{c}_{1,\boldsymbol{\theta}}(\sigma)/\partial\sigma+\alpha\int d\boldsymbol{s}\,\delta(\sigma-\boldsymbol{s}\cdot\boldsymbol{\theta})(s_1+s_2)c(\boldsymbol{s})=0 \tag{27}$$

In order to express the second term in Eq. (27) by means of the ST $\hat{c}_{1,\boldsymbol{\theta}}(\sigma)$ we use a rotated coordinate frame $\boldsymbol{u}=\boldsymbol{R}\boldsymbol{s}$, where $\boldsymbol{R}$ is the orthogonal rotation matrix; $R_{11}=R_{22}=\theta_1$ and $R_{12}=-R_{21}=\theta_2$. In the new system $u_1=\boldsymbol{s}\cdot\boldsymbol{\theta}$ and $u_2=\boldsymbol{s}\cdot\boldsymbol{\theta}_\perp$, where $\boldsymbol{\theta}_\perp\cdot\boldsymbol{\theta}=0$. The Jacobian of the transformation is equal to one. Hence, the Eq. (27) is transformed into the following ODE

$$\partial\hat{c}_{1,\boldsymbol{\theta}}(\sigma)/\partial\sigma+b(\boldsymbol{\theta})\sigma\hat{c}_{1,\boldsymbol{\theta}}(\sigma)=-d(\boldsymbol{\theta})g_{\boldsymbol{\theta}}(\sigma), \tag{28}$$

where $\boldsymbol{\theta}=(\cos\phi,\sin\phi)$, $b(\boldsymbol{\theta})=(\boldsymbol{\vartheta}\cdot\boldsymbol{\theta})^{-1}\alpha$, $d(\boldsymbol{\theta})=(\theta_1-\theta_2)(\boldsymbol{\vartheta}\cdot\boldsymbol{\theta})^{-1}$, and the source term $g_{\boldsymbol{\theta}}(\sigma)$ on the right hand-side is given by

$$g_{\boldsymbol{\theta}}(\sigma)=\int du_2\,u_2\,c'_{\boldsymbol{\theta}}(\sigma,u_2), \tag{29}$$

where $c'_{\boldsymbol{\theta}}(\boldsymbol{u})=c(\boldsymbol{s})$. The source $g_{\boldsymbol{\theta}}(\sigma)$ is not known, because it involves an integral over the unknown function $c(\boldsymbol{s})$. In order to make further progress the source term must be estimated. Invoking separation of variables, i.e., $c'_{\boldsymbol{\theta}}(\sigma,u_2)=c'_{1\boldsymbol{\theta}}(\sigma)c'_{2\boldsymbol{\theta}}(u_2)$, the ODE is

$$\partial\hat{c}_{1,\boldsymbol{\theta}}(\sigma)/\partial\sigma+b(\boldsymbol{\theta})\,\sigma\,\hat{c}_{1,\boldsymbol{\theta}}(\sigma)+v(\boldsymbol{\theta})d(\boldsymbol{\theta})\hat{c}_{1,\boldsymbol{\theta}}(\sigma)=0, \tag{30}$$

where $v(\boldsymbol{\theta})=\int du_2\,u_2\,c'_{2\boldsymbol{\theta}}(u_2)/\int du_2\,c'_{2\boldsymbol{\theta}}(u_2)$. The solution of the above is given by

$$\hat{c}_{1,\theta}(\sigma) = \hat{c}_{1,\theta}(0)\exp[-b(\boldsymbol{\theta})\sigma^2/2 - v(\boldsymbol{\theta})d(\boldsymbol{\theta})\sigma]. \tag{31}$$

The function $v(\boldsymbol{\theta})$ is not known, since $c'_{2\boldsymbol{\theta}}(u_2)$ is not determined. In fact, it is impossible to determine $c'_{2\boldsymbol{\theta}}(u_2)$ based only on the projections along $\boldsymbol{\theta}$. In the simple case that $c'_{2\boldsymbol{\theta}}(u_2)$ is an odd function, it follows that $v(\boldsymbol{\theta}) = 0$. Then Eq. (31) can be inverted either numerically or explicitly by means of

$$c(\boldsymbol{s}) = -\frac{1}{4\pi^2}\mathrm{PV}\int_0^{2\pi}\int_{-\infty}^{\infty}\frac{\mathrm{d}\phi\,\mathrm{d}\sigma}{\sigma - \boldsymbol{s}\cdot\boldsymbol{\theta}}\frac{\partial\hat{c}_{1,\theta}(\sigma)}{\partial\sigma}, \tag{32}$$

where PV denotes the Cauchy principal value of the integral. In the case that $v(\boldsymbol{\theta}) \neq 0$ the function $v(\boldsymbol{\theta})$ must be determined from the initial PDE. This can be accomplished by expressing the PDE in the rotated frame and using the general expression (31) for the ST (Christakos and Hristopulos, 1997). Thus, the following solution is obtained

$$c'_{\boldsymbol{\theta}}(\boldsymbol{u}) = \exp[-b(\boldsymbol{\theta})(u_1^2 + u_2^2)/2 - v(\boldsymbol{\theta})d(\boldsymbol{\theta})u_1 - v(\boldsymbol{\theta})u_2]. \tag{33}$$

The function $v(\boldsymbol{\theta})$ can now be determined from the requirement that in the initial frame the solution $c(\boldsymbol{s})$ be independent of the angle $\boldsymbol{\theta}$, which leads to $v(\boldsymbol{\theta}) = \beta(\cos\theta + \sin\theta)$, where β is an arbitrary constant. The SoV solution (26) then follows. □

In conclusion, the ST method transforms PDEs into ODEs which are simpler to solve. In certain cases, the ODE solution can be solved and directly inverted, explicitly or numerically, thus leading to the solution in n-dimensional space. In other cases, the ODE contains terms that can not be calculated from the projections of the function. Then, the unknown terms should be approximated using physical information or perturbation series expansions. Provided that the ODE can be formulated and solved, use of fast inversion techniques lead to significant computational gains compared with traditional PDE solvers.

Applications of the ST approach to groundwater flow have been investigated (Christakos, 1984a; 1986, 1987b; Christakos and Hristopulos, 1994; 1997; Hristopulos *et al.*, 1998). These studies focus on flow in domains with (i) uniform hydraulic conductivity domains, (ii) deterministic hydraulic conductivity trend, and (iii) random heterogeneity. The results of the ST approach are in good agreement with explicit and numerical solutions. These studies were based on the assumption that fluctuations of the flow velocity vector about the mean direction are negligible. In addition, the solution of the flow ODEs was obtained under the assumption of an infinite domain. Finite size effects were handled by forcing the 3-D solution to satisfy point constraints at the boundaries. Both assumptions can be relaxed: In particular, the velocity fluctuations can be handled perturbatively. A detailed analysis of finite size effects leads to additional boundary terms in the flow ODE.

7.5 Differential Geometric Approach to Multiphase Flow

Another interesting application inspired by the theory of ST is the *stochastic differential geometric* or *stochastic flowpath* approach for multiphase flow analysis proposed by Christakos *et al.* (1998). This approach transforms the multiphase equations using a coordinate system that follows the hydraulic gradients of each phase.

The multiphase flow equations for incompressible fluids can be expressed as

$$\frac{\mu_\alpha}{K_\alpha(\boldsymbol{p})}\frac{\partial}{\partial t}[\phi\, S_\alpha(\boldsymbol{p})] + \nabla\cdot \boldsymbol{j}_\alpha(\boldsymbol{p}) + \nabla \ln K_\alpha(\boldsymbol{p})\cdot \boldsymbol{j}_\alpha(\boldsymbol{p}) = 0, \tag{34}$$

where $\boldsymbol{p}=(\boldsymbol{s},t)$, ϕ denotes the porosity of the medium, K_α the intrinsic permeability, μ_α the dynamic viscosity, $\boldsymbol{j}_\alpha(\boldsymbol{p}) = -\nabla p_\alpha(\boldsymbol{p}) - \rho_\alpha \boldsymbol{g}$ the hydraulic gradient, and ρ_α the mass density of each phase α. Constitutive relations involve the pressure-saturation ($p-S$), and the relative permeability-saturation ($k-S$) models (e.g., Miller *et al.*, 1998). These models couple nonlinearly the saturation, pressure and permeability variables. Hydraulic gradients can be expressed as $\boldsymbol{j}_\alpha(\boldsymbol{p}) = \zeta_\alpha(\boldsymbol{p})\boldsymbol{e}_\alpha(\boldsymbol{p})$, where the (scalar) ζ_α denotes the magnitude and the unit vector $\boldsymbol{e}_\alpha$ defines the local direction of the flowpaths. A phase-α flowpath that passes at time t through the points $\boldsymbol{s}_0$ and $\boldsymbol{s}$ is denoted by $\pi_\alpha(\boldsymbol{s},\boldsymbol{s}_0;t)$. The differential geometric approach leads to a system of 1st-order SODEs for ζ_α

$$\frac{\mu_\alpha}{K_\alpha}\frac{\partial}{\partial t}(\phi\, S_\alpha) + \frac{d}{d\ell_\alpha}\zeta_\alpha + (\nabla\cdot \boldsymbol{e}_\alpha + \frac{d\ln K_\alpha}{d\ell_\alpha})\zeta_\alpha = 0, \tag{35}$$

where $d\ell_\alpha$ denotes the differential length along $\boldsymbol{e}_\alpha = (\varepsilon_{1\alpha},\ldots,\varepsilon_{n\alpha})$. The price paid for replacing the SPDEs by SODEs is that the flowpaths in Eq.(35) are not explicitly known. In fact, the flowpaths have to be determined self consistently with the other variables (i.e., hydraulic gradients and saturation) using, in addition to Eqs. (35), the following $n-1$ independent equations for each phase

$$d(\zeta_\alpha\, \varepsilon_{i,\alpha})/d\ell_\alpha = \partial\zeta_\alpha/\partial s_i - \rho_\alpha g\, \partial\varepsilon_{3,\alpha}/\partial s_i, \tag{36}$$

which represent the rate of change of the hydraulic gradient magnitude along the flowpath. Note that Eqs. (35) and (36) are 1st-order SODEs in the scalar variables ζ_α, while the initial Eqs. (34) are 2nd-order in the hydraulic head $p_\alpha - \rho_\alpha g h$. Since the constitutive equations involve the phase pressures instead of gradients, we need to close the multiphase system with the integral equations

$$p_\alpha(\boldsymbol{p}) = p_\alpha(\boldsymbol{p}_0) + \int_{\pi_\alpha(s,s_0;t)} d\ell'_\alpha \left[\zeta_\alpha(\boldsymbol{p}') - \rho_\alpha g \varepsilon_{\alpha,3}(\boldsymbol{p}') \right] , \tag{37}$$

where $d\ell'_\alpha$ denotes the line integral along the flowpath $\pi_\alpha(s,s_0;t)$. The system of Eqs. (35) through (37) consists of $n+1$ integral and differential equations for each phase in terms of the $n+1$ variables ζ_α, p_α and $\varepsilon_{i,\alpha}$ $(i = 1,\ldots,n-1)$. This system should be augmented by the algebraic constitutive relations, which include the capillary pressure and the relative saturation. The number of the algebraic equations is equal to the number of the additional variables. The SODEs --i.e. Eqs. (35) above-- are coupled via the dependence of the relative permeabilities on the capillary pressure by means of the $p - S$ relations. Also, note that neighboring flowpaths are related. This happens because the rate of change of the pressure gradient along the flowpath depends on the divergence of the direction vector $\boldsymbol{e}_\alpha$.

EXAMPLE 4: For 2-D steady-state flow of two phases we have obtained an iterative numerical solution of the flowpath system that circumvents the interdependence of pressures (Christakos *et al.*, 1998). The numerical method decouples the SODEs of each phase by assuming an initial profile for the capillary pressure; this determines the saturation and relative permeabilities of the two phases. It is then possible to solve the system of flowpath equations numerically. Based on the resulting solution for the pressures the capillary profile is updated; this process is repeated until convergence is obtained. □

7.6 Boundary Effects

In many practical situations the SPDEs must be solved within finite domains. This situation occurs when the solution is not negligible near the boundaries of the domain. In this case the space transformation must account for the position and shape of the boundaries. In addition, as we discussed above, in the case of non-integrable functions finite supports permit practical definitions of the space transformation. The price to be paid is that boundaries complicate the evaluation of space transformations. If the support is anisotropic, e.g., cubic, the space transformation is also anisotropic even for isotropic three dimensional functions. The anisotropy of the transformation reflects the anisotropic variation of the cross-sectional area of the transformation plane cut by the cubic boundary (Fig. 15). In Fig. 16 we show graphs that correspond to the cross-sectional areas for three

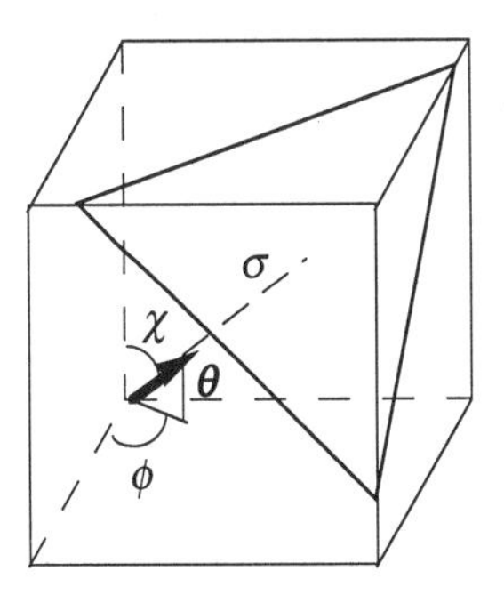

FIGURE 15: Cubic support domain and cross-section of the transform plane with the boundaries.

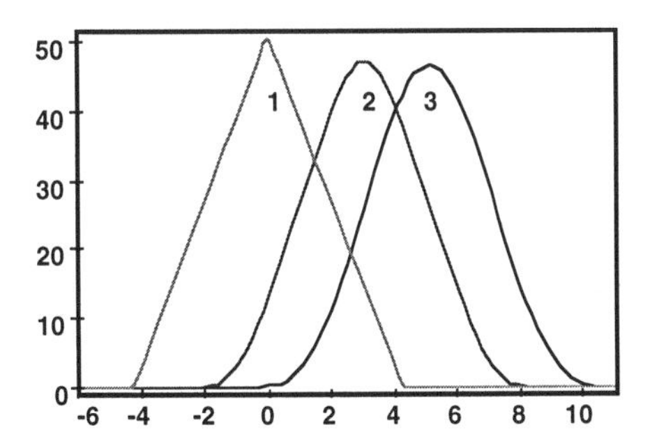

FIGURE 16: Plot of the cross-sectional areas of the ST planes with the cubic support. The ordinate of the graphs corresponds to the projection of the planes in the direction of the transform line. The graph (1) is obtained for $\phi = 0$, $\chi = 3\pi/4$; the graph (2) is obtained for $\phi = 0.65\pi$, $\chi = \pi/4$; and the graph (3) is obtained for $\phi = \chi = \pi/4$.

different orientations of transform lines. The cube is contained within the northeastern hemisphere ($\phi \in [0, \pi/2]$ and $\chi \in [0, \pi/2]$), and its sides have length equal to $L = 6$.

In investigations of groundwater flow in heterogeneous media the hydraulic head is often assumed to have a linear trend, i.e., $H(s) = H_0 - \mathbf{J} \cdot \mathbf{s}$. Since the linear function is not integrable within an infinite domain, the ST approach is applicable only for finite supports. Evaluation of the ST requires the equations that define the edges of the planar cross-sections with the cubic support. These equations determine the integration limits for the ST. The edge equations are obtained by solving simultaneously the equation for the planes, i.e., $\Pi_{\theta}(\mathbf{s}) = \mathbf{s} \cdot \boldsymbol{\theta} - \sigma = 0$ with each of the boundary equations. In the case of cubic support, the six boundary equations are given by $f_{b,i}(\mathbf{s}) = 0$, where $i = 1,..,6$ and $f_{b,i}(\mathbf{s}) = s_i$, $f_{b,i+3}(\mathbf{s}) = s_i - L$ where $i = 1,2,3$. The transform plane crosses the corresponding boundary if the equations $\Pi_{\theta}(\mathbf{s}) = 0$ and $f_{b,i}(\mathbf{s}) = 0$ can be solved simultaneously. Thus, explicit expressions are obtained for the edge equations, which depend on both the direction of the transform lines (Hristopulos *et al.*, 1998) and the projection. Even in the simple linear case the T_3^1 transform can not be evaluated explicitly, due to the implicit dependence of the integration limits on both the direction of the transform line and the projection. The ST of the function $H(\mathbf{s}) = J\,s_2$ within a cubic support is shown in Fig. 17 for four transform lines. For this calculation it has been

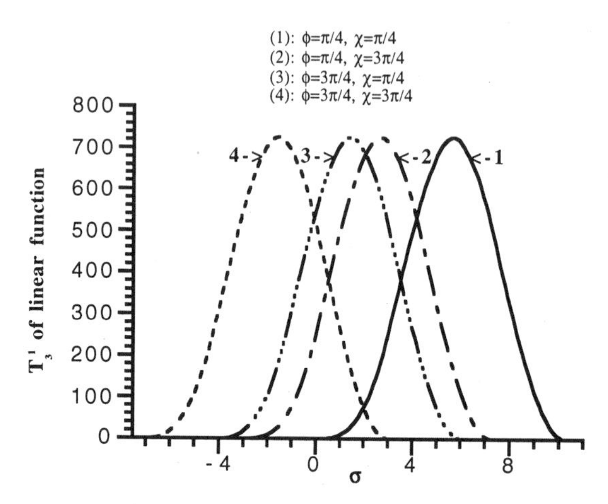

FIGURE 17: Plots of the T_3^1 transform of the linear function $H(s) = J\,s_2$ with a cubic support for four different transform lines.

assumed that $J = -5$ and $L = 6$ (in appropriate length units).

The ST approach for solving PDEs has been shown to work well in the case of simple problems and to have computational advantages due to the accurate reconstruction of multidimensional functions from a finite number of projections. However, a consistent handling of the boundary conditions in the case of finite supports requires more research.

Chapter VIII: Stochastic Physiologically-based Pollutokinetic Modelling

"You are not thinking. You are merely being logical".
N. Bohr to A. Einstein

1. INTRODUCTION

Pollutokinetic analysis is used to predict biomarker distributions in the human organs and tissues (e.g., chemical concentration or metabolites; see Chapter II), resulting from exposure to pollutants. The term pollutokinetics is similar to the pharmacokinetics which is the study of the rate of change in drug and metabolite concentration in the body. The analysis of pollutokinetic data typically involves *compartmental* models (e.g., Piotrowski, 1971; Crawford-Brown, 1997), which are particularly useful in the study of transfer and transformation processes that occur in the body following exposure to a pollutant.

An important development in the compartmental analysis of pollutokinetic data is the advent of *physiological pollutokinetics* (*PPK*), in which each compartment represents a well defined physiological entity (Ramsey and Andersen, 1984; Vinegar *et al.*, 1990; Leung, 1991).

In this chapter, we study pollutokinetic effects that are represented in terms of *stochastic ordinary differential equations* (*SODEs*). These SODEs are based on PPK compartmental models derived from relevant anatomical and physiological studies (e.g., Moore and Agur, 1995; Vander *et al.*, 1998). For the overwhelming majority of pollutants that have the potential to damage the human health, the available knowledge is not sufficient for a deterministic specification of the parameters involved in the PPK compartmental models (e.g., Covello and Merkhoffer, 1993). This indisputable fact makes the use of stochastic analysis a necessary approach.

The SODE representation has two important features: (a) compartmental models are constructed with strict conformity to anatomical and physiological characteristics, and (b) they also account for the uncertainties of the biological processes involved. For example, the biomarker variability obtained from SODE depends not only on exposure variation but also on the variability associated with the biologic characteristics of the individual. Hence, such models offer an adequate quantitative assessment of the actual dose of pollutant absorbed into the body. In this respect, stochastic PPK modelling has considerable advantages over exposure methods based on monitoring pollutant concentration, which do not usually represent the actual dose, or over deterministic pollutokinetics, which do not account for biological uncertainties. Moreover, stochastic PPK leads to closed form analytical expressions of exposure and biologic variabilities that are mathematically rigorous and, also, offer a meaningful representation of reality. Such analytical results may lead to significant improvements over existing empirical statistical expressions.

Just as for SPDE-based physical models (Chapter VII), two main approaches exist for studying SODE of PPK. One method focuses on determining *realizations* of the SODE, while the other concentrates on deriving *stochastic moments* of the PPK models.

2. COMPARTMENTAL ANALYSIS OF POLLUTOKINETICS

After the pollutant enters the body via the exposure routes as discussed in Chapter II, its distribution in the body involves three different groups of biological processes responsible for converting the uptake into burden and dose:

(a) *Transfer* of the pollutant between organs, *transport* within an organ, and *excretion* from the body out to the environment. This is related to the pollutant *biokinetic* properties.

(b) *Transformation* of the pollutant within the body into different chemical, biological or physical forms by means of *activation-deactivation* processes.

(c) *Interaction* with biological structures (e.g., DNA) generating transitions leading to a specific effect. These processes are related to the pollutant *dosimetric* properties.

REMARK 1: It is noteworthy that the above are properties of the pollutant *and* the organism. The same pollutant may have different properties in different organisms (Crawford-Brown, 1997).

EXAMPLE 1: An illustration of chemical distribution within the body and the resulting biological processes are shown in Fig. 1. □

Pollutokinetic modelling is used to describe the fate of pollutants in humans. In particular, pollutokinetics is concerned with the biokinetic and activation-deactivation properties of a pollutant in the body. Traditionally these processes are represented in terms of ODEs of the form (m *th-order kinetics*; usually $m = 0$ or 1)

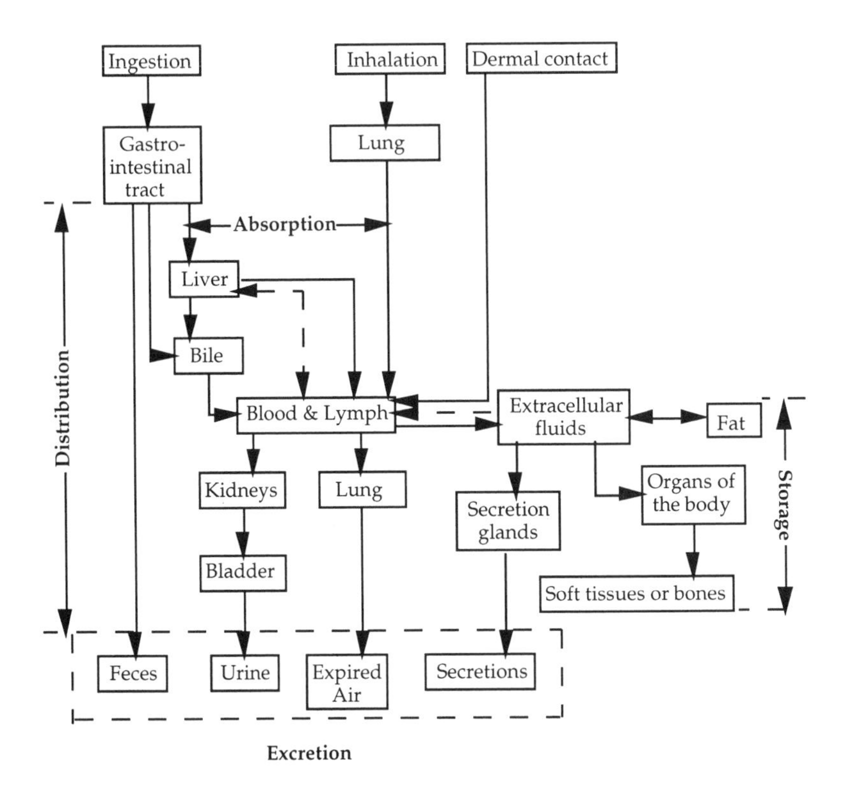

FIGURE 1: Routes of chemical fate within the body (OTA, 1981).

$$d\mathcal{B}(\boldsymbol{p})/dt = \mathcal{U}(\boldsymbol{p}) - \lambda(\boldsymbol{p})\mathcal{B}^m(\boldsymbol{p}), \tag{1}$$

where $\mathcal{B}(\boldsymbol{p})$ and $\mathcal{U}(\boldsymbol{p})$, $\boldsymbol{p} = (\boldsymbol{s},t)$, denote the biomarkers (e.g., uptake, burden) and the exposure processes involved (e.g., pollutant concentrations). The initial condition (IC) on the burden is denoted by $\mathcal{B}(\boldsymbol{s},t=0) = \mathcal{B}_0$. The *temporal* analysis of biomarkers in specified organs is of interest here and, hence, $\boldsymbol{p}$ will occasionally be replaced by t for simplicity. While some biomarkers follow 0th-order kinetics (the exit rate from an organ at time t does not depend on the amount in the organ at time t), most of them follow 1st-order kinetics (the exit rate at time t is proportional to the amount present in the organ at time t).

EXAMPLE 2: In 0th-order kinetics the exit rate of the pollutant from an organ is independent of the amount present in the organ at that time. This happens in conditions of constant infusion, for absorption of pollutants in sustained released formulation, and in enzyme-dependent processes at complete saturation. 1st-order kinetics implies that the exit rate is proportional to the amount present in the organ as, for example, in absorption and excretion by passive diffusion (Leung and Paustenbach, 1988). □

The fate of a pollutant within the body may be described conveniently by means of *compartments* (also called *spaces* or *pools* in anatomy and physiology; Vander *et al.*, 1998) that represent a system of environmental media, organs, or tissues within which the biological processes occur. In literature, a distinction is usually made between *one-compartment* and *multi-compartment* models. The Eq. (1) represents an one-compartment model; a k-compartment model that obeys 1st-order kinetics may be represented by the vector SODE

$$d\boldsymbol{\mathcal{B}}(\boldsymbol{p})/dt = \boldsymbol{\mathcal{U}}(\boldsymbol{p}) + \boldsymbol{\lambda}(\boldsymbol{p})\boldsymbol{\mathcal{B}}(\boldsymbol{p}), \tag{2}$$

where $\boldsymbol{\mathcal{B}}(\boldsymbol{p}) = [\mathcal{B}_1 \ldots \mathcal{B}_k]^T$, $\mathcal{B}_i$ is the amount of material or a biomarker at compartment $i = 1,\ldots,k$; $\boldsymbol{\mathcal{U}}(\boldsymbol{p}) = [\mathcal{U}_1 \ldots \mathcal{U}_k]^T$, $\mathcal{U}_i$ is the entry rate into compartment i from external sources; and $\boldsymbol{\lambda}(\boldsymbol{p})$ is a rate matrix with elements $\lambda_{ij} = -\lambda_i\,\delta_{ij} + \lambda_{ji}(1-\delta_{ij})$, where λ_i is the exit rate from the compartment i and λ_{ij} is the transfer rate from compartment i to compartment j. The ICs are denoted by $\boldsymbol{\mathcal{B}}(\boldsymbol{s}, t=0) = \boldsymbol{\mathcal{B}}_0 = [\mathcal{B}_{1,0} \ldots \mathcal{B}_{k,0}]^T$.

EXAMPLE 2: Consider the three-compartment model: air-lungs-blood (Fig. 2). Let $U_r(\boldsymbol{p})$

FIGURE 2: A three-compartment model.

be the CO uptake for a human at the space-time point $\boldsymbol{p}$. The process of CO transfer from the air to the lungs is 0th-order, and it is governed by

$$dU(\boldsymbol{p})/dt = U_r(\boldsymbol{p}) - \lambda_{\ell a}(\boldsymbol{p}), \tag{3}$$

where $\lambda_{\ell a}$ is the lungs-air transfer rate. Transfer from the lungs to the blood is a 1st-order process. Ignoring back transfer from the blood to the lungs, it is governed by the ODE

$$dB_\ell(\boldsymbol{p})/dt = U_r(\boldsymbol{p}) - \lambda_{\ell b}(\boldsymbol{p})\,B_\ell(\boldsymbol{p}), \tag{4}$$

where $B_\ell(\boldsymbol{p})$ is the burden on the lungs and $\lambda_{\ell b}(\boldsymbol{p})$ is the lungs-blood transfer rate. □

When the transfer rates are constant it is more convenient to work in the *Laplace transform* (*LT*)-domain, in which Eq. (2) is written as

$$\boldsymbol{\lambda}_w \breve{\mathcal{B}}(s,w) = \breve{\mathcal{U}}(s,w) + \mathcal{B}_0, \tag{5}$$

where $\breve{\mathcal{B}}(s,w) = [\breve{\mathcal{B}}_1 \ldots \breve{\mathcal{B}}_k]^T$, $\breve{\mathcal{B}}_i$ denotes the LT of $\mathcal{B}_i$; $\breve{\mathcal{U}}(s,w) = [\breve{\mathcal{U}}_1 \ldots \breve{\mathcal{U}}_k]^T$, $\breve{\mathcal{U}}_i$ denotes the LT of $\mathcal{U}_i$; and $\boldsymbol{\lambda}_w$ is a matrix with diagonal elements $\lambda_{w,ii} = w + \lambda_i$, and off-diagonal elements $\lambda_{w,ij} = -\lambda_{ji}$, where $i \neq j$; or, in general $\lambda_{w,ij} = (w + \lambda_i)\delta_{ij} - \lambda_{ji}(1 - \delta_{ij})$.

PPK compartmental models have recently been advanced that take into consideration the actual physiology, thus offering a more realistic representation of the pollutant behavior within the body than conventional compartment models, which are usually formulated solely on the basis of empirical data. Various PPK models in humans exist in literature, e.g., for tetrachloroethylene (EPA, 1983) and volatile organic chemicals (Leung, 1992).

It is well established that most biomarkers are affected to different extents by random fluctuations and natural variations in exposure, depending on a number of factors which include: (i) their half-lives and initial conditions, (ii) the heterogeneity of the exposure fields in space-time, (iii) the transfer rate parameters, and (iv) the biologic and physiologic characteristics of the individual. Therefore, it is more realistic to represent pollutokinetics relating exposure and health effect in terms of *stochastic* rather than deterministic models.

EXAMPLE 4: Uncertainty in estimates of PPK parameters (metabolic variables, blood-air and tissue-blood partition coefficients, organ weights and blood flow rates, etc.) leads to considerable uncertainty regarding tissue concentrations and resulting risks (Farrar *et al.*, 1989). □

Uncertainty is crucial in the case of exposure to ionizing radiation as illustrated by the following example.

EXAMPLE 5: In the case of ionizing radiation --in which a single particle can kill a cell and a single ionization reaction can decompose a critical molecule-- random fluctuations are important, and they are a major source of uncertainty in radiation action (Rossi and Zaimer, 1996). □

Stochastic pollutokinetics can help to clarify (i) the relationships between exposure variation and biomarkers, and (ii) the influence of critical parameters. The stochastic analysis of the following sections presupposes the validity and appropriate formulation of the relevant PPK compartmental models. Moreover, in order to account for the uncertainties characterizing the fate of pollutants in the body, the biomarkers and other exposure processes will be represented in terms of random fields; as a consequence, the PPK will be modelled as SODEs. To obtain some insight into the stochastic PPK we begin with the analysis of an one-compartment model.

3. ONE-COMPARTMENT STOCHASTIC POLLUTOKINETICS

One-compartment models are usually simplifications of a more complex situation but, given the paucity of data one has to work with in many cases (e.g., plasma or urine levels), it provides a reasonable compromise. Below, we will present a stochastic formulation of the one-compartment pollutokinetics.

Consider the case in which the biomarker of interest is the pollutant burden $B(\boldsymbol{p})$ in an organ. The $B(\boldsymbol{p})$ is represented as a random field which obeys the stochastic 1st-order kinematics

$$dB(\boldsymbol{p})/dt = U_r(\boldsymbol{p}) - \lambda(\boldsymbol{p})B(\boldsymbol{p}), \tag{1}$$

where $\boldsymbol{p} = (\boldsymbol{s}, t)$ are the space/time coordinates of the organ, $U_r(\boldsymbol{p})$ is the random uptake rate and $\lambda(\boldsymbol{p})$ is the random transfer rate out of the organ (these quantities were defined in Chapter II). Since the temporal analysis of the burden in a specified organ is of interest here, the $\boldsymbol{p}$ is replaced by t, and the stochastic Eq. (1) reduces to

$$dB(t)/dt = U_r(t) - \lambda(t)B(t), \tag{2}$$

given the IC $B(0) = 0$. This implies, e.g., that the pollutant was first introduced in another compartment at time $t = 0$, and then transferred to the compartment (2). As we saw in previous sections, there exist two approaches for solving Eq. (2) in a stochastic context: the realization-based approach, and the moment-based approach.

The random field *realization-based* approach leads to the following solution of Eq. (2)

$$B(t) = B(0)\exp(-\zeta_t) + \int_0^t dt' U_r(t')\exp(-\zeta_t + \zeta_{t'}) = \int_0^t dt' U_r(t')\exp(-\zeta_t + \zeta_{t'}), \tag{3}$$

where $\zeta_t = \int_0^t d\tau \lambda(\tau)$. If sufficient information exists so that random field realizations of the uptake $U_r(t)$ and the transfer rate $\lambda(t)$ can be generated (using the methods described in the previous chapters), the corresponding burden $B(t)$ realizations may be obtained from Eq. (3).

In order to obtain the moment-based solution to the stochastic pollutokinetics above, let us consider the following decomposition of the biological fields

$$\left.\begin{array}{l} B(t) = \overline{B}(t) + b(t) \\ U_r(t) = \overline{U_r}(t) + u_r(t) \\ \lambda(t) = \overline{\lambda}(t) + \ell(t) \end{array}\right\}, \tag{4}$$

where $b(t)$, $u_r(t)$ and $\ell(t)$ are random fluctuations around the means $\overline{B}(t)$, $\overline{U_r}(t)$ and $\overline{\lambda}(t)$, respectively. Substituting Eq. (4) into Eq. (2) we find

$$d\overline{B}(t)/dt + db(t)/dt = \overline{U_r}(t) + u_r(t) - \overline{\lambda}(t)\overline{B}(t) - \ell(t)b(t) - \overline{\lambda}(t)b(t) - \ell(t)\overline{B}(t). \tag{5}$$

Subtracting from both sides of Eq. (5) their expected value, we obtain the following two equations for the means and the fluctuations, respectively,

$$d\overline{B}(t)/dt + \overline{\lambda}(t)\overline{B}(t) + \overline{\ell(t)b(t)} = \overline{U_r}(t), \tag{6}$$

and

$$db(t)/dt + \overline{\lambda}(t)b(t) + \ell(t)\overline{B}(t) + [\ell(t)b(t) - \overline{\ell(t)b(t)}] = u_r(t). \tag{7}$$

Next, we introduce the *total time derivative* operator $D_t = [d/dt + \overline{\lambda}(t)]$ and the *modified uptake rate* fluctuation

$$z(t) = u_r(t) - \ell(t)\overline{B}(t). \tag{8}$$

Assuming that fluctuation products are negligible, Eqs. (6) and (7) reduce, respectively, to the following

$$D_t\overline{B}(t) = \overline{U_r}(t), \tag{9}$$

and

$$D_t b(t) = z(t). \tag{10}$$

REMARK 1: Upon reaching *steady state*, i.e., when $\overline{B}(t)$ becomes constant, $d\overline{B}(t)/dt = 0$ and

$$1/\overline{\lambda} = \overline{B}/\overline{U}_r. \tag{11}$$

Since $\overline{B}$ is the average pollutant burden in a compartment and $\overline{\lambda}\,\overline{B}$ is the average exit rate, $1/\overline{\lambda}$ represents physically the average *residence time* of the pollutant in the compartment.

Similarly, based on Eq. (10), the differential equation governing the covariance of the burden is found to be

$$D_t\, D_{t'}\, c_b(t,t') = c_z(t,t'), \tag{12}$$

where the differential operators D_t and $D_{t'}$ are with respect to t and t', respectively.

The Eq. (9) can be solved to obtain the mean burden

$$\overline{B}(t) = \overline{B}(0)\exp(-\overline{\zeta}_t) + \int_0^t dt'\,\overline{U_r}(t')\exp(-\overline{\zeta}_t + \overline{\zeta}_{t'}), \tag{13}$$

where $\overline{B}(0) = 0$ (because of the zero IC assumed above), and $\overline{\zeta}_t = \int_0^t d\tau\,\overline{\lambda}(\tau)$. In many cases experimental information implies a constant mean rate $\overline{\lambda}(t) = \overline{\lambda}$, so that

$$\overline{B}(t) = \exp(-\overline{\lambda}\,t)\int_0^t dt'\,\overline{U_r}(t')\exp(\overline{\lambda}\,t'). \tag{14}$$

The solution of Eq. (12) is obtained most conveniently in the LT-domain. The LT of Eq. (12) is $(w+\overline{\lambda})(w'+\overline{\lambda})\breve{c}_b(w,w') = \breve{c}_z(w,w')$, where the $\breve{c}$ denotes the LT of the corresponding covariance function, or

$$\breve{c}_b(w,w') = \breve{c}_z(w,w')/[(w+\overline{\lambda})(w'+\overline{\lambda})]. \tag{15}$$

To proceed further with the stochastic analysis let us assume the following covariance for the modified uptake rate (Fig. 3) ,

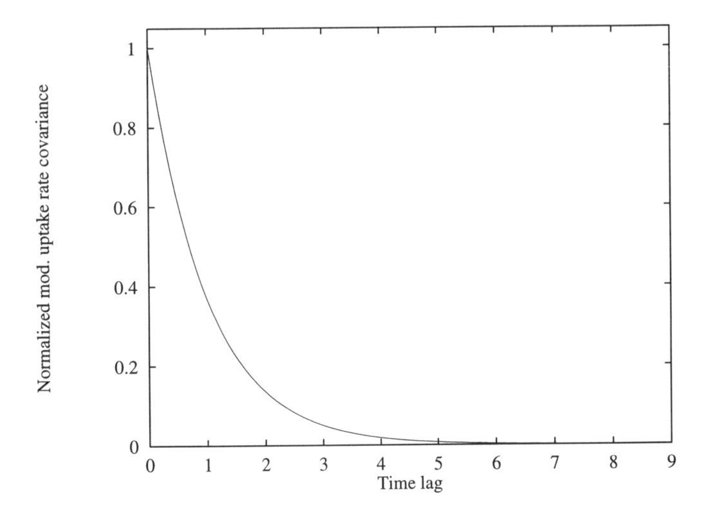

FIGURE 3: Normalized covariance $c_z(t,t')/\sigma_z^2$ of the modified uptake rate fluctuation; Eq. (16), for $\varepsilon = 1$.

$$c_z(t,t') = \sigma_z^2 \exp(-\varepsilon|t' - t|) \tag{16}$$

where ε^{-1} is the *correlation time* of the biological field. The exponential covariance model above is practical, for arbitrary covariances can be approximated with any desired degree of accuracy by functions of the form $\sum_{i=1}^{k} a_i \exp(-\varepsilon_i|t' - t|)$, a_i, $\varepsilon_i > 0$, provided a sufficient number k of terms is used. The LT of the modified uptake rate covariance is $\breve{c}_z(w,w') = \sigma_z^2(w + w' + 2\varepsilon)/[(w+\varepsilon)(w'+\varepsilon)(w+w')]$, and by substituting in Eq. (15) we find

$$\breve{c}_b(w,w') = \sigma_z^2(w + w' + 2\varepsilon)/[(w+\bar{\lambda})(w'+\bar{\lambda})(w+\varepsilon)(w'+\varepsilon)(w+w')]. \tag{17}$$

By taking the inverse LT of Eq. (17) we obtain the burden covariance in the time domain

$$\begin{aligned} c_b(t,t') = \frac{\sigma_z^2}{\bar{\lambda}^2 - \varepsilon^2}[&\exp(-\varepsilon|t'-t|) - \frac{\varepsilon}{\bar{\lambda}}\exp(-\bar{\lambda}|t'-t|) + \frac{\bar{\lambda}+\varepsilon}{\bar{\lambda}}\exp(-\bar{\lambda}t - \bar{\lambda}t') \\ &-\exp(-\varepsilon t - \bar{\lambda}t') - \exp(-\bar{\lambda}t - \varepsilon t')] \end{aligned} \tag{18}$$

As shown in Fig. 4, this is a symmetric function with respect to t and t', and satisfies the ICs $c_b(0,0) = 0$. The burden covariance (18) depends on the absolute time lag $|t' - t|$ as well as on the disposition of both t and t' with respect to the uptake initiation. This shows that the burden is nonstationary, even when the modified uptake rate covariance is

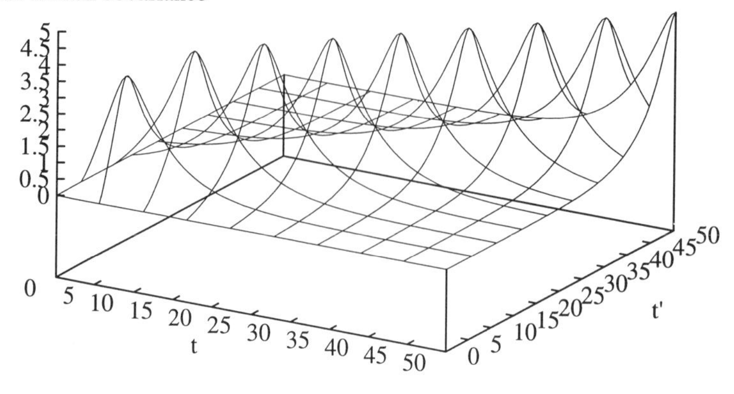

FIGURE 4: The normalized burden covariance $c_b(t,t')/\sigma_z^2$ of Eq. (18) for $\varepsilon = 1$ and $\overline{\lambda} = 0.2$.

stationary. The burden variance is obtained from Eq. (18) by setting $t = t'$ as follows

$$\sigma_b^2(t) = \frac{\sigma_z^2}{\overline{\lambda}^2 - \varepsilon^2}[1 - \frac{\varepsilon}{\overline{\lambda}} + \frac{\overline{\lambda} + \varepsilon}{\overline{\lambda}}\exp(-2\overline{\lambda}t) - 2\exp(-\varepsilon t - \overline{\lambda}t)], \tag{19}$$

FIGURE 5: The normalized burden variance $\sigma_b^2(t)/\sigma_z^2$ of Eq. (19) for $\varepsilon = 1$ and $\overline{\lambda} = 0.2$.

which, as expected, depends on time t (Fig. 5). Note that the parameters in Eq. (19) depend on both the exposure variation and the variability associated with the characteristics of the individual. Some interesting special cases are considered in the following example.

EXAMPLE 1: The burden variance at the asymptotic limit $t \to \infty$ is

$$\sigma_b^2 = \sigma_z^2 / \overline{\lambda}(\overline{\lambda} + \varepsilon). \tag{20}$$

Hence, the burden variance at equilibrium is proportional to the z-variance and inversely proportional to the mean rate and the correlation coefficient. When $t, t' \to \infty$ but $|t' - t| = \tau \neq 0$, the burden covariance (18) tends to the stationary value

$$c_b(t,t') = \frac{\sigma_z^2}{\overline{\lambda}^2 - \varepsilon^2}[\exp(-\varepsilon\tau) - \frac{\varepsilon}{\overline{\lambda}}\exp(-\overline{\lambda}\,\tau)], \tag{21}$$

plotted in Fig. 6. Note that Eq. (20) follows from Eq. (21) by setting $|t' - t| = 0$. □

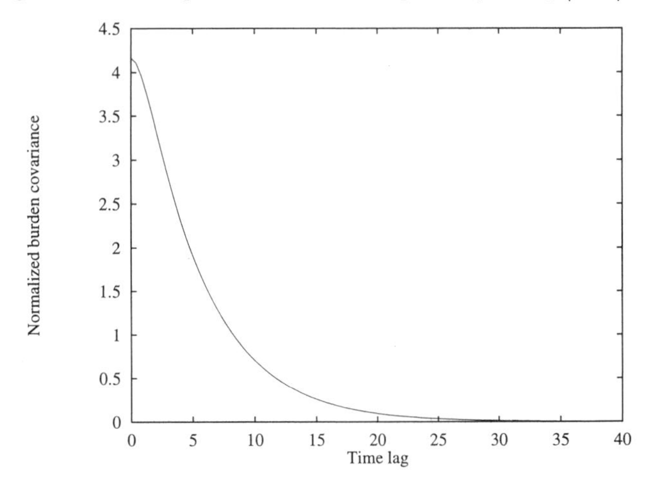

FIGURE 6: The normalized burden covariance $c_b(t,t')/\sigma_z^2$ of Eq. (21) for $\varepsilon = 1$ and $\overline{\lambda} = 0.2$.

In the following we will analyze the physical meaning of the stochastic burden relations obtained above. The *correlation period* τ_c may be defined as the time interval within which there is a significant stochastic dependence between the values of the random biological field $z(t)$. As a criterion for this dependence we may take, e.g., the condition $c_z(\tau) \geq 0.1$ for $\tau \leq \tau_c$ and $c_z(\tau_c) = 0.1\sigma_z^2$ which, in light of Eq. (16), implies $\tau_c \approx 2.3/\varepsilon$.

If the correlation period greatly exceeds the residence time (i.e., $\varepsilon << \overline{\lambda}$ or, equivalently, $\tau_c >> 1/\overline{\lambda}$), then

$$\sigma_b^2 = \sigma_z^2 / \overline{\lambda}^2, \tag{22}$$

$$\rho_b(t,t') = c_b(t,t') / \sigma_b^2 = \exp(-\varepsilon\tau); \tag{23}$$

namely, the burden correlation coincides with that of the modified uptake rate fluctuation, given by Eq. (16). Conversely, if the residence time greatly exceeds the correlation period (i.e., $\varepsilon >> \overline{\lambda}$ or, $\tau_c << 1/\overline{\lambda}$), then

$$\rho_b(t,t') = \exp(-\overline{\lambda}\,\tau), \tag{24}$$

meaning that the burden correlation in time is fully determined by the average transfer rate.

EXAMPLE 2: If we assume constant mean uptake rate, i.e., $\overline{U_r}(t) = \overline{U_r}$, the mean burden given by Eq. (13) reduces to

$$\overline{B}(t) = \overline{U_r}[1 - \exp(-\overline{\lambda}\,t)] / \overline{\lambda}, \tag{25}$$

which is, in fact, the form of the *deterministic* solution for the 1st-order kinetics [see Eq. (II.3.21)]. By letting $t \to \infty$ in Eq. (25) we obtain the steady state mean burden $\overline{B}(t) = \overline{U_r} / \overline{\lambda}$, see Eq. (11) above and Eq. (II.3.23). □

In conclusion, stochastic analysis of pollutokinetic models is a considerably more general and realistic approach than traditional deterministic analysis (the former includes the latter as a special case, if the uncertainty is eliminated). Stochastic modelling rigorously takes into consideration the uncertainty in the biomarkers and exposure processes, and it leads to equations that offer quantitative estimates of possible variations and correlations of these processes in the body. This information is particularly useful in obtaining meaningful estimates of uncertainty for health risk assessment.

4. MULTI-COMPARTMENTAL STOCHASTIC POLLUTOKINETICS

The results provided in the previous section for one-compartment pollutokinetics can be generalized to more complex, multi-compartmental models. While one-compartment models provide useful results in many situations, multi-compartmental models are necessary when a more accurate description of the physiological and anatomical details must be taken into consideration. The body is assumed to consist of several compartments which are built on the basis of physiological knowledge. Each compartment, consisting of tissues or groups of tissues, represents a physical domain in the body through which the chemicals move; intercompartmental coupling is provided by the blood circulation. In developing a multi-compartmental model, various sources of information must be considered including physiological, physicochemical, biochemical and metabolic data.

From a mathematical point of view, the main difference is that the scalar SODE needs to be replaced by a vector SODE. Indeed, in the case of multi-compartmental pollutokinetics the Eq. (3.2) may be generalized as follows [see, also, Eq. (2.2)]

$$d\mathcal{B}(t)/dt = \mathcal{U}(t) + \boldsymbol{\lambda}(t)\,\mathcal{B}(t), \tag{1}$$

given certain ICs $\mathcal{B}_0$; the decomposition of the random vector biological fields is as follows

$$\left.\begin{aligned} \mathcal{B}(t) &= \overline{\mathcal{B}}(t) + \boldsymbol{\mathscr{b}}(t) \\ \mathcal{U}(t) &= \overline{\mathcal{U}}(t) + \boldsymbol{u}(t) \\ \boldsymbol{\lambda}(t) &= \overline{\boldsymbol{\lambda}}(t) + \boldsymbol{\eta}(t) \end{aligned}\right\}, \tag{2}$$

where $\boldsymbol{\mathscr{b}}(t)$ and $\boldsymbol{u}(t)$ are random vector fluctuations around the vector means $\overline{\mathcal{B}}(t)$, $\overline{\mathcal{U}}(t)$; and $\boldsymbol{\eta}(t)$ is a matrix of transfer rate fluctuations around the mean matrix $\overline{\boldsymbol{\lambda}}(t)$. We assume that the fluctuation products are negligible, and we define the vector total time derivative $\boldsymbol{D}_t = [\boldsymbol{I}\,d/dt - \overline{\boldsymbol{\lambda}}(t)]$ and the modified uptake rate fluctuation

$$z(t) = \boldsymbol{u}(t) - \boldsymbol{\eta}(t)\,\overline{\mathcal{U}}(t). \tag{3}$$

Then, Eq. (1) leads to the following equations for the mean, the fluctuation and the covariance of the burden, respectively,

$$\boldsymbol{D}_t\overline{\mathcal{B}}(t) = \overline{\mathcal{U}}(t), \tag{4}$$

$$\boldsymbol{D}_t\boldsymbol{\mathscr{b}}(t) = z(t), \tag{5}$$

and

$$\boldsymbol{D}_t \boldsymbol{D}_{t'} \boldsymbol{c}_{\mathcal{B}}(t,t') = \boldsymbol{c}_z(t,t'), \tag{6}$$

where the operators $\boldsymbol{D}_t$ and $\boldsymbol{D}_{t'}$ are with respect to t and t', respectively; $\boldsymbol{c}_{\mathcal{B}}(t,t')$ and $\boldsymbol{c}_z(t,t')$ are the covariance matrices between the $\mathcal{b}_i$'s and the z_i's. Eqs. (4) and (6) lead to a system of coupled, deterministic differential equations. The degree of difficulty in solving the system depends on the number and the complexity of the physiological compartments involved (numerical rather than analytical procedures may be suitable for PPK models with spatial variability and/or nonlinear dependences).

When studying stochastic multi-compartmental models of PPK, it is usually preferable to work in the LT-domain, in which case Eqs. (4) and (6) become

$$\boldsymbol{\lambda}_w \breve{\overline{\boldsymbol{\mathcal{B}}}}(w) = \breve{\overline{\boldsymbol{\mathcal{U}}}}(w) + \overline{\boldsymbol{\mathcal{B}}}_0, \tag{7}$$

$$\boldsymbol{\lambda}_w \boldsymbol{\lambda}_{w'} \breve{\boldsymbol{c}}_{\mathcal{B}}(w,w') = \breve{\boldsymbol{c}}_z(w,w'), \tag{8}$$

respectively, where the elements of the transfer rate matrix $\boldsymbol{\lambda}_w$ were defined after Eq. (2.5) above. In view of this, note that the solution of Eq. (7) for the $\breve{\overline{\mathcal{B}}}_i$-element of the vector $\breve{\overline{\boldsymbol{\mathcal{B}}}}$ is given by

$$\breve{\overline{\mathcal{B}}}_i = \frac{\overline{\mathcal{B}}_{i,0} + \breve{\overline{\mathcal{U}}}_i}{w + \lambda_i} + \sum_{j \neq i} \frac{\lambda_{ji}}{w + \lambda_i} \breve{\overline{\mathcal{B}}}_j, \tag{9}$$

which leads to a set of linear algebraic equations with respect to the $\breve{\overline{\mathcal{B}}}_i$'s, $i = 1,\ldots,k$.

5. STOCHASTIC POLLUTOKINETICS AND HEALTH EFFECTS

As we saw in Chapter II, exposure generally refers to the contact of a receptor with an environmental pollutant. The amount of the pollutant that exhibits toxic effects on the target organ is denoted by the appropriate pollutant biomarker (e.g., burden) which is not necessarily strictly proportional to the exposure concentration. Stochastic pollutokinetics rigorously take into consideration exposure variations in space-time as well as variabilities linked to the biological and physiological characteristics of the individual. As a result, it can provide information about the exposure-biomarker relationship that is useful in the evaluation of assumptions underlying the dose-response models and the assessment of the actual health effects of exposure.

As we discussed in the preceding sections, there are two approaches for studying important natural (physical, biological, etc.) variabilities characterizing the relationships between exposure and health effect biomarkers: the realization-based, and the moment-based approaches summarized in Table 1.

TABLE 1: Random Field Approaches for Studying Exposure and Biomarker Variabilities

- The random field realization-based approach:

(i) Obtain the exposure $E(t)$-profile (using monitoring and physical modelling).
(ii) Construct a realistic PPK model using physiological, physicochemical, biochemical and metabolic data.
(iii) Obtain the appropriate values for the transfer rates λ_i involved in the PPK model.
(iv) Solve the stochastic PPK equations to obtain the appropriate biomarker $\mathcal{B}(t)$-profile.

- The moment-based approach:

(i) Obtain the exposure mean and covariance (from monitoring data and physical modelling).
(ii) Construct a realistic PPK model as above.
(iii) Obtain the statistics of the transfer rates $\lambda_i(t)$ and the $z(t)$-processes of the PPK model.
(iv) Solve the PPK moment equations to obtain the mean and covariance of the biomarker $\mathcal{B}(t)$ in terms of the exposure statistics and the biological parameters.
(v) Biomarker $\mathcal{B}(t)$-profiles can be generated on the basis of its mean and covariance, using random field methods.

The first approach is most efficient if the transfer rates can be assumed constant or, if random, they are not correlated with the biomarkers and the exposure random fields. The second approach is preferable when the transfer rates are random fields, possibly correlated with the biomarkers and the exposure random fields. Closed-form analytical expressions of the biomarker variability can be obtained in terms of the exposure variability and the biologic variability of the receptors. These expressions are mathematically rigorous and, they also provide a meaningful representation of reality which has significant potential advantages over commonly used statistical expressions (Rappaport, 1991; Droz, 1993). The two stochastic approaches are best explained by means of the following

EXAMPLE 1: Consider a situation in which the uptake rate is proportional to the exposure concentration, i.e.,

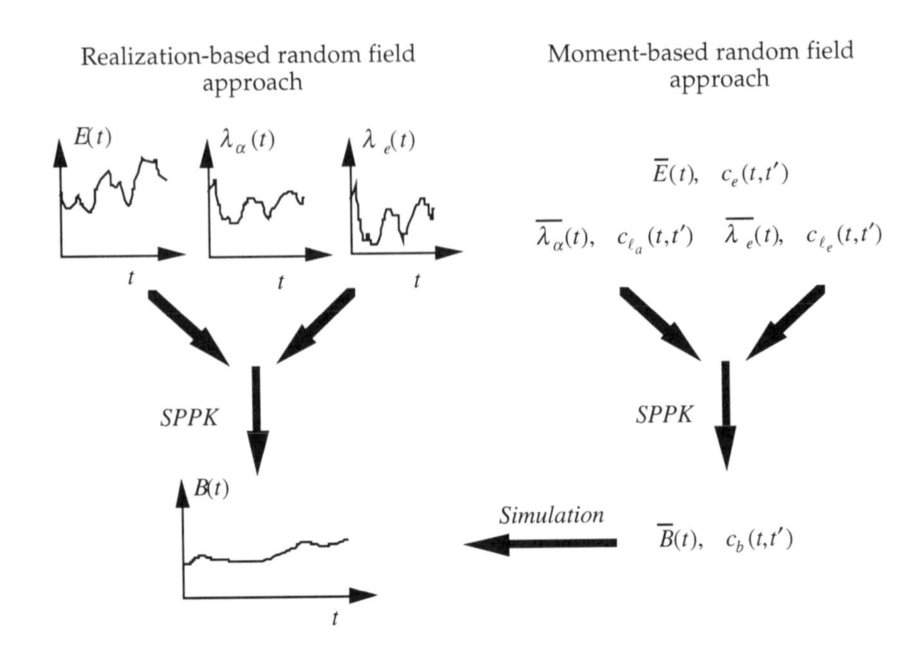

FIGURE 7: Random field approaches; SPPK=stochastic physiological pollutokinetics.

$$U_r(t) = \lambda_\alpha(t)E(t), \tag{1}$$

where $\lambda_\alpha(t)$ is the absorption rate and $E(t)$ is the exposure concentration at time t. Then, the stochastic pollutokinetic Eq. (3.2) becomes

$$dB(t)/dt = \lambda_\alpha(t)E(t) - \lambda_e(t)B(t), \tag{2}$$

where $\lambda_e(t)$ is the elimination rate. The two alternative random field approaches are illustrated in Fig. 7. More specifically, the *realization-based* solution to Eq. (2) is given by

$$B(t) = B(0)\exp(-\zeta_t) + \int_0^t dt'\,\lambda_\alpha(t')E(t')\exp(-\zeta_t + \zeta_{t'}), \tag{3}$$

where $\zeta_t = \int_0^t d\tau\,\lambda_e(\tau)$. Provided that sufficient information exists so that realizations of $E(t)$, $\lambda_\alpha(t)$ and $\lambda_e(t)$ can be generated, the corresponding burden $B(t)$-realizations are obtained from Eq. (3). The burden values can be compared with the exposure values obtained through monitoring. If the $B(t)$-profile follows well the $E(t)$-profile, we may conclude that the latter is a good indicator of the toxic effects on the target organ. For the *moment-based* solutions we consider the decomposition of the random biological fields as before

$$\left.\begin{aligned} &B(t) = \overline{B}(t) + b(t) \\ &E(t) = \overline{E}(t) + e(t) \\ &\lambda_{\psi}(t) = \overline{\lambda_{\psi}}(t) + \ell_{\psi}(t), \quad \psi = \alpha, e \end{aligned}\right\}. \tag{4}$$

The fluctuations in the burden and transfer rates express variations among individuals. Exposure fluctuations represent natural variabilities. The substitution of Eqs. (4) into Eqs. (1) and (2) leads to Eq. (3.10) for the fluctuations $b(t)$, and to the following equations for the burden mean and covariance (fluctuation products are neglected as usually)

$$D_t \overline{B}(t) = \overline{\lambda_{\alpha}}(t)\overline{E}(t), \tag{5}$$

where the time derivative now is $D_t = [d/dt + \overline{\lambda_e}(t)]$ and

$$z(t) = e(t)\overline{\lambda_{\alpha}}(t) + \ell_{\alpha}(t)\overline{E}(t) - \ell_e(t)\overline{B}(t). \tag{6}$$

The covariance of the zero mean z-process is

$$c_z(t,t') = \overline{\lambda_{\alpha}}(t)\overline{\lambda_{\alpha}}(t')c_e(t,t') + \overline{E}(t)\overline{E}(t')c_{\ell_{\alpha}}(t,t') + \overline{B}(t)\overline{B}(t')c_{\ell_e}(t,t'), \tag{7}$$

and it incorporates information about the temporal trends and random fluctuations of exposure, transfer rates and burden. The Eq. (5) is solved explicitly for the burden mean $\overline{B}(t)$, [see Eq. (3.13)] which is then substituted into Eq. (7) to obtain the covariance of the z-process. The cross-covariance of the burden fluctuation with the z-process is

$$D_{t'} c_{zb}(t,t') = c_z(t,t'), \tag{8}$$

which can be solved with the appropriate ICs to give $c_{zb}(t,t')$. Then, the burden covariance $c_b(t,t')$ can be found from

$$D_t c_b(t,t') = c_{zb}(t,t'). \tag{9}$$

To obtain some specific analytical expressions, consider the following situation: The initial burden is zero $B(0)=0$; the mean absorption and elimination rates as well as the mean exposure are constant: $\overline{\lambda_e}(t)=\overline{\lambda_e}$, $\overline{\lambda_\alpha}(t)=\overline{\lambda_\alpha}$ and $\overline{E}(t)=\overline{E}$; the elimination and absorption rate covariances are white noise random processes: $c_{\ell_e}(t,t')=\sigma^2_{\ell_e}\delta(t-t')$ and $c_{\ell_\alpha}(t,t')=\sigma^2_{\ell_\alpha}\delta(t-t')$; and finally that the exposure fluctuations are a Wiener process covariance with covariance $c_e(t,t')=\sigma_e^2\min(t,t')$. Then, the covariance of $z(t)$ is

$$c_z(t,t')=[\overline{E}^2\sigma^2_{\ell_\alpha}+\overline{B}(t)\overline{B}(t')\sigma^2_{\ell_e}]\delta(t-t')+\overline{\lambda_\alpha}^{-2}\sigma_e^2\min(t,t'). \tag{10}$$

For this specific covariance, the solution of Eq. (5) for the mean burden is

$$\overline{B}(t)=\overline{\lambda_\alpha}\,\overline{\lambda_e}^{-1}\,\overline{E}\,[1-\exp(-\overline{\lambda}_e t)]. \tag{11}$$

By integrating Eq. (8) we find that the cross-covariance $c_{zb}(t,t')$ is given by

$$c_{zb}(t,t')=\begin{cases}\overline{\lambda_\alpha}^{-2}\sigma_e^2\, t(1-e^{-\overline{\lambda}_e t'})/\overline{\lambda_e}+e^{-\overline{\lambda}_e(t-t')}[\overline{E}^2\sigma^2_{\ell_a}+\sigma^2_{\ell_e}\,\overline{B}^2(t)], & t\le t'\\ \overline{\lambda_\alpha}^{-2}\sigma_e^2(\overline{\lambda}_e t'+e^{-\overline{\lambda}_e t'}-1)/\overline{\lambda_e}^{3} & t>t'\end{cases}. \tag{12}$$

Next, integrating Eq. (9) over time and in light of Eq. (12) we obtain the following expression for the burden covariance

$$c_b(t,t')=\overline{\lambda_\alpha}^{-2}\sigma_e^2(1-e^{-\overline{\lambda}_e t})(\overline{\lambda}_e t'+e^{-\overline{\lambda}_e t'}-1)/\overline{\lambda_e}^{3} \qquad t>t'. \tag{13}$$

and

$$\begin{aligned}c_b(t,t')=\,&\overline{\lambda_\alpha}^{-2}\sigma_e^2(1-e^{-\overline{\lambda}_e t'})(\overline{\lambda}_e t+e^{-\overline{\lambda}_e t}-1)/\overline{\lambda_e}^{3}+\\ &\overline{\lambda_\alpha}^{-2}\overline{E}^2[e^{-\overline{\lambda}_e(t'-t)}-e^{-\overline{\lambda}_e(t'+t)}]/2\overline{\lambda_e}+\\ &\overline{\lambda_\alpha}^{-2}\sigma^2_{\ell_e}\overline{E}^2[e^{-\overline{\lambda}_e(t'-t)}+(2\overline{\lambda}_e t+3)e^{-\overline{\lambda}_e(t'+t)}-4e^{-\overline{\lambda}_e t'}]/2\overline{\lambda_e}^{3}\end{aligned}, \qquad t\le t' \tag{14}$$

Eqs. (13) and (14) express the burden correlation structure in terms of the exposure correlation and the biologic characteristics of the receptor. Note that at $t=t'$ the burden covariance has a discontinuity proportional to $\overline{E}^2$ which is due to the white noise structure of the exposure correlation. The burden variance is given by

$$
\begin{aligned}
\sigma_b^2(t) = & \frac{\overline{\lambda_\alpha}^2 \sigma_e^2}{\overline{\lambda}_e^3}(1-e^{-\overline{\lambda}_e t})(\overline{\lambda}_e t + e^{-\overline{\lambda}_e t} - 1) + \frac{\overline{\lambda_\alpha}^2 \overline{E}^2}{2\overline{\lambda}_e}(1-e^{-2\overline{\lambda}_e t}) \\
& \frac{\overline{\lambda_\alpha}^2 \sigma_{\ell_e}^2 \overline{E}^2}{2\overline{\lambda}_e^3}[1+(2\overline{\lambda}_e t + 3)e^{-2\overline{\lambda}_e t} - 4e^{-\overline{\lambda}_e t}]
\end{aligned}
\tag{15}
$$

Eq. (15) expresses the burden variability in terms of the exposure uncertainty and the biologic variations. Note that some of the burden statistics can be estimated, even if the burden pdf is not completely known. □

6. SOME GENERALIZATIONS

In the stochastic analysis of §§3 and 5 we have assumed for simplicity that fluctuation products are insignificant. It must be emphasized, however, that this is not necessary, and explicit stochastic expressions for the burden can be obtained even if such products are not ignored. One possible stochastic formulation is suggested in the following example.

EXAMPLE 2: Consider the pollutokinetic Eq. (3.2), and assume that the uptake rate $U_r(t)$ and the transfer rate $\lambda(t)$ are statistically uncorrelated. The realization-based solution of Eq. (3.2) is given by Eq. (3.3). In the following we assume that $B(0)=0$, and that the transfer rate fluctuations follow a normal distribution. By averaging over the fluctuations in Eq. (3.3), the mean and the covariance of the burden can be expressed explicitly in terms of the uptake and transfer rate statistics as follows

$$
\overline{B}(t) = \int_0^t dt' \overline{U_r}(t') \exp[g(t,t')], \tag{16}
$$

and

$$
C_b(t,t') = \overline{B(t)B(t')} = \int_0^t \int_0^{t'} d\tau d\tau' \overline{U_r(\tau) U_r(\tau')} \exp[h(t,t',\tau,\tau')], \tag{17}
$$

where

$$
\begin{aligned}
g(t,t') &= -\overline{(\zeta_t - \zeta_{t'})} + \tfrac{1}{2} Var[(\zeta_t - \zeta_{t'})] \\
&= -\overline{\zeta_t} + \overline{\zeta_{t'}} + \tfrac{1}{2}[\sigma_\zeta^2(t) + \sigma_\zeta^2(t') - 2c_\zeta(t,t')] ,
\end{aligned}
\tag{18}
$$

and

$$h(t,t';\tau,\tau') = -\overline{(\zeta_t + \zeta_{t'} - \zeta_\tau - \zeta_{\tau'})} + \tfrac{1}{2} Var[(\zeta_t + \zeta_{t'} - \zeta_\tau - \zeta_{\tau'})]$$
$$= -\overline{\zeta_t} - \overline{\zeta_{t'}} + \overline{\zeta_\tau} + \overline{\zeta_{\tau'}} + \tfrac{1}{2}[\sigma_\zeta^2(t) + \sigma_\zeta^2(t') + \sigma_\zeta^2(\tau) + \sigma_\zeta^2(\tau') + 2c_\zeta(t,t')$$
$$-2c_\zeta(t,\tau) - 2c_\zeta(t,\tau') - 2c_\zeta(t',\tau) - 2c_\zeta(t',\tau') + 2c_\zeta(\tau,\tau')]. \tag{19}$$

In Eqs. (18) and (19) we use the mean and the covariance function of the integrated transfer process ζ_t; these functions are given by the following integrals

$$\overline{\zeta_t} = \int_0^t d\tau \overline{\lambda}(\tau), \tag{20}$$

and

$$c_\zeta(t,t') = \int_0^t \int_0^{t'} d\tau d\tau' c_\ell(\tau,\tau'). \tag{21}$$

In light of Eq. (21), the variance of the process ζ_t is $\sigma_\zeta^2(t) = c_\zeta(t,t)$. The assumption of uncorrelated $E(t)$ and $\lambda(t)$ can also be relaxed, provided that the cross-covariance between the exposure and the transfer rate processes are known. □

REMARK 1: As a matter of fact, the pollutokinetics equations are very similar to the 1-D flow equation for the hydraulic gradient in a medium with constant gradient in the hydraulic log-conductivity. Explicit expressions for the mean and the covariance of the hydraulic gradient have been obtained in (Christakos *et al.*, 1993b, c; 1995).

Chapter IX: STOCHASTIC EXPOSURE AND HEALTH INDICATORS

"We accept a truth, only after we have initially rejected it from the bottom of our heart".
P. Coelho

1. INTRODUCTION

As we saw in the previous Chapters, environmental models provide pollutant concentrations at any point in space-time, but they do not evaluate the actual effects of these pollutants on human health; this is the goal of human exposure analysis. *Human exposure analysis* involves the study of environmental exposure and its influence on the state of human health. Environmental exposure and health effect are integrated to form a *holistic* system (§I.1 and §II.1), which is of great importance in environmental health risk assessment. The latter focuses on the characterization of adverse health effects on humans of chemical agents and other hazardous substances in the environment. The concept of a holistic environmental health system is, also, supported by politicians and legislators as the most efficient way for ensuring sustainable development. The European Community, e.g., has recently issued a directive on *integrated pollution prevention* and *control (IPPC),* which aims to formulate an integrated framework for industrial activities (Papameletiou, 1995) based on the promotion of *Best Available Techniques* (*BAT*).

An environmental health risk study can be viewed as a six-stage procedure (see, also, §I.7):

(a) *Hazard identification*, which studies whether exposure to a hazardous substance causes a certain disease (§II.2). Does a decrease in the ozone layer, e.g., cause skin cancer? Or, what are the effects of acidic deposition on human health?

(b) *Exposure assessment*, which refers to the contact between a receptor (human organ, skin, etc.) with a pollutant. The exposure to hazards is measured within the space-time domain of interest (§II.3).

(c) *Dose assessment*, which generally predicts chemical concentrations or metabolites in the human organs and tissues resulting from exposure to pollutants. This stage involves a set of *biomarkers* (§II.4 and Chapter VIII).

(d) Determination of a *dose-response relationship*, in which "response" refers to the adverse health effects (disease, death, etc.) caused by the dose (§II.3).

(e) *Health risk assessment*, which involves an evaluation of the state of human health by means of a set of *indicators* evaluating the population exposure and health damage in space-time and providing quantitative measures of risk. The results of stages (b)-(d) above should be rigorously incorporated into the analysis of the present stage.

(f) *Health risk management*: While risk assessment requires scientific and technical expertise, risk management is a policy-oriented as well as a value-oriented activity. Its goals are to evaluate alternative actions for prevention or reduction of unacceptable risks, and to select the best health management approach in view of the available technical knowledge as well as the applicable and relevant state and federal laws and regulations.

Since the basic concepts of hazard identification, exposure, dose and health effect [stages (a)-(c)] were discussed in Chapters II-VIII, in this Chapter we focus mainly on stages (d) and (e). Our intention is to provide a rigorous quantitative formulation and, hence, we refer the reader to the existing literature for a more detailed presentation of the qualitative aspects of human exposure (e.g., Botkin and Keller, 1998). Also, in this book we will not study health risk management (f), which involves important political, economic and social aspects of decision making (CEQ, 1986; Morris, 1990; Mayo and Hollander, 1991).

A predominant feature of environmental health risk analysis is the significant amount of variation and uncertainty inherent in each stage, as well as in the interactions between different stages. Ozone concentration varies considerably over the Eastern U.S. during different time periods; and indoor radon concentration can be larger in the basement than in upper floors; these are just two examples of the several environmental processes discussed in the preceding sections. The salient point here is that any meaningful environmental health risk analysis should be probabilistic in nature. This means that exposure and health effect should be evaluated quantitatively in terms of *stochastic indicators*. Of considerable importance is (a) the derivation of stochastic indicators of human exposure assessment that take into consideration physical space-time variations, and (b) the development of techniques that integrate pollution with dose assessment and health response, and also account for uncertainties in exposure distribution and biological variability.

Exposure and health indicators, as well as the relationships between them, are adequately represented in terms of the theory of *spatiotemporal random fields (S/TRF)* developed in the previous Chapters.

Scholium 1: *An important advantage of the S/TRF-based human exposure indicators over the indicators of classical statistics is that the former rigorously take into consideration composite space-time exposure and biological variabilities.*

This is indeed a considerable improvement over classical statistics indicators which are based on the assumption of independent and identically distributed (iid) random variables (e.g., Reiss and Thomas, 1997). The iid assumption is not realistic, because environmental exposure and health effect exhibit significant spatiotemporal correlations and they involve problems that have common origins. Groundwater contaminant concentrations, e.g., show space-time variations that depend on the location of the source, flow direction, time interval and local permeability fluctuations. Intake, effect and susceptibility biomarkers exhibit significant intersubject and intrasubject variabilities.

Stochastic exposure indicators are discussed in §2, including functions quantifying extreme exposure conditions, connectivity (correlation) measures of space-time pollutant distributions, and statistical measures of level-crossing contours. These indicators provide a rigorous foundation for measuring the effectiveness of technological advances in decreasing the environmental impact on human health. Exposure indicators, e.g., may provide the basis for a quantitative assessment of the uncertainties, risks and costs faced by health administration and management. Estimates of cleanup costs can be directly formulated in terms of exposure indicators which, thus, become significant tools for the effective implementation of remedial strategies.

Health indicators generally provide a quantitative measure of exposure effects on human populations. Building on the fundamental biomarkers and population health indicators of Chapter II, §3 introduces cell-based stochastic indicators of health effects. Traditional population health indicators such as incidence, prevalence, and mortality ratios are revisited in §4 in the light of S/TRF analysis. Health and exposure indicators are closely related and offer powerful tools for the analysis of causal links between the natural environment and human health. Relationships between exposure and health indicators are studied in §§5-7. Some examples are briefly presented below to whet your interest.

EXAMPLE 1: The cleanup of hazardous waste sites has proved to be a formidably difficult task, largely due to the natural heterogeneities in the space-time distribution of contaminants (e.g., Loaiciga, 1989; Christakos and Killam, 1993; Asante-Duah, 1996). Material properties affecting contaminant movement in the subsurface --such as permeability, porosity and geochemical conditions-- can vary significantly over short distances. Hence, rigorous contaminant level analysis should take into consideration the heterogeneities and uncertainties that characterize the contaminant processes. □

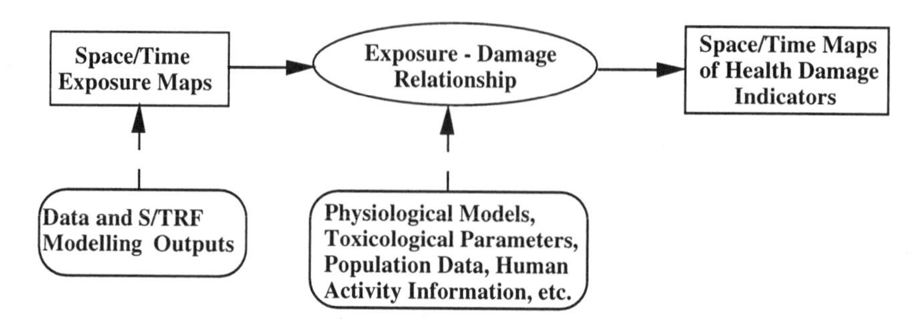

FIGURE 1: An environmental exposure-health damage approach.

EXAMPLE 2: Space-time exposure maps serve as input to exposure-health effect relations and the outputs are health damage indicator maps (Fig. 1). This approach leads to practical advances, such as techniques for mapping environmental impact on populations. At the same time, theoretical advances are made by developing new basic tools to better analyze and understand exposure and health data. □

EXAMPLE 3: *Spatiotemporal information systems* (*S/TIS*) provide valuable tools for epidemiologic data analysis and health management purposes by integrating exposure and health information in the form of maps. Numerous disease case and control residences can be incorporated into an environmental S/TIS with medical and personal information obtained by field interviews. This information is then used to produce disease maps, which can reveal disease hotspots and potential links to the environmental features of the area (Christakos and Lai, 1997; Christakos, 1998a). The approach can help to identify the mechanisms of hotspot incidence and to visualize potential relationships between environment and disease risk. □

2. STOCHASTIC EXPOSURE INDICATORS

The U.S. EPA and other environmental agencies all over the world have documented that a wide spectrum of hazardous substances exist in the environment in alarming quantities (e.g., Shields, 1990). Despite significant progress, an accurate assessment of the relation between exposure and health effects are still very difficult to make. Some of the difficulties involved in such an effort are summarized in Table 1.

TABLE 1: Difficulties faced by exposure-health effect studies

- Lack of knowledge bases (limited data sets, inadequate understanding of physical-chemical-biological processes, etc.).
- Significant uncertainties involved at all stages of the exposure-health effect continuum.
- Poor modelling of spatiotemporal exposure and biological variabilities.
- Multi-media exposures and exposures acting in synergy.
- Confounding variables.
- Hazard identification.
- Inadequate monitoring procedures.
- Exposed organisms are removed from polluted areas.
- Latency periods involved.
- Delayed response in applying the appropriate environmental health procedures and measures.
- Financial restrictions and political issues.

S/TRF analysis offers powerful theoretical background and techniques for improving the situation in certain of the areas identified in Table 1. The basic S/TRF theory and methods were discussed in the preceding Chapters. Applications of S/TRF analysis in human exposure assessment are discussed in the following sections.

2.1 The Exposure S/TRF-Pair

Humans are exposed to chemicals and other hazardous substances primarily through air and water. The impact of chemicals on the atmospheric compartment is classified under two main categories according to scale: (i) *Macro* effects occur when the introduction of chemicals leads to large scale disruption of natural patterns, as is the case for global warming and stratospheric ozone depletion. (ii) *Micro* effects occur at a local scale, usually near the pollution source.

EXAMPLE 1: Local weather patterns can lead to high concentrations of pollutants. In many urban areas, when the pollutant concentrations exceed the maximum allowed thresholds, policy measures that aim to reduce emission of pollutants (e.g., constraints on traffic patterns) are automatically put into effect. Unlike the atmosphere, in the case of drinking water micro effects are more important. This difference is due to the constrained distribution and movement of water in the hydrosphere --with the exception of oceans-- that leads primarily to local effects. □

The goal of exposure analysis is to obtain simplified models of complex environmental processes that allow predictions. Meaningful simplifications should account for variabilities in the spatiotemporal exposure distributions (intrasubject variations in individual exposure, and intersubject exposure variations for different individuals), uncertainties in the available data, and natural laws. S/TRFs play a central role in all stages of environmental risk assessment, which includes collecting, analyzing and communicating information for use in health risk management and policy making.

Consider a space-time domain $\Omega = D \times T$ where exposure to a pollutant takes place (D is the spatial domain and T is the time interval of interest), and let $E(\boldsymbol{p})$, $\boldsymbol{p} = (\boldsymbol{s}, t) \in \Omega$, be an S/TRF modelling the exposure. We associate with the exposure field $E(\boldsymbol{p})$ the binary-valued S/TRF

$$I_E(\boldsymbol{p}, \zeta) = \begin{cases} 1 & \text{if exposure } E(\boldsymbol{p}) \geq \zeta \\ 0 & \text{if exposure } E(\boldsymbol{p}) < \zeta \end{cases}, \tag{1}$$

where ζ is a threshold chosen on the basis of environmental and health requirements; possible forms of $E(\boldsymbol{p})$ were discussed in Chapter II. Environmental exposure analysis can be performed by means of an appropriate combination of the two "generic" random fields $E(\boldsymbol{p})$ and $I_E(\boldsymbol{p}, \zeta)$. This idea gives rise to the following fundamental definition.

Definition 1: *An S/TRF-pair is expressed as*

$$\{E(\boldsymbol{p}), I_E(\boldsymbol{p}, \zeta)\}, \tag{2}$$

where the stochastic characteristics of the component S/TRFs $E(\boldsymbol{p})$ *and* $I_E(\boldsymbol{p}, \zeta)$ *are mutually dependent.*

The S/TRF-pair (1) plays a central role in the study of exposures exceeding certain threshold levels.

EXAMPLE 2: A waste site can be modelled in terms of (2), where $E(\boldsymbol{p})$ represents the contaminant concentration at point $\boldsymbol{p}$ in a space-time domain Ω and $I_E(\boldsymbol{p},\zeta)$ identifies points of *excess* exposure. □

While the exposure field $E(\boldsymbol{p})$ is generally a continuous function, the excess exposure field $I_E(\boldsymbol{p},\zeta)$ is a discrete function. The mutual dependence of $E(\boldsymbol{p})$ and $I_E(\boldsymbol{p},\zeta)$ is an important feature of model (2) that is expressed in various ways, like,

$$I_E(\boldsymbol{p},\zeta) = \mathrm{P}[E(\boldsymbol{p}) \geq \zeta | \varepsilon(\boldsymbol{p})] \tag{3}$$

(i.e., =1 if $\varepsilon(\boldsymbol{p}) \geq \zeta$, =0 otherwise; $\varepsilon(\boldsymbol{p})$ is a realization of the exposure S/TRF),

$$E(\boldsymbol{p}) \geq \zeta\, I_E(\boldsymbol{p},\zeta); \tag{4}$$

and

$$E(\boldsymbol{p}) = \int_0^\infty [1 - I_E(\boldsymbol{p},\zeta)]\mathrm{d}\zeta \tag{5}$$

for all $\boldsymbol{p} \in D$. Also, the equality $E(\boldsymbol{p}) = E(\boldsymbol{p}')$ implies $I_E(\boldsymbol{p},\zeta) = I_E(\boldsymbol{p}',\zeta')$, but the converse is not true. This property reflects the fact that $I_E(\boldsymbol{p},\zeta)$ is uniquely determined from $E(\boldsymbol{p})$, but not vice versa. Hence, the $I_E(\boldsymbol{p},\zeta)$ and its statistics can be expressed in terms of probability functions of $E(\boldsymbol{p})$, and this has significant consequences in exposure analysis. Another interesting consequence of the mutual dependence of the components of the S/TRF-pair (2) is that if the $E(\boldsymbol{p})$ is stochastically homogeneous-stationary, the $I_E(\boldsymbol{p},\zeta)$ is homogeneous-stationary, as well. The complementary *deficit exposure* field $I_E^c(\boldsymbol{p},\zeta)$ is defined as $I_E^c(\boldsymbol{p},\zeta) = 1 - I_E(\boldsymbol{p},\zeta)$, which represents exposures below the threshold value ζ.

2.2 The Stochastic Exposure Indicator Concept

Exposure indicators are defined in terms of the S/TRF-pair (2), which provide a rigorous characterization of exposure levels for heterogeneous pollutant distributions. They also offer valuable information for a cost-effective waste site cleanup analysis, health management, etc.

Various stochastic exposure indicators can be established by means of the following general expression

$$\mathcal{Y}_E = \mathcal{F}[E, I_E, \zeta], \tag{6}$$

where $\mathcal{F}[\cdot]$ is a functional involving the quantity and the domain of extreme exposure as well as the permissible threshold. Specific examples of Eq. (6) will be given below. Generally, stochastic indicators can be *one-point* or *multi-point* functions. While the former evaluate point or local averages of extreme values, the latter provide a measure of their connectivity in space-time. In certain cases, the exposure threshold also involves some degree of *uncertainty*, and it can be considered as a random variable z (§2.5).

In practical applications, stochastic exposure indicators are estimated based on the available knowledge bases $\mathcal{K}$ (data, models, etc.; §I.5). Functional relationships between indicators are useful, because they allow direct evaluation of other indicators based on few that are initially estimated from the data. The relative importance and usefulness of exposure indicators depends on the exposure characterization problem under consideration.

2.3 One-Point Stochastic Exposure Indicators

One-point exposure indicators are useful tools for arriving at optimal remediation and health management decisions. These indicators are functions of the threshold level ζ which is subject to change based on scientific evidence, available information, environmental policy, and public awareness of health or ecological risks. If the exposure S/TRF has a continuous probability law, the expectations of exposure indicators are also continuous functions in ζ.

There are various ways to present the one-point exposure indicators. In theory, these indicators can be defined in terms of the pdf of the exposure field (Christakos and Hristopulos, 1996a and b). A disadvantage of this approach is that in practice the pdf may not be available. A different approach is to start with a fundamental one-point exposure indicator, based on which other useful indicators can be derived using a set of constitutive equations. Here we will use the second approach. The fundamental one-point exposure indicator defined below is justified by the fact that in many situations the assessment of extreme exposure values can be of outstanding importance, since these values present higher health risk.

Definition 2: *The expected excess differential exposure at point* $\boldsymbol{p}$ *is given by*

$$L_E(\boldsymbol{p},\zeta) = \overline{[E(\boldsymbol{p}) - \zeta] I_E(\boldsymbol{p},\zeta)} \, . \tag{7}$$

Eq. (7) is a convex and decreasing function in ζ; its range is between $L_E(\boldsymbol{p},0) = \overline{E(\boldsymbol{p})} = m_E$ and $L_E(\boldsymbol{p},\infty) = 0$. It can be also written as $L_E(\boldsymbol{p},\zeta) = \overline{\max[E(\boldsymbol{p}) - \zeta, 0]}$, which has the form of a linear cost function [i.e. the expected cost is zero if $E(\boldsymbol{p})$ is less than ζ, and it is equal to the expected difference between exposure and threshold if $E(\boldsymbol{p})$ exceeds ζ]. The linear cost function plays an important role in certain site cleanup optimization problems (Christakos and Killam, 1993).

The *constitutive equations* relate L_E to some other useful one-point exposure indicators as follows (Christakos and Hristopulos, 1998; here we drop the $\boldsymbol{p}$ and ζ, for simplicity)

$$P_E = R_E U_E, \tag{8}$$

$$L_E = R_E (U_E - \zeta), \tag{9}$$

$$dL_E / d\zeta = -R_E, \tag{10}$$

$$dP_E / d\zeta = \zeta \, dR_E / d\zeta, \tag{11}$$

$$O_E = R_E / (1 - R_E). \tag{12}$$

Given L_E, all other indicators can be found from the above equations. These equations also establish restrictions on the relative changes of the exposure indicator values.

Let us now illustrate the physical meaning of the exposure indicators above. The $R_E = R_E(\boldsymbol{p},\zeta)$ is, simply, the *expected exposure indicator* at the space-time point $\boldsymbol{p}$, i.e.,

$$R_E(\boldsymbol{p},\zeta) = \overline{I_E(\boldsymbol{p},\zeta)}, \tag{13}$$

which is a decreasing continuous function in ζ with values between 1 and 0. The R_E is related to the exposure cdf, i.e., $R_E(\boldsymbol{p},\zeta) = 1 - F_E(\boldsymbol{p},\zeta)$, but it acquires a specific physical interpretation in the context of exposure characterization.

EXAMPLE 3: Consider a homogeneous exposure distribution with $R_E(\boldsymbol{p},\zeta) = R_E(t,\zeta)$ in a space-time domain $\Omega = D \times T$ (Fig. 2; $D \subset R^2$ is the spatial region and T is the time period of interest). If the exposure S/TRF is also ergodic, $R_E(t,\zeta)$ is equal to the over-exposed fraction of the spatial domain at time t, i.e., $R_E(t,\zeta) = |\Theta_D(t,\zeta)| / |D|$, where

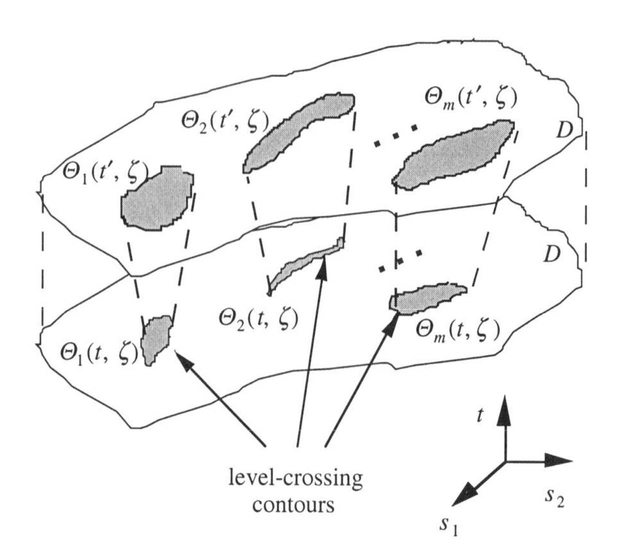

FIGURE 2: An example of R_E-calculation in $\Omega = D \times T \subset R^2 \times T$.

$|\Theta_D(t,\zeta)| = \bigcup_{i=1}^{m} |\Theta_i(t, \zeta)|$ at any t is the area of the over-exposed subdomain $\Theta_D(t,\zeta) \subset D$ (also, §2.7 below). Moreover, $L_E(\boldsymbol{p},\zeta) = L_E(t,\zeta)$, which provides an assessment of the expected excess differential exposure in $\Theta_D(t,\zeta)$. □

The $P_E = P_E(\boldsymbol{p},\zeta)$ is the *expected excess exposure* above the threshold ζ at point $\boldsymbol{p}$, i.e.,

$$P_E(\boldsymbol{p},\zeta) = \overline{E(\boldsymbol{p}) I_E(\boldsymbol{p},\zeta)}. \tag{14}$$

P_E is a measure of the average exposure at points where the threshold is exceeded. For each space-time point $\boldsymbol{p}$, P_E is a positive and decreasing function in ζ. Its values range between mean exposure m_E at zero threshold and zero at maximum threshold.

EXAMPLE 4: In the stochastically homogeneous case, $P_E(\boldsymbol{p},\zeta) = P_E(t,\zeta)$ provides an assessment of the expected exposure in the over-exposed subdomain $\Theta_D(t,\zeta)$. □

The $U_E = U_E(\boldsymbol{p},\zeta)$ is the *conditional mean excess exposure* over a threshold ζ, i.e.,

$$U_E(\boldsymbol{p},\zeta) = \overline{E(\boldsymbol{p}) | E(\boldsymbol{p}) \geq \zeta}. \tag{15}$$

U_E is an increasing function in ζ that involves only the over-exposed area of the domain Ω. The $O_E = O_E(\boldsymbol{p},\zeta)$ is the *exposure exceedance odds* indicator which expresses the probability ratio of excess over below-threshold exposure at point $\boldsymbol{p}$. The range of O_E is 0 to ∞. In addition to their useful physical meaning, some interesting mathematical results are valid for the one-point indicators.

Proposition 1 (Christakos and Hristopulos, 1998): *Consider the exposures $E_1(\boldsymbol{p})$ and $E_2(\boldsymbol{p})$ with univariate cdf F_{E_1} and F_{E_2}, respectively, mean values m_{E_1} and m_{E_2}, and bivariate cdf $F_{E_1E_2}$. (a) It is valid that*

$$F_{E_2}(\zeta) \leq F_{E_1}(\zeta) \text{ if and only if } R_{E_1}(\boldsymbol{p},\zeta) \leq R_{E_2}(\boldsymbol{p},\zeta); \tag{16}$$

and (b) if $m_{E_1} = m_{E_2} = m$, then

$$\overline{E_1|E_2} = E_2 \text{ if and only if } L_{E_2}(\boldsymbol{p},\zeta) \leq L_{E_1}(\boldsymbol{p},\zeta) \tag{17}$$

for all $\zeta \geq 0$.

The one-point exposure indicators above are not the whole story. The mathematically rich S/TRF theory provides the means to derive new indicators, depending on the requirements of the specific problem.

The one-point indicators provide *local* exposure measures. They depend on the space-time point $\boldsymbol{p}$ and can be defined as expected values of the S/TRF-pair (2) with respect to the exposure distribution and the threshold level. In the case of a homogeneous-stationary exposure distribution, the one-point indicators evaluate *global* averages of extreme exposure values. Stochastic averages are equal to the sample averages (i.e., the average exposure over the space-time domain) if the exposure S/TRF is ergodic (§2.7 below).

2.4 Two-Point Stochastic Exposure Indicators

The one-point stochastic indicators provide no information regarding the spatiotemporal correlation (connectivity) of exposure. The latter is established by means of two-point exposure indicators. We first consider two-point stochastic indicators that evaluate the correlations of extreme exposure concentrations in space-time.

Definition 3: *The non-centered exposure indicator covariance between the space-time points* $\boldsymbol{p}$ *and* $\boldsymbol{p}'$ *is defined as*

$$C_I(\boldsymbol{p},\boldsymbol{p}';\zeta) = \overline{I_E(\boldsymbol{p},\zeta) I_E(\boldsymbol{p}',\zeta)}. \tag{18}$$

C_I may be viewed as a risk indicator expressing the probability that exposure reaches or exceeds ζ concurrently at the space-time points $\boldsymbol{p}$ and $\boldsymbol{p}'$. If the exposure distribution is homogeneous-stationary, the C_I depends only on the spatial $\boldsymbol{r}$ and time τ lags, i.e., $C_I(\boldsymbol{p},\boldsymbol{p}';\zeta) = C_I(\boldsymbol{r},\tau;\zeta)$. The *centered indicator covariance* is given by

$$c_I(\boldsymbol{p},\boldsymbol{p}';\zeta) = C_I(\boldsymbol{p},\boldsymbol{p}';\zeta) - R_E(\boldsymbol{p};\zeta) R_E(\boldsymbol{p}';\zeta), \tag{19}$$

which expresses spatiotemporal connectivity between exposure values at $\boldsymbol{p}$ and $\boldsymbol{p}'$. For homogeneous-stationary exposure distributions, $c_I(\boldsymbol{r},\tau;\zeta) = C_I(\boldsymbol{r},\tau;\zeta) - R_E^2(\zeta)$. Then, $c_I(\boldsymbol{r},\tau;\zeta)$ represents fluctuation correlations about $R_E(\zeta)$. Furthermore, the *exposure indicator semi-variogram* is defined as

$$\gamma_I(\boldsymbol{p},\boldsymbol{p}';\zeta) = \overline{\tfrac{1}{2}[I_E(\boldsymbol{p},\zeta) - I_E(\boldsymbol{p}',\zeta)]^2}. \tag{20}$$

The indicator semi-variogram contains the same connectivity information about exposures at points $\boldsymbol{p}$ and $\boldsymbol{p}'$ as the covariance functions; however, it is easier to estimate the semi-variogram from data. The semi-variogram increases as the connectivity between the exposures at points $\boldsymbol{p}$ and $\boldsymbol{p}'$ decreases. In the homogeneous-stationary case, the indicator semi-variogram is given by

$$\gamma_I(\boldsymbol{r},\tau;\zeta) = R_E(\zeta) - c_I(\boldsymbol{r},\tau;\zeta). \tag{21}$$

Physically, Eq. (21) represents the difference between the probability that exposure exceeds the threshold at a single point, and the probability that it exceeds the threshold at two points separated by the lags $\boldsymbol{r}$ and τ. Next, we define three additional two-point exposure indicators which involve both $E(\boldsymbol{p})$ and $I_E(\boldsymbol{p},\zeta)$.

Definition 4: *The excess exposure covariance is defined as*

$$P_{EE}(\boldsymbol{p},\boldsymbol{p}';\zeta) = \overline{E(\boldsymbol{p})E(\boldsymbol{p}') I_E(\boldsymbol{p},\zeta) I_E(\boldsymbol{p}',\zeta)}; \tag{22}$$

Eq. (22) measures exposure connectivity in space-time. The *excess differential exposure covariance* is given by

$$L_{EE}(\boldsymbol{p},\boldsymbol{p}';\zeta)=\overline{[E(\boldsymbol{p})-\zeta][E(\boldsymbol{p}')-\zeta]I_E(\boldsymbol{p},\zeta)I_E(\boldsymbol{p}',\zeta)}; \tag{23}$$

and the *conditional excess exposure covariance* is

$$U_{EE}(\boldsymbol{p},\boldsymbol{p}';\zeta)=\overline{E(\boldsymbol{p})E(\boldsymbol{p}')|E(\boldsymbol{p}),E(\boldsymbol{p}')\geq\zeta}. \tag{24}$$

Eq. (23) assesses correlations between excess exposure values, and Eq. (24) is a marker of conditional exposure connectivity.

In the following we consider an exposure field $E(\boldsymbol{p})=E(s_1,s_2,t)$ in two spatial dimensions (i.e., $D\subset R^2$). The *level-crossing contours* are lines along which $E(\boldsymbol{p})=\zeta$ at a fixed time t. The *continuum specific length* $\ell_s(\zeta;t)$ of the ζ-crossing contours at time t is calculated by

$$\ell_s(\zeta;t)=|D|^{-1}\,\overline{\Pi_E}(\zeta;t), \tag{25}$$

where $\overline{\Pi_E}(\zeta;t)$ denotes the expectation of the total length of the ζ-crossing contours over the spatial domain D at time t. In the case of ergodic exposure,

$$\ell_s(\zeta;t)=\lim_{|D|\to\infty}|D|^{-1}\,\Pi_E(\zeta;t), \tag{26}$$

where $\Pi_E(\zeta;t)$ is the length of the level-crossing contours for a given sample of $E(\boldsymbol{p})$ at time t.

EXAMPLE 5: For an isotropic Gaussian S/TRF $E(\boldsymbol{p})$, it is valid

$$\ell_s(\zeta;t)=[2\,\lambda_E(t)]^{-1}\exp\{-[\zeta-m_E(\mathrm{t})]^2/2\sigma_E^2(t)\}; \tag{27}$$

where $m_E(t)=\overline{E(\boldsymbol{p})}$, $\sigma_E^2(t)=c_E(\boldsymbol{p},\boldsymbol{p})$, and the length $\lambda_E(t)$ is given by

$$\lambda_E(t)=\sqrt{-c_E(r,t)/c_E''(r,t)}\Big|_{r=0}, \tag{28}$$

where $c_E(r,t) \equiv c_E(\boldsymbol{s}+\boldsymbol{r},t;\boldsymbol{s},t)$, and $c''_E(r,t)$ denotes the 2nd-order derivative with respect to the space lag. The $\lambda_E(t)$ is proportional to the spatial correlation length of $E(\boldsymbol{p})$. Note that Eq. (27) is valid provided that $c'_E(r,t)=0$. This condition is not satisfied by the spherical covariance model used in geostatistics (Deutsch and Journel, 1992). In some cases, the exposure indicator covariance may include a nugget term. However, what matters in the specific length analysis is the slope at the origin rather than the nugget discontinuity. □

Simulations of exposure indicators provide the means for studying excess exposures. Simulation procedures generate realizations of exposure fields that honor the statistics of the data. Each realization can be conditioned on the data at the sampling points; at the same time, it provides possible representations of exposure at unobservable locations. Hence, simulation procedures are numerical tools that offer valuable information for risk assessment and remediation strategies. Both one-point and two-point indicators can be estimated from simulations (Christakos and Hristopulos, 1996a and b). Exposure S/TRFs are simulated on lattices (numerical grids). Continuous quantities are discretized, due to the fact that on a lattice the smallest length that can be resolved is the distance α between lattice sites (nodes). The level-crossing contours on a lattice, e.g., are approximated by the bonds that join neighbor sites with unequal indicator values; the $\ell_s(\zeta;t)$ is approximated by the *lattice specific length* $\ell_b(\zeta;t)$.

EXAMPLE 6: Consider a square lattice with length L and spacing α (in appropriate length units). The sites are labeled by their integer Cartesian coordinates so that $\boldsymbol{s}_{ij}$ represents a site with coordinates $(i\alpha, j\alpha)$, $i,j=1,\dots,N$. Let $\boldsymbol{s}_{ij}$ and $\boldsymbol{s}_{kl}$ represent neighbor lattice sites; the bond function is then defined by

$$T(\boldsymbol{s}_{ij},\boldsymbol{s}_{kl};\zeta) = \begin{cases} 1, & \text{if } I_E(\boldsymbol{s}_{ij},\zeta) + I_E(\boldsymbol{s}_{kl},\zeta) = 1 \\ 0, & \text{otherwise} \end{cases} . \tag{29}$$

The $\ell_b(\zeta)$ measures the expectation of the bond function over the entire lattice as follows

$$\ell_b(\zeta) = \alpha \sum_{i,j} \sum_{\sigma,\sigma'=\pm 1} \overline{[T(\boldsymbol{s}_{ij},\boldsymbol{s}_{i+\sigma,j};\zeta) + T(\boldsymbol{s}_{ij},\boldsymbol{s}_{i,j+\sigma'};\zeta)]} \big/ 2L^2 , \tag{30}$$

where the first sum is over all lattice sites and the second sum is over the nearest neighbors of $\boldsymbol{s}_{ij}$. For an isotropic and Gaussian exposure field, Eq. (30) is expressed as

$$\ell_b(\zeta) = 4\,\alpha^{-1} \int_{-\infty}^{\zeta - m_E} d\chi_2 \int_{\zeta - m_E}^{\infty} d\chi_1\, f_E(r = \alpha; \chi_1, \chi_2), \tag{31}$$

where

$$f_E(\chi_1, \chi_2; \boldsymbol{r}) = \frac{2\pi}{[c_E^2(0) - c_E^2(\boldsymbol{r})]^{1/2}} \exp\{-\frac{c_E(0)(\chi_1^2 + \chi_2^2) - 2c_E(\boldsymbol{r})\chi_1\chi_2]}{2[c_E^2(0) - c_E^2(\boldsymbol{r})]}\}, \tag{32}$$

the bivariate pdf of the exposure fluctuations. □

2.5 Uncertainty in the Exposure Threshold Level

In certain cases there is uncertainty regarding the permissible exposure threshold ζ. This uncertainty may be due to inconclusive scientific evidence, limited available information, or changes in environmental guidelines. The definitions of the stochastic exposure indicators above allow the flexibility of considering a range of threshold levels ζ. If the threshold is modelled as a random variable z two useful parameters, the representative threshold and the exposure indicator dispersion, can be defined as follows. Let $\mathcal{Y}_E$ denote a general exposure indicator as in Eq. (6) above. The *representative exposure threshold,* $\hat{\zeta}_E$, with respect to $\mathcal{Y}_E$ is found by solving the equation

$$\overline{\mathcal{Y}_E(z)} = \mathcal{Y}_E(\hat{\zeta}_E), \tag{33}$$

where the expectation is with respect to the pdf of z.

Eq. (33) determines $\hat{\zeta}_E$ as the threshold level for which the exposure indicator is equal to the mean exposure indicator value over all possible ζ-thresholds. The representative threshold $\hat{\zeta}_E$ depends on the functional form of the indicator $\mathcal{Y}_E$. Consider an exposure indicator $\mathcal{Y}_E$ that is a monotonically increasing function: if $\hat{\zeta}_E > \bar{z} = m_z$, the $\mathcal{Y}_E$ is concave; if $\hat{\zeta}_E < m_z$, the $\mathcal{Y}_E$ is convex; and if $\hat{\zeta}_E = m_z$, the $\mathcal{Y}_E$ has a linear form.

EXAMPLE 7: If $\mathcal{Y}_{E,1}(z) = z > 0$, then $\hat{\zeta}_{E,1} = m_z$; if $\mathcal{Y}_{E,2}(z) = \ell n\, z$, $\hat{\zeta}_{E,2} = \exp[\overline{\ell n\, z}]$; and if $\mathcal{Y}_{E,3}(z) = z^{-1}$, $\hat{\zeta}_{E,3} = [\overline{z^{-1}}]^{-1}$. Clearly, $\hat{\zeta}_{E,1} > \hat{\zeta}_{E,2} > \hat{\zeta}_{E,3}$. □

Using a Taylor series expansion it can be shown that a low order approximation of $\hat{\zeta}_E$ is given by

$$\hat{\zeta}_E = m_z + \sigma_z^2 \mathcal{Y}_E''(m_z)/2\mathcal{Y}_E'(m_z), \tag{34}$$

where $\mathcal{Y}_E'$ and $\mathcal{Y}_E''$ denote the first and the second order derivatives of $\mathcal{Y}_E$, respectively. As a result of Eq. (34), the $\hat{\zeta}_E$-value implies certain restrictions on the behavior of the exposure indicator $\mathcal{Y}_E$.

Definition 5: *The coefficient of exposure indicator dispersion is defined as*

$$\Psi_E(\boldsymbol{p}) = \overline{L_E(\boldsymbol{p},z)}/m_E(\boldsymbol{p}) \ , \tag{35}$$

where $\overline{L_E(\boldsymbol{p},z)}$ *is the average of* L_E *over the pdf of* z.

The Ψ_E is a measure of the ratio of (a) the average exposure for all possible threshold values ζ over the mean exposure and (b) excess exposure dispersion over the mean exposure. Clearly, higher Ψ_E-values indicate a higher level of exposure. The Ψ_E can be evaluated in terms of either the one-point L_E or the two-point (non-centered) indicator covariance by means of the following

Corollary 1 (Christakos and Hristopulos, 1998)**:** *It is valid that*

$$\Psi_E(\boldsymbol{p}) = \int_0^\infty C_I(\boldsymbol{0},\zeta)d\zeta \Big/ m_E(\boldsymbol{p}). \tag{36}$$

EXAMPLE 8: The Ψ_E's for the Gaussian and log-normal pdf's are functions of the coefficient of variation $\mu_E(\boldsymbol{p}) = \sigma_E(\boldsymbol{p})/m_E(\boldsymbol{p})$, i.e., $\Psi_E(\boldsymbol{p}) = \mu_E(\boldsymbol{p})/\sqrt{\pi}$ and $\Psi_E(\boldsymbol{p}) = 2\Phi_E[\mu_E(\boldsymbol{p})/\sqrt{2}]-1$, respectively; the exponential pdf is characterized by a constant $\Psi_E(\boldsymbol{p}) = 0.5$. □

The Ψ_E exists and is finite if and only if the mean exposure is $m_E(\boldsymbol{p}) = \overline{L_E(\boldsymbol{p},0)} < \infty$. Also, the corollary below is a direct consequence of the definitions above.

Corollary 2 (Christakos and Hristopulos, 1998)**:** *If* $L_{E_1}(\boldsymbol{p},\zeta) \geq L_{E_2}(\boldsymbol{p},\zeta)$ *for all* ζ, *then* $\Psi_{E_1}(\boldsymbol{p}) \geq \Psi_{E_2}(\boldsymbol{p})$.

A summary of the one- and two-point exposure indicators is given in Table 2. It is worth emphasizing that these stochastic exposure indicators are general tools that can be used to

TABLE 2: One- and Two-Point Exposure Indicators

Indicator	Definition
R_E	Expected exposure indicator (trend) at point $\boldsymbol{p}$
L_E	Expected excess differential exposure (trend) at point $\boldsymbol{p}$
P_E	Expected excess exposure (trend) at point $\boldsymbol{p}$
U_E	Conditional expected excess exposure (trend) at point $\boldsymbol{p}$
O_E	Exposure exceedance odds at point $\boldsymbol{p}$
C_I	Non-centered exposure indicator covariance (connectivity) between points $\boldsymbol{p}$ and $\boldsymbol{p}'$
c_I	Centered exposure indicator covariance (connectivity) between points $\boldsymbol{p}$ and $\boldsymbol{p}'$
γ_I	Exposure indicator semi-variogram (connectivity) between points $\boldsymbol{p}$ and $\boldsymbol{p}'$
P_{EE}	Non-centered excess exposure covariance (connectivity) between points $\boldsymbol{p}$ and $\boldsymbol{p}'$
L_{EE}	Excess differential exposure covariance (connectivity) between points $\boldsymbol{p}$ and $\boldsymbol{p}'$
U_{EE}	Conditional excess exposure covariance (connectivity) between points $\boldsymbol{p}$ and $\boldsymbol{p}'$
ℓ_s	Continuum specific length
ℓ_b	Lattice specific length
Ψ_E	Coefficient of exposure indicator dispersion

study human exposure at a macroscopic (waste site, geographical regions, etc.), as well as a microscopic (cellular) level (see later, §3).

2.6 Other Properties of Exposure Indicators

Relationships analogous to these between one-point exposure indicators can be derived for two-point indicators, as well.

EXAMPLE 9: It is true that

$$P_{EE} = C_I U_{EE}; \tag{37}$$

Eq. (37) states that, due to the mutual dependence property of the S/TRF pair (2), the 4th-order moment of $E(\boldsymbol{p})$ and $I_E(\boldsymbol{p},\zeta)$ is equal to the product of the 2nd-order moment of

$I_E(\boldsymbol{p},\zeta)$ and the conditional 2nd-order moment of $E(\boldsymbol{p})$ given that $I_E(\boldsymbol{p},\zeta)I_E(\boldsymbol{p}',\zeta)=1$. Also, for a spatially isotropic-temporally stationary Gaussian exposure, it is valid

$$L_{EE}(r,\tau;\zeta)=P_{EE}(r,\tau;\zeta)+\zeta^2 C_I(r,\tau;\zeta)-2\zeta\,\alpha(r,\tau,\zeta)P_{EE}(\zeta), \tag{38}$$

where $r=|\boldsymbol{r}|$ and the function $\alpha(r,\tau,\zeta)$ satisfies $\alpha(\boldsymbol{0},0;\zeta)=1$ and $\alpha(\infty,0;\zeta)=R_E(\zeta)$. □

In the case of homogeneous-stationary exposure, the $C_I(\boldsymbol{r},\tau;\zeta)$ and $c_I(\boldsymbol{r},\tau;\zeta)$ are expressed in terms of the $R_E(\zeta)$ in the limiting cases of 0 and ∞ lags; i.e.,

$$\left.\begin{array}{l} C_I(\boldsymbol{0},0;\zeta)=R_E(\zeta) \\ \lim_{r\to\infty} C_I(\boldsymbol{r},0;\zeta)=R_E^2(\zeta) \end{array}\right\}, \tag{39}$$

where $r=|\boldsymbol{r}|$, and

$$c_I(\boldsymbol{0},0;\zeta)=R_E(\zeta)-R_E^2(\zeta), \tag{40}$$

The $C_I(\boldsymbol{r},\tau;\zeta)$ and $O_{EE}(\boldsymbol{r},\tau;\zeta)$ are related to the covariance $C_{I'}(\boldsymbol{r},\tau;\zeta)$ of the complementary deficit random field $I_E'(\boldsymbol{p},\zeta)$ by

$$C_I(\boldsymbol{r},\tau;\zeta)=O_{EE}(\boldsymbol{r},\tau;\zeta)C_{I'}(\boldsymbol{r},\tau;\zeta). \tag{41}$$

The $\ell_s(\zeta)$ is related to the indicator covariance $C_I(\boldsymbol{r},\tau;\zeta)$ in $R^2\times T$ by

$$\lim_{r\to 0}\ \partial[\textstyle\int_0^{2\pi} C_I(\boldsymbol{r},t;\zeta)\mathrm{d}\phi]/\partial r=-2\,\ell_s(\zeta;t), \tag{42}$$

where $C_I(\boldsymbol{r},t;\zeta)\equiv C_I(\boldsymbol{s}+\boldsymbol{r},t;\boldsymbol{s},t;\zeta)$ and ϕ is the polar angle of the lag vector $\boldsymbol{r}$. This result is valid for homogeneous exposure fields, in general. Eq. (42) is a generalization of the isotropic relationship

$$\lim_{r\to 0}\partial C_I(\boldsymbol{r},t;\zeta)/\partial r=-\pi^{-1}\ell_s(\zeta;t). \tag{43}$$

In Christakos and Hristopulos (1998) it is also proven that in the case of an isotropic Gaussian exposure field, Eq. (43) is equivalent to Eq. (27).

REMARK 1: Consider the $\ell_s(\zeta)$ and $\ell'_s(\zeta')$ associated with the Gaussian fields $E(\boldsymbol{p}) \sim N(m_E, \sigma_E^2, b)$ and $E'(\boldsymbol{p}) \sim N(m'_E, \sigma_E'^2, b')$, respectively. Then, the *scaling* relationship below is true

$$b' \ell'_s(\zeta') = b\, \ell_s(\zeta), \tag{44}$$

where $\zeta' = (\zeta - m_E)\sigma'_E / \sigma_E + m'_E$.

Assuming that the contaminant observation network follows a random Poisson pattern, the $c_I(\boldsymbol{r},0;\zeta)$ is expressed as follows

$$c_I(\boldsymbol{r},0;\zeta) = c_I(\boldsymbol{0},0;\zeta)\exp\{-\ell_s(\zeta)/\pi R_E(\zeta)[1 - R_E(\zeta)]\}. \tag{45}$$

In general, N-point exposure indicators ($N \geq 3$) can be established in terms of N-order moments of the random field $I_E(\boldsymbol{p},\zeta)$.

EXAMPLE 10: The three-point exposure indicator moment is defined by

$$\beta_I(\boldsymbol{p},\boldsymbol{p}',\boldsymbol{p}'';\zeta) = \overline{I_E(\boldsymbol{p},\zeta)I_E(\boldsymbol{p}',\zeta)I_E(\boldsymbol{p}'',\zeta)}. \tag{46}$$

In the case of spatial homogeneity and if ζ is equal to the median of $E(\boldsymbol{p})$, the three-point spatial indicator can be expressed in terms of $R_E(\zeta)$ and the semi-variogram as

$$\begin{aligned}\beta_I(\boldsymbol{s}-\boldsymbol{s}',\boldsymbol{s}-\boldsymbol{s}'';\zeta_{1/2}) = R_E(\zeta_{1/2}) - \tfrac{1}{2}[\gamma_I(\boldsymbol{s}-\boldsymbol{s}',\zeta_{1/2}) + \gamma_I(\boldsymbol{s}-\boldsymbol{s}'',\zeta_{1/2}) \\ +\gamma_I(\boldsymbol{s}'-\boldsymbol{s}'',\zeta_{1/2})],\end{aligned} \tag{47}$$

where $R_E(\zeta_{1/2}) = 0.5$. □

REMARK 2: Certain exposure indicators can be expressed explicitly in terms of probabilities to substantial exposure (either at a global scale or spatially distributed over the region of concern) and, thus, they can be considered as risk indicators. Risk assessment procedures can then incorporate the information provided by these indicators in risk management and health policy-making (Wentz, 1989).

Finally, one-point indicators may be also parametrized with respect to $R_E(\boldsymbol{p},\zeta)$; see Christakos and Hristopulos (1996b).

2.7 Ergodicity and Sample Averaging of Exposure Indicators

Let us pause at this point and consider a fundamental hypothesis in the modelling of uncertainty and variability in environmental systems. The S/TRF theory involves an infinity of random field realizations, which satisfy the constraints imposed by the knowledge bases (§I.5). and they may also be conditioned to the data at the measurement locations. One can argue that there is only one realization that represents the state of the environmental system. However, this realization is for all practical purposes non-observable. Hence, the S/TRF representation provides acceptable estimates in the face of uncertainty. We have seen that useful S/TRF characteristics are provided by the stochastic moments (Chapter III). Indeed, the indicators above involve moments of the exposure S/TRF. Evaluation of the moments involves calculating averages over an infinite ensemble of realizations, which are very expensive computationally. Significant gains are obtained if the moments can be estimated based on spatial or temporal averages --sample averages-- of a single S/TRF realization. The S/TRFs that possess this property are called *ergodic* or *self-averaging*.

Definition 6: *A homogeneous-stationary exposure random field $E(\boldsymbol{p})$ within a space-time domain Ω is ergodic in the mean if the following is satisfied*

$$\lim_{|\Omega|\to\infty} E_\Omega = \lim_{|\Omega|\to\infty} |\Omega|^{-1} \int_\Omega E(\boldsymbol{p})\, d\boldsymbol{p} = m_E. \tag{48}$$

The E_Ω is a random variable that depends on the domain Ω, such that $\overline{E_\Omega} = m_E$. The limit in the above is considered in the sense of mean square (m.s.) or almost sure (a.s.) convergence (Christakos, 1992). Ergodicity requires that at the limit of an infinite domain the fluctuations in E_Ω tend to zero. The above definition can be modified in the case of S/TRF that are ergodic only in time or space. It is clear that ergodicity in space (time) requires that the S/TRF be homogeneous (stationary), because the sample average is space (time) independent. If the exposure is ergodic in the mean, the one-point exposure indicators of effectively infinite domains can be estimated using sample means. If an S/TRF is ergodic in both the mean and the covariance, it is called second-order ergodic.

Rigorous mathematical theorems that establish sufficient and necessary conditions for ergodicity in the mean are given by Adler (1981). However, these conditions can not be verified in practice in most environmental health applications, where a single S/TRF sample is the only available information and the domain size is finite. While the ergodic hypothesis can not be rigorously established in practice, it is often a reasonable *working hypothesis* (Christakos, 1992). In such cases ergodicity in the mean and the covariance is assumed, based on which the exposure indicators are approximated in terms of sample averages over the exposure domain. The stochastic mean of $E(\boldsymbol{p})$, e.g., is approximated by

$$m_E \cong \varepsilon_\Omega = |\Omega|^{-1} \int_\Omega d\boldsymbol{p}\, \varepsilon(\boldsymbol{p}); \tag{49}$$

$|\Omega|$ represents a practically infinite domain, and $\int_\Omega d\boldsymbol{p}$ denotes the Riemann integral of the continuous sample function. A domain is considered practically infinite if its characteristic spatial and temporal scales are significantly larger than the correlation lengths of the S/TRF. Hence, an exposure $E(\boldsymbol{p})$ with long range correlations is not ergodic. This requirement leads to certain necessary conditions given below. Analogous empirical expressions in terms of space-time sample averages are obtained for the binary S/TRF $I_E(\boldsymbol{p},\zeta)$.

The following proposition formulates a necessary ergodicity condition in terms of indicator parameters for exposure random fields.

Proposition 2 (Christakos and Hristopulos, 1998): *A necessary condition for the validity of the ergodic hypothesis is that*

$$R_E(\zeta) = \lim_{|\Omega|\to\infty} [|\Omega|^{-1} \int_\Omega d\boldsymbol{r}\, d\tau\, c_I(\boldsymbol{r},\tau;\zeta)]^{1/2}. \tag{50}$$

REMARK 3: Condition (50) is not sufficient for ergodicity, but it provides an explicit test for ergodicity breaking --in the exposure mean or covariance-- that could be implemented approximately on the basis of available measurements. For a Gaussian pdf, it is possible to extend the analysis as is shown by the following corollary.

Corollary 3 (Christakos and Hristopulos, 1996b): *Consider a normally distributed, homogeneous-stationary exposure field. Then, the necessary condition (50) for the ergodic hypothesis can be expressed as*

$$\lim_{|\Omega|\to\infty} |\Omega|^{-1} \int_\Omega d\boldsymbol{r}\, d\tau\, \Phi(\boldsymbol{r},\tau;\zeta) = 0, \tag{51}$$

where $\Phi(\boldsymbol{r},\tau;\zeta)=\int_0^{\sin^{-1}\rho_E(\boldsymbol{r},\tau)} d\theta \exp[-\frac{(\zeta-m_E)^2}{\sigma_E^2(1+\sin\theta)}]$ *and* $\rho_E(\boldsymbol{r},\tau)= c_E(\boldsymbol{r},\tau)/c_E(\boldsymbol{0},0)$.

Note that $\Phi(\boldsymbol{r},\tau;\zeta)$ becomes essentially equal to zero as the spatial and time lags exceed 4-5 times the correlation lengths ξ_r and ξ_τ respectively. Hence, provided that the correlation lengths are finite, $\int_\Omega d\boldsymbol{r}\, d\tau\, \Phi(\boldsymbol{r},\tau;\zeta) \propto \xi_r^n\, \xi_\tau$ and Eq. (51) is satisfied. Long range correlations, on the other hand, break ergodicity. Finally, the $\lim_{|\boldsymbol{r}|\to\infty} c_E(\boldsymbol{r}) = 0$ is a sufficient condition for ergodicity in the mean for homogeneous Gaussian exposure fields.

Useful spatiotemporal representations of the exposure indicators can be established if ergodicity holds. In general, an indicator $\mathcal{Y}_E$ may be expressed in terms of space-time averaging,

$$\mathcal{Y}_E = |\Omega|^{-1} \int_\Omega \mathfrak{F}[E, I_E, \zeta]\, ds\, dt, \tag{52}$$

where $\mathfrak{F}[\cdot]$ is a function of S/TRF-pair realizations and the threshold level ζ. Hence, the indicators defined above in terms of stochastic averages can be conveniently calculated by means of space-time averages. This has considerable implications in practice, where often a single sample (realization) is available. For illustration, let us consider space-time R_E-representations.

EXAMPLE 11: The $R_E(\zeta)$ represents the fraction of the total domain Ω where the exposure exceeds the threshold level ζ, i.e.,

$$R_E(\zeta) = |\Omega|^{-1} \int_\Omega I_E(\boldsymbol{p},\zeta) \mathrm{d}\boldsymbol{p} = |\Omega|^{-1} |\Theta_\Omega(\zeta)|, \tag{53}$$

where $\Theta_\Omega(\zeta) \subset \Omega$ is the over-exposed subdomain. □

Similar spatiotemporal expressions can be obtained for the two-point indicators.

EXAMPLE 12: The

$$C_I(\boldsymbol{p}-\boldsymbol{p}';\zeta) = |\Omega|^{-1} \int_\Omega d\boldsymbol{p}\, d\boldsymbol{p}'\, I_E(\boldsymbol{p},\zeta)\, I_E(\boldsymbol{p}',\zeta) \tag{54}$$

expresses space-time correlations between pollutant concentrations exceeding the threshold level. □

Representations of the stochastic exposure indicators can be established by means of *space transformation* operators (Chapter VII; also, Christakos, 1992; Christakos and Panagopoulos, 1992).

EXAMPLE 13: For a spatially isotropic exposure field the statistics should be the same in all directions in space. If we consider a line segment Λ across the site with direction determined by the unit vector $\boldsymbol{\theta} = (\theta_1, \ldots, \theta_n)$, the R_E can be expressed as

$$R_E(\zeta) = \lim_{|\Lambda| \to \infty} |\Lambda|^{-1} \int_\Lambda d\eta \, \hat{I}_{E,\boldsymbol{\theta}}(\eta, \zeta), \tag{55}$$

where

$$\hat{I}_{E,\boldsymbol{\theta}}(\eta, \zeta) = \int_\Omega ds \, I_E(\boldsymbol{p}, \zeta) \, \delta(\boldsymbol{s} \cdot \boldsymbol{\theta} - \eta) \tag{56}$$

is a space transformation of $I_E(\boldsymbol{p}, \zeta)$; $\delta(\cdot)$ is a delta function and $|\Lambda|$ is the length of the line segment Λ ($\eta \in \Lambda \subset R^1$). □

In view of the preceding analysis, depending on the data available in practical applications one may distinguish between calculated (on the basis of the available data) exposure indicators and derived exposure indicators (from constitutive equations, etc.). The exposure indicators can be estimated robustly from the data and other kinds of available knowledge using the S/TRF techniques discussed in the previous Chapters. Stochastic exposure indicators provide a set of quantitative tools with useful environmental health applications including: human exposure assessment and control in space-time; evaluation of the effectiveness of site restoration strategies; identification of damaged ecosystems; estimation of the potential for pollutant transport from smaller to larger regions; and policy and decision making; and health management.

REMARK 4: The stochastic exposure indicators have important consequences in the financial cost analysis of environmental health management projects such as, e.g., waste site cleanup cost analysis. The issue will not be pursued here, but the interested reader is referred to Christakos and Hristopulos (1996b) and references therein.

3. STOCHASTIC CELL-BASED HEALTH EFFECT INDICATORS

3.1 Cellular Structures

Cells are considered as the primary structural and functional units of all living organisms. The human body is composed of trillions of cells (a good start for an effective application of random field methods!). Most cells in a human have diameters in the range of 10 to 20 μm; cells as small as 2 μm and as large as 120 μm also exist .

Cells are the simplest structural units that retain essentially all of the functions characteristic of life. A human organism begins as a single cell, which is divided into two cells, each of which is again divided into two cells (thus resulting in four cells), etc.. During this process each cell undergoes certain changes, such as: (i) Cells become specialized for the performance of specific functions -this is called *cell differentiation.* (ii) Cells migrate to new locations forming a hierarchy of multicellular structures (tissues, organs, organ systems and total organism), as shown in Fig. 3. A detailed discussion of the biological and physiological properties of the cells can be found in physiology texts like Vander *et al.* (1998).

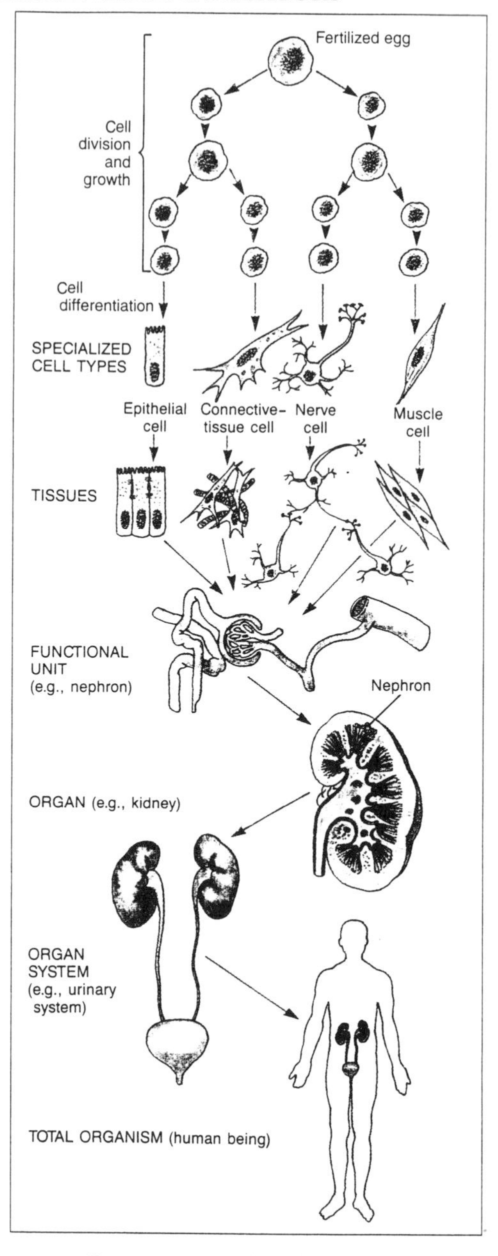

FIGURE 3: Hierarchy of cellular structures (Vander *et al.*, 1998; used with permission of McGraw-Hill Co.)

3.2 The Threshold Postulate

Let N be a number of normal cells distributed in a target organ. For purposes of quantitative description, each cell is identified by its spatiotemporal location $\boldsymbol{p} = (\boldsymbol{s}, t)$ within a domain $\Omega = V \times T$, where V denotes the volume of the organ occupied by the cells and T is the exposure time under consideration (the volume can be replaced by the area D when appropriate). In the following the basic cell indicators are introduced and theoretically justified. Our analysis will be based on the following postulate.

Postulate 1: *A cell is considered affected (e.g., it can develop a tumor) if a biomarker $\mathcal{B}(\boldsymbol{p})$ reaches or exceeds a certain threshold ζ, i.e.,*

$$I_c(\boldsymbol{p}, \zeta) = \begin{cases} 1 & (\textit{i.e., the cell p is affected}), \quad \textit{if} \quad \mathcal{B}(\boldsymbol{p}) \geq \zeta \\ 0 & (\textit{i.e., the cell p is normal}), \quad \textit{if} \quad \mathcal{B}(\boldsymbol{p}) < \zeta \end{cases}, \tag{1}$$

where $\boldsymbol{p} = (\boldsymbol{s}, t)$ denotes the spatial location $\boldsymbol{s}$ of the cell and t the time of occurrence of the biomarker.

The threshold ζ is usually selected on the basis of biological and toxicological data (§II.4). Note that the random cell field is related to the biomarker field by

$$I_c(\boldsymbol{p}, \zeta) = \mathrm{P}[\mathcal{B}(\boldsymbol{p}) \geq \zeta | \beta(\boldsymbol{p})] \tag{2}$$

for any cell $\boldsymbol{p} \in \Omega$, where $\beta(\boldsymbol{p})$ is a realization of the random biomarker field [say, a dose rate $D_r(\boldsymbol{p})$]. Eqs. (1) and (2) establish some very important relationships: The stage of a cell $\boldsymbol{p}$ (normal vs. affected) is expressed in terms of extreme biomarker values. Other useful random cell fields can be derived by combining the biomarker and the binary random cell fields.

Definition 1: *The expected cell indicator is defined by*

$$R_c(\boldsymbol{p}, \zeta) = \overline{I_c(\boldsymbol{p}, \zeta)} = 1 - F_{\mathcal{B}}(\boldsymbol{p}, \zeta) \tag{3}$$

for each cell $\boldsymbol{p}$.

To obtain an idea of the physical meaning of the expected cell indicator consider a stochastically homogeneous and ergodic biomarker distribution $\mathcal{B}(\boldsymbol{p})$; then $R_c(\boldsymbol{p},\zeta) = R_c(t,\zeta)$ at all spatial cell locations $\boldsymbol{s}$, and the expected cell indicator at time t is equal to the fraction of the organ area covered by affected cells over the total cell area (or, equivalently, the fraction of affected cells). As we shall see in §3.5 below, equations governing the temporal evolution of the fraction of affected cells can be used to calculate $R_c(t,\zeta)$.

Definition 2: *The non-centered cell covariance between the cells* $\boldsymbol{p}$ *and* $\boldsymbol{p}'$ *is given by*

$$C_c(\boldsymbol{p},\boldsymbol{p}';\zeta) = \overline{I_c(\boldsymbol{p},\zeta)\, I_c(\boldsymbol{p}',\zeta)} = \int_\zeta^\infty \int_\zeta^\infty d\beta\, d\beta'\, f_{\mathcal{B}}(\boldsymbol{p},\boldsymbol{p}';\beta,\beta'), \tag{4}$$

where $f_{\mathcal{B}}(\boldsymbol{p},\boldsymbol{p}';\beta,\beta')$ *is the bivariate pdf of the biomarker between any two cells.*

Eq. (4) expresses the probability $P[\mathcal{B}(\boldsymbol{p}) \geq \zeta \wedge \mathcal{B}(\boldsymbol{p}') \geq \zeta]$ that both cells $\boldsymbol{p}$ and $\boldsymbol{p}'$ are affected. It provides a measure of the spatiotemporal correlation (connectivity) between affected cells, which is important in the assessment of health effects. The *centered cell covariance* is expressed in terms of (3) and (4) as

$$c_c(\boldsymbol{p},\boldsymbol{p}';\zeta) = C_c(\boldsymbol{p},\boldsymbol{p}';\zeta) - R_c(\boldsymbol{p};\zeta) R_c(\boldsymbol{p}';\zeta). \tag{5}$$

For a homogeneous-stationary biomarker distribution, $c_c(\boldsymbol{r},\tau;\zeta) = C_c(\boldsymbol{r},\tau;\zeta) - R_c^2(\zeta)$, which represents cell fluctuation correlations around $R_c(\zeta)$. The analysis above suggests the following approach for studying cell distributions in space-time (see, also, Fig. 4).

Cellular approach: *Given the biomarker bivariate pdf and the critical threshold, we can calculate the cell indicators from Eqs. (3) -(5), and then use the random field simulation methods to generate several possible cell field realizations for a target organ.*

The biomarker pdf may be estimated from experimental animal data or experience with similar situations of exposed organs. The cellular approach takes into consideration certain important factors of cell evolution: (i) the spatial location of each cell, (ii) the biomarker level and time, (iii) spatiotemporal correlations between cells, and (iv) spatiotemporal correlations between biomarker levels. As we shall see in §3.5 below the temporal cell evolution may, also, involve repaired cells --i.e. formerly affected cells that have been repaired by systems of enzymes. We illustrate the above approach by means of a numerical example, as follows.

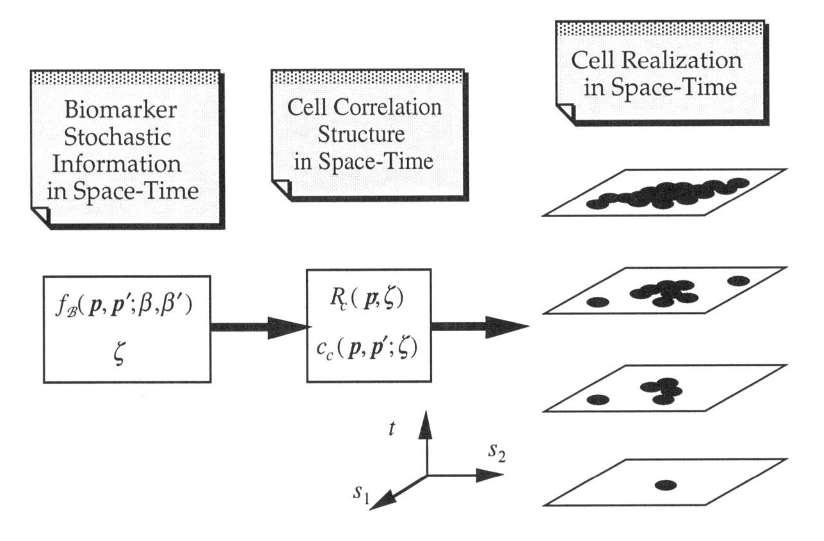

FIGURE 4: The cellular approach; affected cells are denoted as ⬬.

EXAMPLE 1: A realization of the cell distribution in space-time is shown in Fig. 5. It is

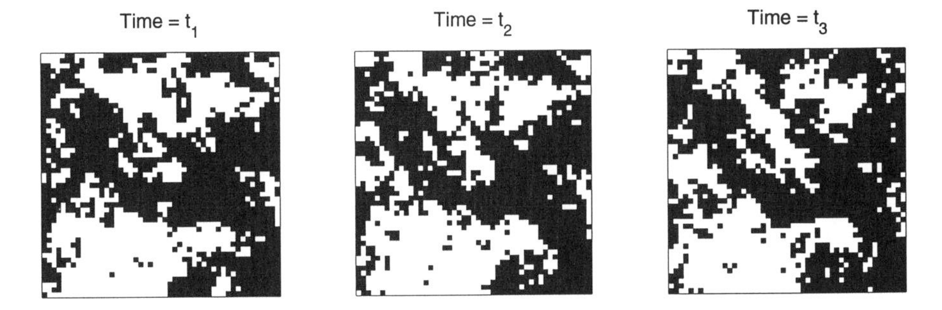

FIGURE 5: A simulated spatiotemporal cell field for a target organ; the affected cells are white and the normal cells are black (some repair is taking place, as well).

assumed that $R_c(\boldsymbol{p},\zeta)=0.4\,units$ and the cell correlation is $c_c(\boldsymbol{r},\tau;\zeta)=(2\pi)^{-1}\int_0^{\sin^{-1}\rho(\boldsymbol{r},\tau)} d\theta \exp[-0.0625/(1+\sin\theta)]$, with $\rho_c(\boldsymbol{r},\tau)=\exp[-(r^2+\tau^2)/36]$. Notice the change in the number of normal vs. affected cells in space-time (some of the affected cells are repaired in time). An advantage of the S/TRF-based cellular approach is that, unlike the purely statistical methods, it takes into consideration spatial and temporal correlations between cells. □

We can proceed further with the cellular approach. Let the random field $n^a(t)$ denote the number of affected cells in the target organ at time t. Given a spatial realization of the binary cell field $I_c(\boldsymbol{p},\zeta)$ at time t, $n^a(t)$ can be expressed in the continuous space-time domain as

$$n^a(t)=\upsilon_c^{-1}\int_V ds\, I_c(\boldsymbol{p},\zeta); \tag{6}$$

υ_c is a uniform normalization volume per cell, and $|V|=\upsilon_c N$, where N is the total number of cells. On the basis of the last equation, the temporal mean and correlation functions of $n^a(t)$ are expressed in terms of the cell indicators (3)-(5), i.e.,

$$\overline{n^a}(t)=\upsilon_c^{-1}\int_V ds\, \overline{I_c}(\boldsymbol{p},\zeta)=\upsilon_c^{-1}\int_V ds\, R_c(\boldsymbol{p},\zeta), \tag{7}$$

and

$$c_n^a(t,t')=\upsilon_c^{-2}\int_V\int_V ds\, ds'\, c_c(\boldsymbol{p},\boldsymbol{p}',\zeta). \tag{8}$$

On the basis of the above functions, an interesting expression for the health effect indicator $H_p(t)$ of Eq. (II.4.9) can be obtained as follows. Let us define the random field $W(t)$ that represents the fraction of affected cells as

$$W(t)=|V|^{-1}\int_V ds\, I_c(\boldsymbol{p},\zeta) \tag{9}$$

with mean and variance, respectively,

$$\overline{W}(t)=\overline{n^a}(t)/N=|V|^{-1}\int_V ds\, R_c(\boldsymbol{p},\zeta), \tag{10}$$

and

$$\sigma_w^2(t) = |V|^{-2} \int_V \int_V ds\, ds'\, c_c(s,s',t,\zeta). \tag{11}$$

Then, the following expression is obtained for the health effect indicator at time t

$$H_p(t) \in [\overline{W}(t) - \eta\,\sigma_w(t),\ \overline{W}(t) + \eta\,\sigma_w(t)], \tag{12}$$

where the value of the coefficient η depends on the desired confidence and the shape of the $W(t)$-pdf. If the $W(t)$ is assumed to have a Gaussian shape, $\eta = 1.96$ implies a 95% confidence interval, etc.. By taking into consideration spatial and temporal correlations between cells, the health indicator (12) provides a more realistic health effect probability than purely statistical methods --e.g. Poisson (Crawford-Brown, 1997)-- which do not account for these correlations.

EXAMPLE 2: In a discrete representation, cells are counted by summations

$$n^a(t) = \sum_{j=1}^N I_c(s_j,t,\zeta), \tag{13}$$

where N is the total number of cells of the target organ,

$$\overline{n^a}(t) = \sum_{j=1}^N \overline{I_c}(s_j,t,\zeta) = \sum_{j=1}^N R_c(s_j,t,\zeta), \tag{14}$$

$$c_n^a(t,t') = \sum_{j=1}^N \sum_{j'=1}^N c_c(s_j,t,s_{j'},t',\zeta), \tag{15}$$

and

$$H_p(t) \in [\overline{n^a}(t)/N - \eta\, c_n^a(t,t')/N^2,\ \overline{n^a}(t)/N + \eta\, c_n^a(t,t')/N^2]; \tag{16}$$

the $\overline{n^a}(t)/N$ may be considered as a 1st-order approximation of the health indicator. □

REMARK 1: The classical statistical expression at time t

$$H_P(t) \approx n^a(t)/N \tag{17}$$

is an approximation to the above general representations that does not take into consideration spatiotemporal correlations between cells; the expected number of affected cells is approximated by the number of affected cells at time t, $\sum_{j=1}^{N} I_c(s_j,t,\zeta) = n^a(t)$.

Furthermore, the analysis above may be modified appropriately to assign greater weights ξ_j to cells s_j that form clusters or groups. This is important, for there is evidence that many health effects require that the cells be in close proximity or even in contact. As the number of the affected cells in a cluster increases, the weight ξ_j should also increase; i.e.,

$$W_\xi(t) = \sum_{j=1}^{N} \xi_j I_c(s_i,t,\zeta); \tag{18}$$

then, the health effect indicator is written as

$$H_p(t) \in [\overline{W_\xi}(t) - \eta\sigma_{w,\xi}(t),\ \overline{W_\xi}(t) + \eta\sigma_{w,\xi}(t)], \tag{19}$$

where $\overline{W_\xi}(t) = \sum_{j=1}^{N} \xi_j R_c(s_j,t,\zeta)$ and $\sigma_{w,\xi}^2(t) = \sum_{j=1}^{N} \sum_{j'=1}^{N} \xi_j \xi_{j'} c_c(s_j,s_{j'},t,\zeta)$. Eq. (16) is a special case of Eq. (19) for $\xi_j = N^{-1}$. Another modification is suggested next.

EXAMPLE 3: Assume that the number of affected cells $n^a(t)$ at time t can be decomposed into M clusters, i.e., $n^a(t) = \sum_{k=1}^{M} n_k^a(t)$, where $n_k^a(t)$ is the number of affected cells in the kth group. An approximate expression of the health effect indicator is given by

$$H_P(t) = N^{-1} \sum_{k=1}^{M} \xi_k(n_k^a) n_k^a(t), \tag{20}$$

where $\xi_k(n_k^a)$ are weight functions that depend on the number of affected cells $n_k^a(t)$ in each group. The $\xi_k(n_k^a)$ should satisfy certain constraints; e.g., a permissible set of weights satisfies

$$\sum_{k=1}^{M} \xi_k(n_k^a) = 1, \tag{21}$$

$$\xi_k(n_k^a) > \sum_{i=1}^{p} \xi_i(n_i^a)\,\delta(n_k^a - \sum_{i=1}^{p} n_i^a) \tag{22}$$

for any integer p, which ensures that larger clusters of cells are assigned larger weights than smaller clusters. A possible choice of a weight function is

$$\xi_i(n_i^a) = be^{n_i^a(t)}, \tag{23}$$

where $b^{-1} = \sum_{j=1}^{k} e^{n_{i,j}^a(t)}$ is a normalization coefficient; Eq. (23) satisfies both conditions (21) and (22). If several realizations of the random cell field are available at time t, Eq. (20) becomes

$$H_P(t) = (K\,N)^{-1} \sum_{j=1}^{K} \sum_{k=1}^{M} \xi_k(n_k^a) n_k^{a(j)}(t), \tag{24}$$

where j denotes the realization, and K is the number of realizations. The aggregate (statistical) insight gained from Eq. (24) can be decisive in health risk assessment. □

There is sufficient evidence that the same total biomarker (e.g., exposure or dose) produces a higher probability of health effect when delivered at a higher rate. Useful cell indicators, in this respect, are the *expected excess cell biomarker* above the threshold ζ at cell $\boldsymbol{p}$,

$$P_c(\boldsymbol{p},\zeta) = \overline{\mathcal{B}(\boldsymbol{p}) I_c(\boldsymbol{p},\zeta)}, \tag{25}$$

which is a measure of the average biomarker at cells where the threshold is exceeded; and the *expected excess differential biomarker* at cell $\boldsymbol{p}$,

$$L_c(\boldsymbol{p},\zeta) = \overline{[\mathcal{B}(\boldsymbol{p}) - \zeta] I_c(\boldsymbol{p},\zeta)}, \tag{26}$$

which is a measure of the average amount by which the biomarker exceeds the threshold at the affected cell. P_c-values range between mean biomarker at zero threshold and zero at maximum threshold. The range of L_c is between $L_c(\boldsymbol{p},0) = \overline{\mathcal{B}}(\boldsymbol{p})$ and $L_c(\boldsymbol{p},\infty) = 0$. The *conditional mean excess biomarker* over the threshold ζ is given by

$$U_c(\boldsymbol{p},\zeta) = \overline{\mathcal{B}(\boldsymbol{p}) | \mathcal{B}(\boldsymbol{p}) \geq \zeta}, \tag{27}$$

which refers only to the affected cells conditional to the excess exposure values.

3.3 Relationships Between the Basic Cell Indicators

The basic cell indicators above may be parametrized with respect to $R_c(\boldsymbol{p},\zeta)$, as follows

$$dP_c / dR_c = \zeta, \tag{28}$$

$$dL_c / dR_c = -R_c \, d\zeta / dR_c, \tag{29}$$

and

$$R_c \, dU_c / dR_c + U_c = \zeta, \tag{30}$$

where in the above ζ is viewed as a function of R_c. Eq. (28) implies that $P_c(R_c)$ is an increasing and concave function in R_c. Similarly, $L_c(R_c)$ is an increasing function in R_c. In other words, the expected values of both the excess cell biomarker and the excess differential cell biomarker increase as the relative area of affected cells increases.

EXAMPLE 4: For illustration, some plots of the cell indicators obtained from the solution of

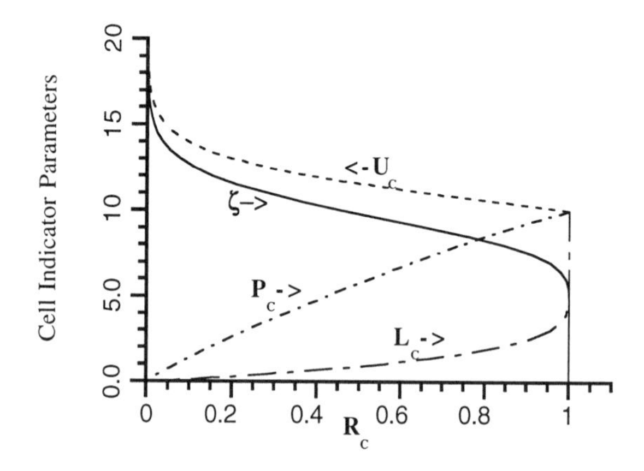

FIGURE 6: Plot of the exposure threshold and the indicators P_c, L_c and U_c vs. R_c for a lognormal burden distribution with $\overline{\mathcal{B}} = 10$ and $\sigma_{\mathcal{B}} = 2$ (in appropriate units).

the above system of constitutive equations are given in Fig. 6. □

The cell indicators can be, also, parametrized with respect to the threshold ζ, leading to the set of *constitutive* Eqs. (51) - (54) below. Hence, if one of them is known, say L_c, the others can be found from these equations. Furthermore, if the health effect threshold is uncertain, threshold-averaged cell indicators can be obtained such as in

Corollary 1 (Christakos and Hristopulos, 1996b): *The threshold-averaged excess differential biomarker can be expressed as*

$$L_c(\boldsymbol{p}) = \overline{L_c(\boldsymbol{p},z)} = \int_0^{\infty} d\zeta\, F_{\mathcal{B}}(\zeta)\,[1 - F_{\mathcal{B}}(\zeta)]\,. \tag{31}$$

In practice, threshold-averaging may be evaluated over a finite domain $[\zeta_{\min}, \zeta_{\max}]$ instead of $[0, \infty)$. The *coefficient of cell indicator dispersion* is given by

$$\Psi_c(\boldsymbol{p}) = \overline{L_c(\boldsymbol{p},z)}/\overline{\mathcal{B}(\boldsymbol{p})} = \int_0^{\infty} C_c(\boldsymbol{0},\zeta)\,d\zeta/\overline{\mathcal{B}(\boldsymbol{p})}, \tag{32}$$

where $\overline{L_c(\boldsymbol{p},z)}$ is the average of L_c over the pdf of the random threshold z. Eq. (32) is a measure of the ratio of (a) the average biomarker for all possible threshold values over the mean biomarker, and (b) the excess cell biomarker dispersion over the mean biomarker. Higher Ψ_c-values indicate a higher biomarker level.

3.4 Derived Cell Indicators

Using the above basic cell indicators, additional two-point indicators can be derived. The *cell interaction ratio*, e.g., is given by

$$\mathcal{G}_c(\boldsymbol{p},\boldsymbol{p}';\zeta) = C_c(\boldsymbol{p},\boldsymbol{p}';\zeta)/[R_c(\boldsymbol{p},\zeta) + R_c(\boldsymbol{p}',\zeta) - C_c(\boldsymbol{p},\boldsymbol{p}';\zeta)]; \tag{33}$$

$\mathcal{G}_c$ is the ratio of the probability that both biomarker values of cells separated by lags $\boldsymbol{r}$ and τ reach or exceed the threshold over the probability that either one of them does.

The *cell biomarker exceedance odds* on $\mathcal{B}(\boldsymbol{p})$ such that $\mathcal{B}(\boldsymbol{p}) > \zeta$ and $\mathcal{B}(\boldsymbol{p}') > \zeta$ is

$$O_c(\boldsymbol{p},\boldsymbol{p}';\zeta) = C_c(\boldsymbol{p},\boldsymbol{p}';\zeta)[1 - R_c(\boldsymbol{p},\zeta) - R_c(\boldsymbol{p},\zeta) + C_c(\boldsymbol{p},\boldsymbol{p}';\zeta)], \tag{34}$$

which expresses the ratio of the probability that biomarker values of cells separated by lags $\boldsymbol{r}$ and τ exceed ζ over the probability that both biomarker values are below ζ.

3.5 Scale and Modelling Effects

Cell biomarker characterization depends on the analysis of biomarker values at various *scales*, and on the formulation of suitable quantitative connections between the results obtained at each one of these scales. This is important, for considerable differences exist between the behavior and the statistics of the same biological process when analyzed at different scales. The appropriate choice of a scale should lead to a cell biomarker model that contains sufficient detail to reproduce accurately all the relevant biological and health effects without rendering the computations impractical.

A change of cell scale operation leads to changes in the cell indicators and the underlying pdf. Consider, e.g., an *upscaling operation* $\Im[\cdot]$ that maps the biomarker random field $\mathcal{B}(\boldsymbol{p})$ into a new field $\mathcal{B}_u(\boldsymbol{p})$, i.e.,

$$\mathcal{B}_u(\boldsymbol{p}) = \Re[\mathcal{B}(\boldsymbol{p})], \tag{35}$$

where the vector $\boldsymbol{u}$ determines the change of scale operation in space. The upscaling operation can be better understood in the context of lattice S/TRF --although it can be extended to continuous S/TRF as well. For lattice random fields, the upscaling functional replaces a group of variables $(\beta_1,\ldots,\beta_N)$ that represent S/TRF values at cells $\boldsymbol{p}_i$ $(i=1,\ldots,N)$ with a single variable; the latter is given by a linear or non-linear combination of the individual variables determined by the operator $\Re[\cdot]$. That is, the upscaled cell biomarker pdf is

$$f^{\boldsymbol{u}}_{\mathcal{B}}(\beta^*) = \int d\beta_1 \ldots d\beta_N \, \delta[\beta^* - \Re(\beta_1,\ldots,\beta_N)] f_{\mathcal{B}}(\beta_1,\ldots,\beta_N). \tag{36}$$

In Eq. (36), the upscaling operator assigns the value β^* to a specified space-time cell. Hence, the coarse grained S/TRF depends on the form of $\Re[\cdot]$.

EXAMPLE 5: Consider a square lattice with L^2 cells and spacing α. An upscaling transformation coarse grains the exposure $\mathcal{B}(\boldsymbol{p})$ by replacing a cluster of $N=n^2$ cells with a single cell (n must be a divisor of L) so that $\boldsymbol{u}=(\alpha n, \alpha n)$. The value of the upscaled cell biomarker at the center of the large cell is given by a functional of the biomarker values in the initial cells. A specific case of the upscaling transformation is given in Fig. 7.

When the upscaling functional is given by the arithmetic average we obtain

$$f^{\boldsymbol{u}}_{\mathcal{B}}(\beta^*) = N \int d\beta_1 \ldots d\beta_{N-1} \, f_{\mathcal{B}}(\beta_1,\ldots,\beta_{N-1},\beta_N)\Big|_{\beta_N = N\beta^* - \sum_{i=1}^{N-1} \beta_i} \tag{37}$$

for the corresponding pdf. □

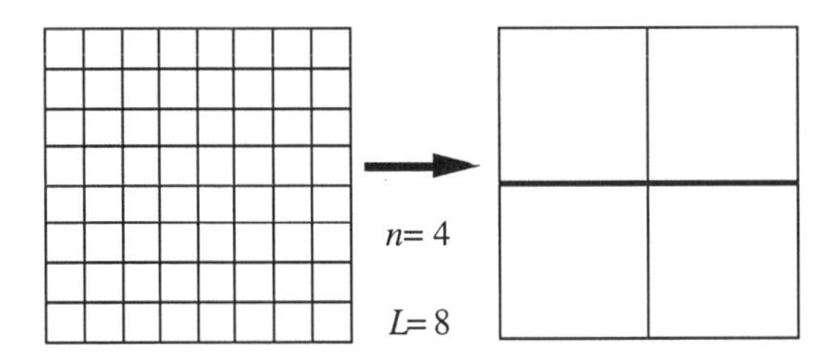

FIGURE 7: A simple example of cell upscaling transformation. The initial square lattice that contains 64 cells is replaced by a lattice containing 4 cells. The initial random biomarker field has values on all 64 cells. The coarse-grained biomarker field in the upscaled lattice is represented by 4 values at the centers of the cells.

Eqs. (36) and (37) are useful in analytical upscaling studies, in which the multivariate pdf can be assumed known. In certain cases, explicit expressions can be obtained for the changes in the biomarker pdf parameters, e.g. the variance, under upscaling. The arithmetic average upscaling transformation leads to a reduction in the variance of the pdf. The reduction of the variance can be intuitively understood as follows: For a large class of pdf models the pdf of the arithmetic average tends asymptotically towards a Gaussian form (Gnedenko and Kolmogorov, 1954). A sum of N Gaussians is also a Gaussian with variance equal to N times the variance of the individual Gaussians. Hence, the variance of the arithmetic average is $O(1/N)$, and thus it is reduced with increasing N.

For practical purposes, the biomarker pdf change due to upscaling can be calculated in terms of the cell indicators as follows (Christakos and Hristopulos, 1996a and b)

$$\delta f_u(\beta) \equiv f^u_{\mathcal{B}}(\beta) - f_{\mathcal{B}}(\beta) = \begin{cases} -\mathrm{d}\, \delta R_c(\beta)/d\beta, & \delta R_c = R^u_c - R_c \\ -\mathrm{d}^2\, \delta L_c(\beta)/d\beta^2, & \delta L_c = L^u_c - L_c \\ -\beta^{-1}\, \mathrm{d}\, \delta P_c(\beta)/d\beta, & \delta P_c = P^u_c - P_c \end{cases}, \tag{38}$$

where R^u_c, L^u_c and P^u_c are the upscaled exposure indicators. Given the form of the operator $\Re[\cdot]$, the upscaling transformation can be applied iteratively and the pdf change at each step is calculated by means of Eq. (38) above. Under certain conditions, scale effects can be quantified in terms of the cell indicators, which are calculated from analytical expressions or numerical simulations.

EXAMPLE 6: Consider a target organ Ω of size $|\Omega|$ and consisting of N_1 subregions υ_1 of the same size $|\upsilon_1|$, each subregion including a number of cells. Let $\mathcal{B}_\Omega = |\Omega|^{-1} \int_\Omega ds\, \mathcal{B}(s)$ be a spatial linear average of the biomarker field $\mathcal{B}(\boldsymbol{p})$ over the organ Ω and let $\mathcal{B}_1$ be a similar average biomarker over υ_1 considered as an estimator of $\mathcal{B}_\Omega$ (the size of the subregion may vary depending on the heterogeneity of the biomarker, biological constraints, health and toxicological data, etc.). We repeat the partitioning to smaller subregions n times, i.e., $|\upsilon_n| < |\upsilon_{n-1}| < \ldots < |\upsilon_2| < |\upsilon_1|$; the final subregion υ_n may be, e.g., a cell. The excess differential cell indicators of biomarker $\mathcal{B}(\boldsymbol{p})$ over the subregions υ_n ($n = 1, 2, \ldots$) and that of the entire region Ω satisfy

$$\left.\begin{array}{l} L_c^n(\zeta) \text{ decreases in } |\upsilon_n| \le |\Omega| \\ \lim_{|\upsilon_n| \to |\Omega|} L_c^n(\zeta) = L_c(\zeta) \end{array}\right\} \tag{39}$$

for all $\zeta \ge 0$. I.e., the larger the subregion size, the smaller the excess differential cell indicator; as the subregion size increases --including more cells-- its cell indicator approximates more accurately the actual $L_c(\zeta) = L_c^\Omega(\zeta)$. As a consequence, the following is also true for the corresponding coefficient of cell indicator dispersion

$$\left.\begin{array}{l} \Psi_c^n \text{ decreases in } |\upsilon_n| \le |\Omega| \\ \lim_{|\upsilon_n| \to |\Omega|} \Psi_c^n = \Psi_c \end{array}\right\}. \tag{40}$$

Since the coefficient of indicator dispersion is a function of the standard deviation for each scale level considered, Eq. (40) also provides a quantitative measure of the change in the standard deviation due to change-of-scale. □

Other scale relations of interest concern total and marginal biomarker quantities related to the stochastic cell indicators.

EXAMPLE 7: Let $\vartheta_\upsilon(\zeta) = |\upsilon| R_c^\upsilon(\zeta)$ be the total affected volume (or area) of an organ υ of size $|\upsilon|$, where $R_c^\upsilon(\zeta)$ is the expected cell indicator for a threshold level ζ --assuming homogeneous and stationary biomarker distribution. Then, a scale effect involving a change $d|\upsilon|$ in υ leads to a marginal value of $\vartheta_\upsilon(\zeta)$ expressed analytically by

$$d\vartheta_\upsilon(\zeta)/d\zeta = R_c^\upsilon(\zeta) + |\upsilon| dR_c^\upsilon(\zeta)/d|\upsilon|. \tag{41}$$

Note that $dR_c^{\upsilon}(\zeta)/d|\upsilon| > 0$ (< 0) implies $d\vartheta_{\upsilon}(\zeta)/d\zeta > R_c^{\upsilon}(\zeta)$ $[< R_c^{\upsilon}(\zeta)]$. Hence, whenever R_c^{υ} decreases as a function of $|\upsilon|$, the marginal value of the affected volume (area) $d\vartheta_{\upsilon}/d\zeta$ lies below it, and vice versa. □

In many practical applications, analytical evaluations of the scale effect require complex mathematical tools. Numerical estimates can be obtained in terms of random field simulations, which generate values of the biomarker on a discrete lattice covering the whole region of interest Ω (an organ, tissues, etc.); these are then averaged over any subregion υ $(\subset \Omega)$ to give the upscaled biomarker values. Cell indicators are evaluated at scales of interest and the corresponding scale effects can be quantified directly.

Exposure characterization also depends on *modelling choices* used to describe the biological processes (pdf, spatial correlation functions). A change in the pdf model can lead to significant changes in the stochastic cell indicators and, consequently, the estimated health risk. Modelling choices should be guided, when possible, by the available data. As we saw above, if the coefficient of variation and the coefficient of cell indicator dispersion can be determined from the data, the latter can be used to distinguish between the underlying probability models. Predictions of biomarker distributions are based on approximations of the actual biological processes. The impact of these approximations on exposure characterization can be expressed in terms of health indicators. The adequacy of model assumptions can, thus, be quantified in terms of differences between the actual and the model-predicted health indicators. A poor estimation model can lead to incorrect health management decisions.

3.6 The Multistage Postulate

In this section we are interested in the temporal evolution of biological processes and, thus, the spatial position of the cells (though implicitly present) will be dropped for notational convenience. As we saw above, the number of affected cells $n^a(t)$ varies in time. Thus, at this point we will examine the following postulate (compare with Postulate 1 above).

Postulate 2: *A cell can be considered affected (e.g., it can develop a tumor) only after it has undergone a series of changes in a prescribed order.*

The temporal cell evolution may involve various stages (Fig. 8): normal, initiated, promoted, progressed (e.g., cancerous), dead, and repaired cells. In particular, the following sequence of cell stages or mutations are assumed.

Stage 0: All N cells begin as normal cells at time $t = 0$.

Stages 1 through K: As a result of a biomarker reaching or exceeding certain threshold, some of the cells are affected going through a series of stages at later times $t>0$.

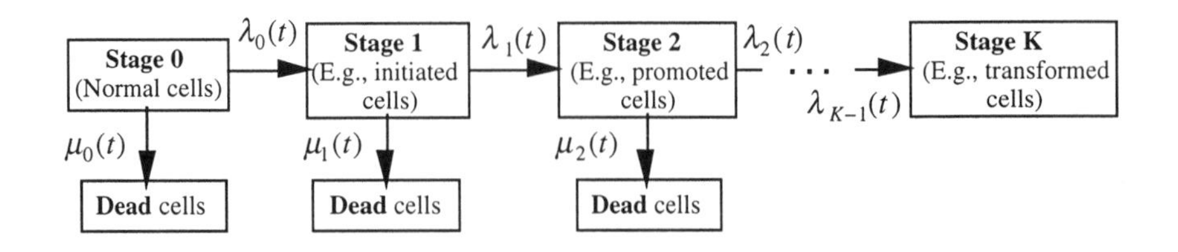

FIGURE 8: Multistage representation of cell transitions.

As a consequence, the health stages are placed into a sequence that is typified by the above series of cell stages or mutations. Mathematically, the temporal cell evolution may be represented in terms of the SODEs discussed in Chapter VIII, i.e.,

$$\left.\begin{aligned} &dn^0(t)/dt = -[\lambda_0(t) + \mu_0(t)]n^0(t) \\ &dn^a(t)/dt = -[\lambda_a(t) + \mu_a(t)]n^a(t) + \lambda_{a-1}(t)n^{a-1}(t), \quad a = 1,\ldots,K-1 \\ &dn^K(t)/dt = -\lambda_{K-1}(t)n^{K-1}(t) \\ &dn^d(t)/dt = \textstyle\sum_{a=0}^{K-1} \mu_a(t)n^a(t) \end{aligned}\right\}, \tag{42}$$

where $n^0(t)$ is the number of normal cells at time t, $n^a(t)$ is the number of cells in stage a at time t, $n^d(t)$ is the number of dead cells at time t, $\lambda_a(t)$ are transition rates between stages, and $\mu_a(t)$ is the transition rate from stage a to the dead cell state (Fig. 8).

REMARK 2: Note that unlike Postulate 1, Postulate 2 does not consider the spatial variation of affected cells. This is an important drawback if there is spatial correlation between cells in the final stage. Indeed, as it has been reported in Crawford-Brown (1997), there is evidence that many health effects require not only a certain number of affected cells, but also that the cells be in close proximity or even in contact.

Depending on the situation, some modifications may be necessary for Eqs. (42). To indicate the fact that transitions between stages are due to the effect of a pollutant biomarker $\mathcal{B}(t)$, the transition rates $\lambda_a(t)$ and $\mu_a(t)$ may be replaced by $\lambda_a(t)\mathcal{B}(t)$ and $\mu(t)\mathcal{B}(t)$.

Cell damage can be repaired by enzymes. To include this event, a repair rate $\zeta_{r,a}(t)$ may be introduced in Eqs. (42) that moves cells back to previous stages. Let us explain some of the main ideas of this approach by means of a simple example.

EXAMPLE 8: Consider the case in which there exist only two stages: normal cells and cells affected by a pollutant biomarker $\mathcal{B}(t)$ --Fig. 9. The evolution of the number of cells at each stage is expressed in terms of the following SODEs,

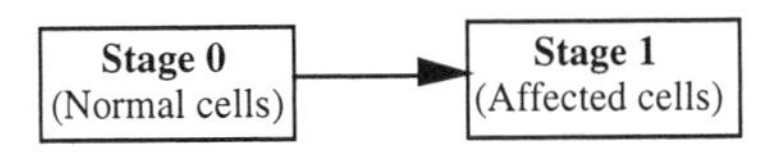

FIGURE 9: Two-stage cell transitions.

$$dn^0(t)/dt = -\lambda(t)\mathcal{B}(t)n^0(t), \tag{43}$$

and

$$dn^a(t)/dt = \lambda(t)\mathcal{B}(t)n^0(t), \tag{44}$$

where $n^0(t)$ and $n^a(t)$ are, respectively, the numbers of normal and affected cells at time t, and $\lambda(t)$ is the transition rate from normal to affected cells. The random field *realization-based* solution to Eq. (43) is

$$n^0(t) = n^0(0)\exp[-\int_0^t d\tau\,\lambda(\tau)\mathcal{B}(\tau)], \tag{45}$$

where $n^0(0) = N$ is the total number of cells in the target organ before the pollutant causes any transitions. Assuming that sufficient information exists so that random field realizations of $\mathcal{B}(t)$ and $\lambda(t)$ can be generated (using random field methods; Christakos, 1992), realizations of the time evolution of the corresponding normal cell number $n^0(t)$ may be obtained from Eq. (45). Given these realizations, Eq. (44) can be solved as

$$n^a(t) = \int_0^t dt'\,\lambda(t')\mathcal{B}(t')n^0(t') = n^0(0)\int_0^t dt'\,\lambda(t')\mathcal{B}(t')\exp[-\int_0^{t'} d\tau\,\lambda(\tau)\mathcal{B}(\tau)]. \tag{46}$$

At this point, the probability of a health effect may be approximated by the probability of a cell being affected, i.e.,

$$H_P(t) \approx n^a(t)/n^0(0) = \int_0^t dt' \lambda(t') \mathcal{B}(t') \exp[-\int_0^{t'} d\tau \lambda(\tau) \mathcal{B}(\tau)]; \tag{47}$$

in the special case of constant biomarker and transition rate, Eq. (47) reduces to $H_P(t) \approx n^a(t)/n^0(0) = 1 - \exp(-\lambda \mathcal{B} t)$. □

Since we are dealing with random fields an improved estimate of the health effect probability may be evaluating the average over a large number K of $n^a(t)$-realizations

$$H_P(t) = \overline{n^a}(t)/n^0(0) \approx \sum_{i=1}^K n_i^a(t)/[Kn^0(0)]. \tag{48}$$

The reader may have noticed that Eq. (48) looks like a 1st-order approximation to Eq. (16) above. There is, however, an important difference: while Eq. (16) takes into consideration spatial correlations between cells, Eq. (48) does not in its current form.

Under certain conditions, the stochastic techniques of §3.2 can be combined with the present analysis to produce useful results.

EXAMPLE 8(cont.): Assume a stochastically homogenous and ergodic biomarker distribution $\mathcal{B}(\boldsymbol{p})$. According to the threshold analysis in §3.2, the expected cell indicator at time t, $R_c(t,\zeta)$, may be estimated by the fraction of the affected cells over the total number of cells, i.e.,

$$R_c(t,\zeta) \approx n^a(t,\zeta)/N, \tag{49}$$

where ζ denotes that the number of affected cells depends on the threshold. By comparing Eqs. (47) and (49) we see that

$$R_c(t,\zeta) \approx \int_0^t dt' \lambda(t') \mathcal{B}(t') \exp[-\int_0^{t'} d\tau \lambda(\tau) \mathcal{B}(\tau)], \tag{50}$$

assuming, as before, $n^0(0) = N$. It is noteworthy that Eq. (50) establishes a connection between the threshold of Postulate 1 and the SODE solution of Postulate 2. Given $R_c(t,\zeta)$, other useful cell indicators can be found by solving the following set of constitutive equations

$$dL_c/d\zeta = -R_c, \tag{51}$$

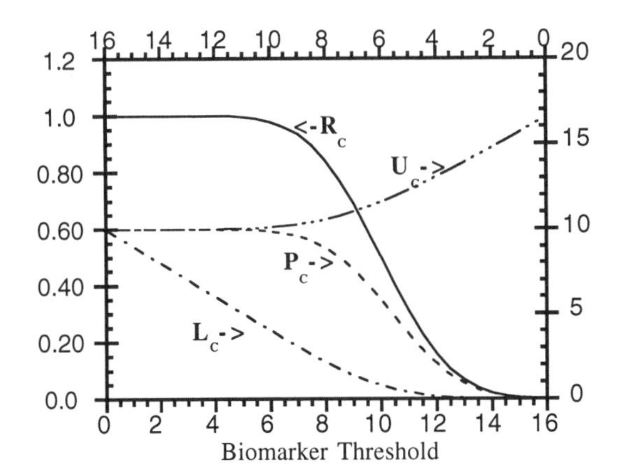

FIGURE 10: Plot of cell indicators R_c, L_c, P_c and U_c vs. threshold ζ for a Gaussian cell biomarker distribution. The values of the R_c are given on the left vertical axis and its threshold on the bottom horizontal axis. The values of the other indicators are given on the right vertical axis and their thresholds on the top horizontal axis.

$$dP_c/d\zeta = \zeta\, dR_c/d\zeta, \tag{52}$$

$$P_c = R_c U_c, \tag{53}$$

$$L_c = R_c(U_c - \zeta). \tag{54}$$

Fig. 10 presents plots of the cell indicators obtained using the system of equations above. □

Scholium 1: *The threshold cell analysis in §3.2 can be generalized to include several cell transition stages, that is, a combination of Postulates 1 and 2 is possible.*

This combination may involve an extension of the definition of the cell indicator field in Eq. (1) to incorporate the multistage cell evolution. In this case, the $I_c(\boldsymbol{p},\zeta)$ will take its values in the interval [0, 1] associated with the various stages, instead of the binary set {0, 1}.

REMARK 3: Due to the presence of cell *repair* mechanisms, however, the resulting system of differential equations may be coupled. Then, analytical solutions can be derived only in a limited number of cases, provided that the system can be linearized. In most other cases, a numerical solution must be obtained.

4. STOCHASTIC FORMULATION OF TRADITIONAL POPULATION HEALTH EFFECT INDICATORS

With regard to geographically and temporally distributed populations, the study of health effects and the associated levels of health needs are of interest to environmental epidemiology and medical geography (Briggs and Elliott, 1995). The investigation goals are threefold: (i) the study of disease distributions in populations, (ii) the analysis of the factors that influence these distributions, and (iii) the evaluation of health management services.

The message of the book is that environmental epidemiology can benefit considerably by incorporating in its methodology the rigorous concepts and tools of modern stochastics. The latter can offer valuable assistance in making appropriate epidemiologic hypotheses and developing suitable models for their evaluation.

As we saw in Chapter II, environmental health science is concerned with two main scales: The individual receptor scale, and the population scale. In the first case the health effect is expressed in terms of measurable transitions between health states of the individual receptor. In the second case the health effect is expressed in terms of the number of receptors affected by the exposure within the population of interest. The population scale is the focus of this section. Hence, the effects of environmental exposure on human health are estimated by means of population health indicators.

Due to the randomness and space-time (intrasubject and intersubject) variabilities and biological uncertainties (in intake routes, physiology, etc.), the health effects are represented in terms of S/TRFs. An important attraction of the S/TRF model is that it provides a unified formalism for all three kinds of epidemiologic studies, viz., cohort, case-control and ecologic studies. Then, one speaks of the probability of an individual contracting a disease, given information about geographical, temporal, exposure, intake routes, and other health state parameters.

S/TRFs representing health effects have some interesting properties. While in the case of natural processes the S/TRF representation is a function of the spatiotemporal coordinates, in the case of health processes the S/TRF is also a function of "internal coordinates" such as health condition, exposure intensity, age, race and sex (Chapters I and II). The state of health of a population can be adequately described in terms of S/TRF-based *health effect indicators.* A variety of health indicators, including rates and ratios, can be derived on the basis of two fundamental indicators as follows.

4.1 Two Fundamental Health Effect Indicators

Consider a population of N receptors facing a health risk during the time period T. Let the population N be located within a domain $\Lambda(\boldsymbol{p})$ surrounding a reference point $\boldsymbol{p}$ in space-time. In a dynamic population (i.e., a population that gains and loses members), N is the average size, often the estimated population at mid-period. A health-related event e (a specific disease, death, etc.) may occur at any point/individual within $\Lambda(\boldsymbol{p})$.

EXAMPLE 1: A population domain $\Lambda(\boldsymbol{p})$ in $R^2 \times T$ is shown in Fig. II.10. Each point/individual ○ corresponds to an event e. □

Next, we give the definition of two fundamental health effect indicators.

Definition 1: *The individual health effect indicator is defined as the event-dependent S/TRF*

$$I_H(\boldsymbol{p},e) = \begin{cases} 1, & \text{if the event } e \text{ occurs at point } \boldsymbol{p} \\ 0, & \text{otherwise} \end{cases} . \tag{1}$$

Definition 1 associates each individual with a point $\boldsymbol{p} = (\boldsymbol{s},t)$ in space-time and the occurrence or absence of an event e. The mean value $\overline{I_H}(\boldsymbol{p},e)$ of the health effect indicator (1) gives the probability of occurrence of the event e at point $\boldsymbol{p}$ in space-time. Furthermore, it gives rise to the following

Definition 2: *The population health effect indicator per* 10^m *person-years is defined as*

$$P_H(\boldsymbol{p},e) = \overline{I_H}(\boldsymbol{p},e) \times 10^m, \tag{2}$$

where m *is an integer.*

In practice, $\overline{I_H}(\boldsymbol{p},e)$ can be estimated by

$$\overline{I_H}(\boldsymbol{p},e) \approx \sum_{\boldsymbol{p}' \in \Lambda(\boldsymbol{p})} I_H(\boldsymbol{p}',e) \big/ N = n(\boldsymbol{p},e)/N, \tag{3}$$

where $n(\boldsymbol{p},e) \subset N$ denotes the number of events within $\Lambda(\boldsymbol{p})$ (i.e., within the population N at risk during the time period T). The estimate (3) is based on a *population ergodicity* assumption, which requires that the ensemble average over all S/TRF-realizations be equal to the sample average over the space-time domain $\Lambda(\boldsymbol{p})$. Using Eq. (3), Eq. (2) gives

$$P_H(\boldsymbol{p},e) \approx n(\boldsymbol{p},e) \times 10^m \big/ N, \tag{4}$$

which expresses the population health effect indicator as a ratio (proportion) of the total population. The term *population rate* is used for the health effect indicator.

Sometimes, studies focus on a specific population factor, such as age, sex, race, or marital status. The indicator $P_H(\boldsymbol{p},e)$ in Eq. (4) is called *age-specific* if $n(\boldsymbol{p},e)$ and N refer to the same age group. Similarly, the $P_H(\boldsymbol{p},e)$ is called *cause-specific* if it represents health effects due to a specific cause.

A population is called *heterogeneous* with respect to a specific factor if the value of the latter is not constant for the entire population. A heterogeneous population N can be divided into K strata and λ_i ($i = 1,\ldots,K$) are the ratios of all members of N that belong to the respective stratum (e.g., age intervals or race groups). Clearly, $\sum_{i=1}^{K} \lambda_i = 1$.

EXAMPLE 2: The data in this example are borrowed from Fleiss (1981). Let $N =$ population of mothers that gave birth to mongoloid infants in the state of Michigan during the time period from 1950 to 1964. The population is divided into $K = 6$ strata according to maternal age interval (column 1 of Table 3). The age-specific distribution λ_i ($i = 1,\ldots,K$) is shown in column 2 of Table 3. □

When the distribution λ_i ($i = 1,\ldots,K$) of the specific population factor is unknown, Eq. (4) provides the *crude* or *total health indicator* for the population N.

EXAMPLE 2 (cont.): The total population of infants in the state of Michigan, from 1950 to 1964, was $N = 2{,}825{,}173$. The mongoloids were $n(\boldsymbol{p},e) = 2{,}529$. Hence, from Eq. (4) the crude health indicator (rate of mongolism) for the state as a whole is $P_H(\boldsymbol{p},e) = 2{,}529 \times 10^5 \big/ 2{,}825{,}173 = 89.5$ per 100,000 live births. □

TABLE 3: Data on mongoloid infants, state of Michigan, 1950-1964 (from Fleiss, 1981)

(1) Maternal age	(2) Distribution for all Michigan λ_i	(3) Specific rates for all Michigan per 10^5 $P_{H,i}$	(4) $\lambda_i P_{H,i}$
<20	.113	42.5	4.80
20-24	.330	42.5	14.03
25-29	.278	52.3	14.54
30-34	.173	87.7	15.17
35-39	.084	264.0	22.18
>40	.022	864.4	19.02
Total	1.000		89.74

When the distribution λ_i is known, a *factor-weighted* health indicator may be defined as follows.

Definition 3: *Assume that $P_{H,i}(\boldsymbol{p},e)$ are indicators specific to the stratum i of the population with known distribution λ_i $(i=1,...,K)$. The factor-weighted health effect indicator is defined as*

$$P_H^{\lambda}(\boldsymbol{p},e) \approx \sum_{i=1}^{K} \lambda_i \, P_{H,i}(\boldsymbol{p},e). \tag{5}$$

EXAMPLE 2 (cont.): The indicators (rates of mongolism) specific to each population interval are shown in Table 3 (column 3). It follows from Eq. (5) that the weighted rate of mongolism in the state of Michigan is $P_H^{\lambda}(\boldsymbol{p},e) = 89.74$ per 100,000 live births. The rates per stratum are shown in column 4 of Table 3. □

In some cases, *factor-adjustments* (mainly age-adjustments) are performed on health indicators. The age-adjustment procedure, e.g., replaces the indicators for a given population with values that represent the age distribution observed in a standard population N^s. Adjustment is typically performed *standardization* (Mahon and Pugh, 1970; Wassertheil-Smoller, 1996; also, §4.3 below). *Indirect* standardization is used when the specific indicators $P_{H,i}(\boldsymbol{p},e)$ are either statistically unstable or unknown. *Direct* standardization is applicable only when the $P_{H,i}(\boldsymbol{p},e)$ are known. Related to these two kinds of standardization are the following definitions.

Definition 4: *Assume that* $P^s_{H,i}(\boldsymbol{p},e)$ *are strata-specific indicators of a selected standard population* N^s *with population strata-distribution* λ^s_i $(i = 1,...,K)$*. The indirectly standardized health effect indicator is defined as*

$$P_{H,indir}(\boldsymbol{p},e) = P^s_H(\boldsymbol{p},e)\,P_H(\boldsymbol{p},e)\,/\,P^{\lambda,s}_H(\boldsymbol{p},e), \tag{6}$$

where $P^s_H(\boldsymbol{p},e) = \sum_{i=1}^K \lambda^s_i P^s_{H,i}(\boldsymbol{p},e)$, $P^{\lambda,s}_H(\boldsymbol{p},e) = \sum_{i=1}^K \lambda_i P^s_{H,i}(\boldsymbol{p},e)$ *and* $P_H(\boldsymbol{p},e)$ *is the crude health indicator.*

EXAMPLE 2 (cont.): We continue with the analysis of the data on mongoloid infants. In the state of Michigan, from 1950 to 1964, 731,177 infants were first-born to their mothers; of these, 412 were mongoloids giving a crude rate $P_H(\boldsymbol{p},e) = 56.3$ per 100,000 first-borns. The selected standard population was all the live births during the years 1950-1964 so that $P^s_H(\boldsymbol{p},e) = 89.5$ per 100,000 live births. In column 4 of Table 4 the specific rates

TABLE 4: Data on mongoloid infants, state of Michigan, 1950-1964 (from Fleiss, 1981)

(1) Maternal age	(2) Distribution for all Michigan	(3) Specific rates for all Michigan per 10^5	Birth order (4) First		(5) Fifth or more	
	λ^s_i	$P^s_{H,i}$	λ_i	$\lambda_i P^s_{H,i}$	λ_i	$\lambda_i P^s_{H,i}$
<20	.113	42.5	.315	13.4	.001	0.0
20-24	.330	42.5	.451	19.2	.069	2.9
25-29	.278	52.3	.157	8.2	.279	14.6
30-34	.173	87.7	.054	4.7	.339	29.7
35-39	.084	264.0	.019	5.0	.235	62.0
>40	.022	864.4	.004	3.5	.078	67.4
Total	1.000			54.0		176.6

per 100,000 first-borns are shown. The crude rate that would follow if the population of first-borns were subject to the standard population schedule of rates $P^s_{H,i}(\boldsymbol{p},e)$ is $P^{\lambda,s}_H(\boldsymbol{p},e) = 54.0$ per 100,000 live births (see column 4 of Table 4). Hence, from Eq. (6) the indirectly age-standardized rate of mongolism is $P_{H,indir}(\boldsymbol{p},e) = 93.3$ mongoloids per 100,000 first-borns. This value makes sense, because the age distribution for mothers of first-borns is skewed towards the younger age groups that have lower rates of mongolism. Hence, the value of the rate increases when it is adjusted to the standard population distribution. □

In certain cases one needs to compare the rate for some event across different populations. If the populations are not similarly constituted with respect to factors associated with the event (e.g., age, sex or race), the direct comparison of the overall health indicators may be misleading. It is more reasonable to compare the factor-adjusted indicators obtained by means of standardization.

EXAMPLE 2 (cont.): In the same 15-year period, 442,811 infants were the result of fifth or higher births; of these, 740 were mongoloid, giving a crude rate of $P_H(\boldsymbol{p},e) = 167.1$ per 100,000 fifth-born or more. A similar adjustment analysis gives $P_{H,indir}(\boldsymbol{p},e) = 84.7$ per 100,000 infants born fifth or later. The conclusions drawn from crude rates are different from these obtained on the basis of the age-adjusted rates. Indeed, the crude rates for first-borns are $P_H(\boldsymbol{p},e) = 56.3$ per 100,000 first-borns, and for fifth or later births $P_H(\boldsymbol{p},e) = 167.1$ per 100,000. This may lead to the conclusion that there is an almost threefold increase in risk for mongolism as the order of birth increases from one to five or higher. However, the corresponding age-adjusted rates are $P_{H,indir}(\boldsymbol{p},e) = 93.3$ mongoloids per 100,000 for first births, and $P_{H,indir}(\boldsymbol{p},e) = 84.7$ per 100,000 for fifth or higher births. Hence, after adjustment there is no substantial difference in risk of mongolism as the birth order increases. The apparent greater risk for later births, suggested by the comparison of the crude rates, follows from differences in maternal age distribution. Age is the actual risk factor, and proportionately more mothers of later-born infants are in older age groups, where the specific rates are higher, than are mothers of first-born infants. After adjustment for differences in the maternal age distribution, it appears that, if anything, the rate of mongolism in later-born infants is reduced compared to the rate in first-born infants. □

Direct standardization is applicable only when the $P_{H,i}(\boldsymbol{p},e)$ are known. This standardization is defined as follows:

Definition 5: *Let $P_{H,i}(\boldsymbol{p},e)$ be strata-specific indicators of the population N of interest, and let λ_i^s ($i = 1,\ldots,K$) be the population strata-distribution of a selected standard population N^s. The directly standardized health indicator is defined as*

$$P_{H,dir}(\boldsymbol{p},e) = \sum_{i=1}^{K} \lambda_i^s \, P_{H,i}(\boldsymbol{p},e). \tag{7}$$

EXAMPLE 2 (cont.): Let us calculate the directly standardized rates for first-birth as well as 5th- or higher-order birth mongolism from the Michigan data. The detailed calculations involved in Eq. (7) are shown in Table 5. They give $P_{H,dir}(p,e) = 92.5$ mongoloids per

TABLE 5: Data on mongoloid infants, state of Michigan, 1950-1964 (from Fleiss, 1981)

(1) Maternal age	(2) Distribution for all Michigan λ_i^s	Birth order (3) First $P_{H,i}$	$\lambda_i^s P_{H,i}$	(4) Fifth or more $P_{H,i}$	$\lambda_i^s P_{H,i}$
<20	.113	46.5	5.3	0.0	0.0
20-24	.330	42.8	14.1	26.1	8.6
25-29	.278	52.2	14.5	51.0	14.2
30-34	.173	101.3	17.5	74.7	12.9
35-39	.084	274.5	23.1	251.7	21.1
>40	.022	819.1	18.0	857.8	18.9
Total	1.000		92.5		75.7

100,000 first-borns, and $P_{H,dir}(p,e) = 75.7$ per 100,000 fifth-born or more. The conclusion is that there are no essential differences in mongolism risk between first-birth and fifth-birth infants, in agreement with the conclusion reached based on the indirectly standardized rates . □

The usefulness of standardization is the subject of an ongoing debate (Mausner and Bahn, 1974; Fleiss, 1981; Last, 1995). Specific health indicators (Definition 2) provide a complete characterization of the population health conditions. Standardized health indicators (Definitions 3 through 5) are used in certain cases, for they provide a single summary indicator for a population instead of a set of specific indicators. However, a single summary indicator fails to capture the variability of specific indicators across the population strata. Nevertheless, standardized health indicators are useful if specific indicators are not available, or if the population strata include only a few individuals, thus making the calculation of specific health indicators unreliable.

4.2 Person-Time Rates

On the basis of the general definitions above, several health indicators can be derived in the form of *person-time rates.*

Definition 6: *If e=death occurrence in $\Lambda(\boldsymbol{p})$, the population health effect indicator (4) is called the crude mortality rate (or death rate).*

EXAMPLE 3: Consider the event e = death due to breast cancer and the space-time domain $\Lambda(\boldsymbol{p})$ in which $n(\boldsymbol{p},e) = 505$ deaths, $N = 2{,}093{,}000$ people, $T = 5$ years and $m = 6$. Then, the crude mortality rate is $P_H(\boldsymbol{p},e) = 505 \times 10^6/2{,}093{,}000 = 241.3$ per million per 5 years. Alternatively, $P_H(\boldsymbol{p},e) = 505 \times 10^6/2{,}093{,}000 \times 5 = 48.3$ per million per year. □

Definition 7: *If e=death due to a specific cause in $\Lambda(\boldsymbol{p})$, the population health effect indicator (4) is called the cause-specific mortality rate.*

EXAMPLE 4: The data in this example are from MacMahon and Pugh (1970). In Table 6 data on white population with acute leukemia in $\Lambda(\boldsymbol{p})$ = Brooklyn, New York, between 1948 and 1952 are presented. In 1950, the white population was $N = 2{,}525{,}000$ people. For the calculation of the acute leukemia-specific mortality rate we have $n(\boldsymbol{p},e) = 395$ deaths (column 3 of Table 6), $T = 5$ years and let $m = 6$. Then, the crude mortality rate is

TABLE 6: Data on patients with acute leukemia, Brooklyn, New York, whites, 1948-1952 (from MacMahon and Pugh, 1970)

Year	(1) Patients alive at beginning of year	(2) New cases diagnosed in year	(3) Deaths in year	(4) Patients lost to trace
1948	7	69	54	7
1949	15	91	86	3
1950	17	83	73	3
1951	24	99	101	1
1952	21	68	81	1
Total	84	410	395	15

$P_H(\boldsymbol{p},e) = 395 \times 10^6/2{,}525{,}000 = 156.4$ per million per five years, or $P_H(\boldsymbol{p},e) = 156.4/5 = 31.3$ per million per year. □

Definition 8: *If e=death occurring in the specific age group in $\Lambda(\boldsymbol{p})$, the population health effect indicator (4) is called the age-specific mortality rate.*

EXAMPLE 4 (cont.): Consider the data of Table 6. Assume that among the deaths $n(\boldsymbol{p},e) = 215$ were people of 40 years of age. Then, the age-specific mortality rate is

$P_H(\boldsymbol{p},e) = 215 \times 10^6/2{,}525{,}000 = 85.1$ per million per 5 years, or $P_H(\boldsymbol{p},e) = 215 \times 10^6/2{,}525{,}000 \times 5 = 17$ per million per year. □

The following two indicators (rates) -incidence and prevalence- provide measures of morbidity (illness).

Definition 9: *If* e *=new case of a disease in* $\Lambda(\boldsymbol{p})$*, the population health indicator (4) is called the incidence rate.*

EXAMPLE 4 (cont.): To calculate the incidence rate for the data of Table 6, we consider the $n(\boldsymbol{p},e) = 410$ new cases diagnosed with acute leukemia during the $T = 5$ year period (column 2) and derive the incidence rate $P_H(\boldsymbol{p},e) = 410 \times 10^6/2{,}525{,}000 = 162.4$ per million per 5 years, or $P_H(\boldsymbol{p},e) = 410 \times 10^6/2{,}525{,}000 \times 5 = 32.5$ per million per year. □

REMARK 1: Even though it is believed that many forms of cancer are preventable, most causes of cancer have yet to be identified with a sufficient degree of certainty. In the case of breast cancer, e.g., a variety of risk factors are known, but it remains unclear which factors influence most actively cancer incidence. Incidence rates give a clear picture of the burden of breast cancer in a population. They often give an accurate specification of anatomical site where the cancer occurs and an indication of histological type. Also, as is discussed in Boyle (1989), cancer incidence data have a number of advantages over cancer mortality data for describing both the nature and extent of the cancer burden in a community, and for affording international comparisons.

Definition 10: *If* $e =$ *occurrence of a disease at a particular time* t *(or during period* T*), the population health indicator (4) is called the prevalence rate.*

EXAMPLE 4 (cont.): To calculate the prevalence rate for the data of Table 6, we consider the $n(\boldsymbol{p},e) = 84$ existing cases of acute leukemia during the $T = 5$ year period (column 1) and derive the prevalence rate by $P_H(\boldsymbol{p},e) = 84 \times 10^6/2{,}525{,}000 = 33.5$ per million per 5 years, or $P_H(\boldsymbol{p},e) = 84 \times 10^6/2{,}525{,}000 \times 5 = 6.7$ per million per year. □

Note that the prevalence rate refers to existing cases of disease, while the incidence rate refers to new cases of disease developed during a particular time period. For chronic diseases the prevalence rate is larger than the incidence rate, because prevalence accounts for new as well as existing cases. For rapidly fatal or quickly curable diseases the prevalence rate may be approximately equal to the incidence rate.

4.3 Rate Ratios

Another useful group of population health indicators is expressed in terms of rate ratios. One of the reasons for studying population health indicators is the subsequent investigation of etiological disease factors. In order to compare indicators obtained under different conditions (geographical regions, populations, year, etc.), the effects of *confounding* factors, which potentially lead to differences should first be removed. Removing these effects is the job of *standardization*. We start with a general definition.

Definition 11: *A standardized rate ratio is a rate ratio in which the numerator and denominator rates have been standardized to the same (standard) population distribution.*

The most common standardization methods use weighted averaging of rates specific for age (mainly), sex, marital status, or other potential confounding factors based on a specified distribution of these factors (Fleiss, 1981). *Standardized mortality ratio (SMR)* is the ratio of the number of deaths observed in the study population to the number of deaths that would be expected if the study population had the same specific rates as the standard population; SMR is usually expressed as a percentage. *Standardized incidence ratio (SIR)* is the ratio of the incident number of cases of a specified condition in the study population to the incident that would be expected if the study population had the same --known-- incidence rate as a standard population. The SIR is usually expressed as a percentage.

4.4 Spatiotemporal Variation and Mapping of Population Health Indicators - The case of the North Carolina Breast Cancer Incidence Distribution

The spatiotemporal variation analysis of a population health indicator $P_H(\boldsymbol{p}, e)$ can provide clues to disease etiology, thus allowing the implementation of preventive measures and stimulating fruitful investigation (Burkitt, 1962). A cancer incidence atlas, e.g., constitutes a useful basis for the investigation of cancer causation and the efficient organization of anti-cancer campaigns (Hutt and Burkitt, 1986; Pickle, *et al.*, 1989).

Since population health indicators are modelled as S/TRFs, the stochastic methods presented in the previous Chapters can be used to study them. There is significant epidemiologic evidence that supports the use of the S/TRF model in disease indicator analysis. For example, the well-established fact that commonly-occurring cancers show considerable geographical and time variations in incidence (Doll, 1980; Higginson, 1979) provides a rationale for the analysis and mapping of breast cancer incidence variation by means of the S/TRF model. The analysis can produce a sequence of space-time maps of incidence estimates and an assessment of the accuracy of these estimates at any unsampled location/time (i.e., where incidence has not been registered). Incidence maps are obtained using data at the same time period as well as at other time periods and spatial locations. Maps of disease incidence in human populations may be utilized to quantify the variations of disease risk in space and time, elucidate causal mechanisms, explain the occurrence of disease at local or global scales, describe the natural history of disease, or provide guidance in the administration of health services (adoption of cancer prevention measures and control strategies). In recent years, the overall U.S. breast cancer rates (incidence and mortality) among women has remained rather stable (Pisani, 1992; Sturgeon *et al.*, 1995), while rising rates have been observed in many areas of the South, including the State of North Carolina (e.g., in the poor rural areas of Appalachia). Below, we will use the S/TRF model to present a cartographic picture of the "breast cancer landscape" of North Carolina.

EXAMPLE 5: The breast cancer dataset for the years 1990, 1991 and 1992 (incidence per 100,000 population) includes age-adjusted data (on the basis of the 1970's U.S. population) at 100 counties throughout the state (Christakos and Lai, 1997). Fig. 11 shows the map of the 100 North Carolina counties. Many cancer cases from the NE

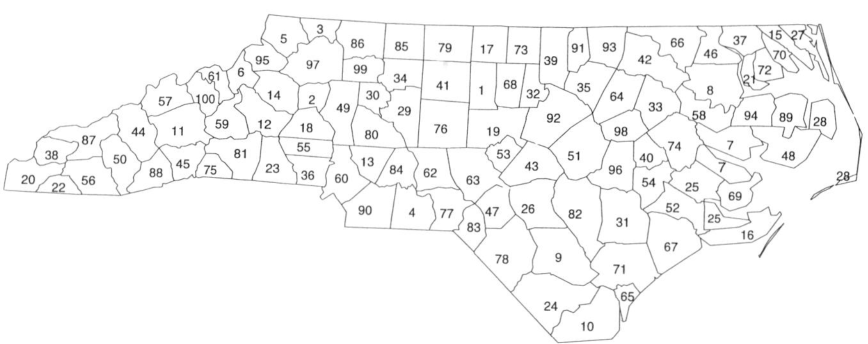

FIGURE 11: Map of the counties of the State of North Carolina.

counties go to Norfolk (Virginia) for treatment and, thus, data for these counties are not reliable (and, hence, are not considered in this analysis). Let the S/TRF $P_H(\boldsymbol{p},e)$ be the

e =breast cancer incidence at an area denoted by $\boldsymbol{p}$. Summary statistics of the breast cancer dataset are shown in Table 7. There is a significant spatial variability in incidence

TABLE 7: Summary statistics of breast cancer incidence (per 100,000 cases).

	Mean	Standard Deviation	Median	Highest value	Lowest Value
1990	92.04	28.48	90.15	189.9	20.9
1991	97.19	32.41	99.20	196.5	4.5
1992	101.16	21.76	100.35	140.3	41.8

distribution. Breast cancer variation analysis using the generalized space-time covariance model of Eq. (IV.8.32) showed that both the spatial and temporal continuity orders ν and μ varied in space and time. This variation was a result of the composite spatiotemporal structure of the breast cancer incidence P_H. A map of the order difference $\nu - \mu$ for the year 1992 was plotted in Fig. IV.3. The structure of the map depends on the scale of analysis (compare Fig. IV.3 with Fig. 14 below). Incidence maps for the years 1990, 1991 and 1992 (Fig. 12) were constructed using the MMSE space-time mapping technique of Chapter VI. This technique generates P_H estimates at several points within each county. Estimates of possible incidence patterns at these points would otherwise be unavailable. An important property of the mapping method is that the estimates are identical with the data at sampling points. At the unsampled points, the mapping technique produces P_H-estimates by using information from data sampled at neighboring locations or years. In some cases, this may lead to some smoothing of the P_H-distribution within each county, as well as in between counties. The temporal variations observed in the limited 3-year period were less significant than spatial variations. The population changes that occurred at certain counties during these years possibly had a contribution to the temporal incidence changes observed. Although there are quantitative differences between the three incidence maps of Fig. 12, they all seem to have the same qualitative features: the incidence contours show a downhill slope in the Western counties and ridges in certain North-Central counties. The space-time breast cancer incidence maps are very accurate, as the mapping error standard deviation histograms testified (Fig. 13). □

Composite space-time analysis is a considerable improvement over purely spatial methods (e.g., the former can provide more accurate and informative incidence maps than the latter). Moreover, spatiotemporal analysis can use temporal data to compensate for the lack of spatial data. Therefore, it may offer accurate estimates based on fewer disease registries, which gives the method a considerable economic advantage over purely spatial techniques.

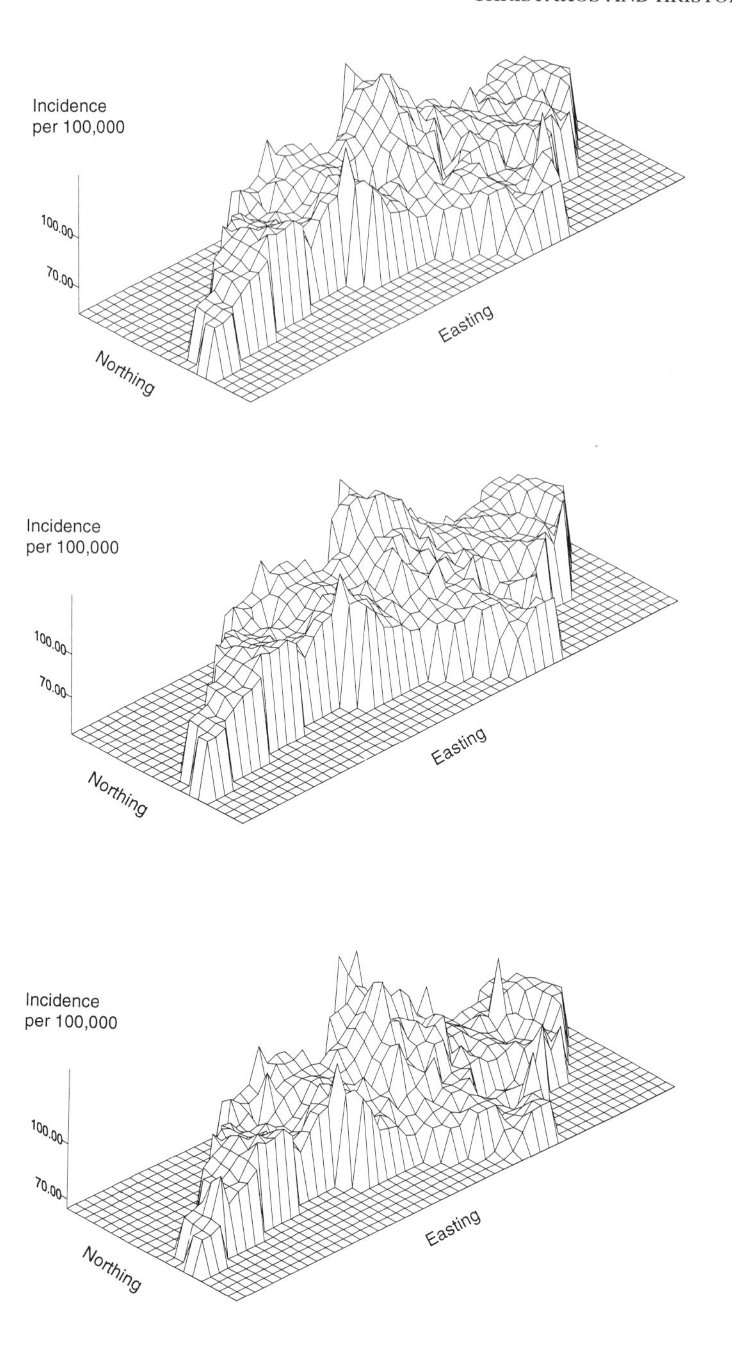

FIGURE 12: Maps of the breast cancer incidence estimates at the year (a) 1990, (b) 1991, and (c) 1992 (Λ =60 miles/3 years).

EXAMPLE 6: As was shown in Christakos and Lai (1997), breast cancer incidence estimates obtained by the spatiotemporal method are in all cases more accurate than

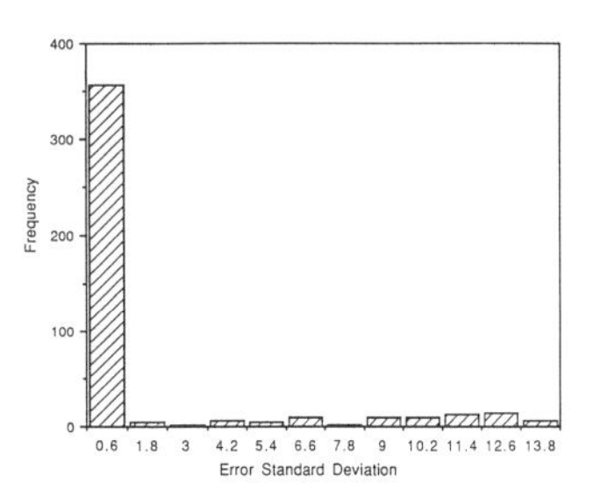

FIGURE 13: Histogram of the error standard deviations of the breast cancer incidence estimates of Fig. 12c at the year 1992.

estimates produced by a purely spatial (kriging) method; the ratio of the spatial estimation error over the spatiotemporal error lied in the interval [1.5, 56.1]. The spatiotemporal method is more accurate even when it used fewer data points in space than kriging. □

Breast cancer incidence maps may show different variations depending on the scale at which the estimates are obtained (counties, socio-economic areas or random county-aggregates). At fine scales, the local space-time fluctuations of the incidence are resolved. Coarser scales may lead to smoothing of high or low incidence areas. Conversely, mapping incidence on dense grids (e.g., using a number of nodes within each county) may identify patterns or high risk areas that would not be noticed if larger geographic units were used. The observation scale also plays an important role regarding the detail of local incidence variation, the detection of large scale trends, and the accuracy of the incidence estimates. Studying incidence at an hierarchy of scales offers information about connections between fine-grain variation and the larger context. This kind of information is important for complex systems requiring the consideration of several levels of analysis.

EXAMPLE 7: The map of Fig. 12c was generated at a local observation scale $\Lambda=$ 60 miles/T=3 years (the scale definition was given in Fig. V.3) that led to small average incidence estimation errors and, at the same time, it provided sufficient detail of the incidence variation. If a larger observation scale (D= 83 miles, T=3 years) is used, the smoothed map of Fig. 14 is obtained; the corresponding $\nu-\mu$ is plotted in Fig. 15 (compare with the map of Fig. IV.3). The map of Fig. 14 captures fewer local details of the breast cancer incidence variability than the map of Fig. 12c, and it has the undesirable

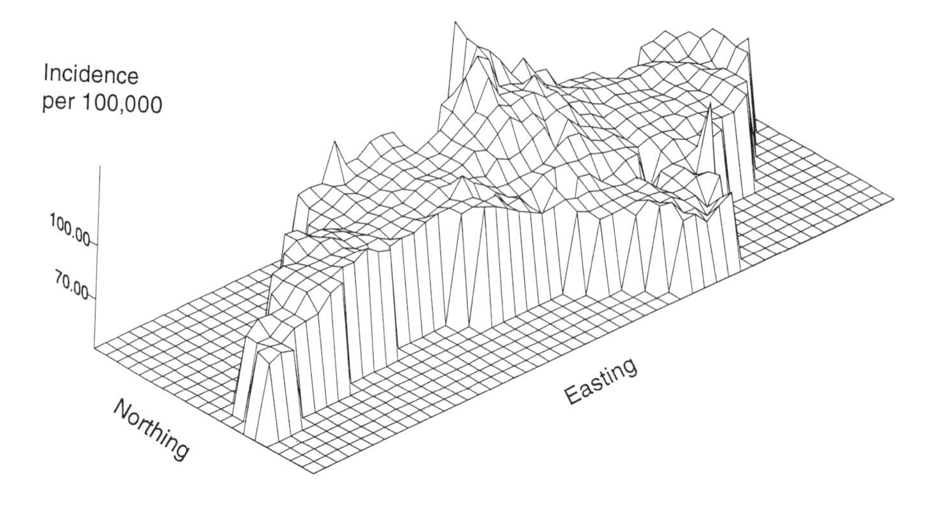

FIGURE 14: Map of the estimates of breast cancer incidence at the year 1992 using the observation scale Λ =86 miles/3 years.

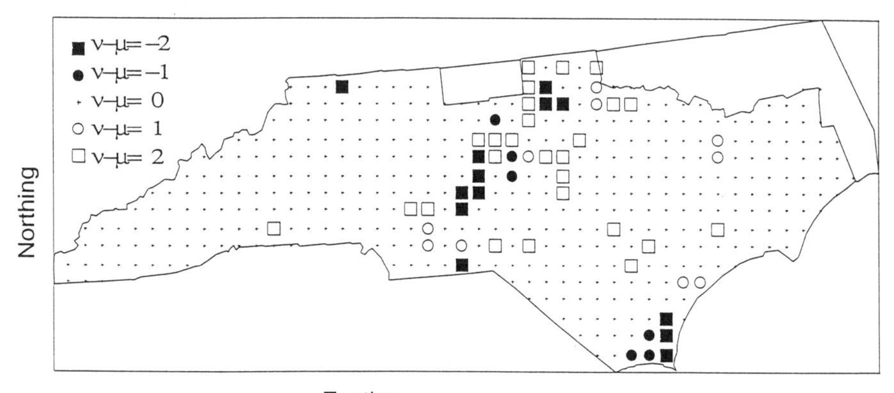

FIGURE 15: Map of the difference $v - \mu$ at the year 1992 (Λ =86 miles/3 years).

effect of reducing the variance of the actual incidence data. □

In conclusion, space-time mapping is an important component of environmental health studies. Mapping can generate realistic descriptions of incidence variation, the study of which can help to identify high incidence areas and to motivate field studies aiming to discover the underlying environmental or life-style factors. Thus, incidence maps constitute a useful research tool for associative studies in analytical epidemiology and medical geography. Certainly, due to the numerous factors involved in breast cancer causation, incidence maps alone do not usually lead directly to the discovery of disease causes. However, they can help in health management and policy by providing quantitative measures of risk. In addition, they offer the means for generating hypotheses and designing analytical studies that can improve understanding of breast cancer etiology.

5. CORRELATION ANALYSIS

The purpose of *correlation analysis* --also called *ecological analysis*-- is to detect associations between response and exposure levels, to estimate the former as a function of the latter, and to suggest --or preferably verify-- explanatory hypotheses.

In the field of toxicology the relationship between the dose of a chemical agent taken into the receptor and the response of the receptor to this dose is of fundamental importance (e.g., Loomis, 1978). In general, receptors show no response to doses below a minimum threshold. As the dose is increased from the minimum threshold to the maximum level a graded response is observed, instead of a sharp transition from the state of no response to the state of complete response (death). We can all appreciate the fact that the dose-response curve shows a graded slope instead of a discontinuous jump.

EXAMPLE 1: Several studies have reported associations between ozone concentration and health effect. Exposure of humans to ozone alters spirometric and permeability functions of the lung (Miller *et al.*, 1985; Lippmann, 1989; Horvath and McKee, 1994). The impacts of ozone concentration to humans are usually analyzed by means of an impact pathway process (McJilton *et al.*, 1972; McDonnell *et al.*, 1985; McCurdy, 1994). Certain studies have indicated significant receptor's response for first-day ozone exposures followed by attenuation of response on subsequent days (e.g., Horvath *et al.*, 1981). □

5.1 Population Health Damage Indicators

Our discussion will focuses on a community wide basis. As a consequence, the health effect (or response or impact) $H_P(\boldsymbol{p})$ on a population due to exposure $E(\boldsymbol{p})$ at the space-time point $\boldsymbol{p} = (s_1, s_2, t)$ is expressed as a frequency distribution or percentage of the total population.

EXAMPLE 2: The health impact $H_P(\boldsymbol{p})$ may denote the frequency of pulmonary function decrements --measured in forced expiratory volume FEV_1-- due to exposure to ozone concentration $E(\boldsymbol{p})$; Whitfield *et al* (1995). □

In addition to chemical agents and hazardous substances, the term "exposure" sometimes refers to living habits (e.g., dietary and reproductive) that present risk factors, because they lead to exposure to certain chemicals or biological agents.

EXAMPLE 3: Known risk factors for breast cancer include dietary and reproductive habits, hormonal patterns and family history (Berrino *et al.*, 1989; De Waard and Trichopoulos, 1988; Parkin, 1989). The interpretation of the spatiotemporal breast cancer incidence maps produced in the previous sections should involve carefully designed analytical studies aiming to elucidate the role of the above factors. Incidence maps can be generated for different ethnic, sex and age groups, and they may alert clinicians to strong trends or unusual clustering of rare tumors in certain population groups. This will in turn help to identify responsible exposures, motivate scientific research, and stimulate public and political concern (e.g., the conclusions of many existing etiologic analyses are consistent with geographic disease patterns; Glattre, 1989). □

Since exposure $E(\boldsymbol{p})$ is represented as an S/TRF, so is the population health effect $H_P(\boldsymbol{p})$. The two are generally related by a *probabilistic exposure-response relationship* (*PER*) of the form --see Eq. (II.4.8)--

$$H_P(\boldsymbol{p}) = \mathcal{F}_{PER}[E(\boldsymbol{p})]. \tag{1}$$

Eq. (1) accounts implicitly for the absorption of the pollutant from the air into the bodies of the receptors. This is typically the case of classical air pollutants (ozone, sulfate, nitric oxides). The relationship between air quality and health status in a population is usually complicated (McCurdy, 1994). Typical $F_{PER}[\cdot]$-curves were plotted in Fig. II.8. The PER relationship (1) is subject to a considerable amount of uncertainty and, thus, any decisions based upon it are also subject to uncertainty. Other factors, in addition to the exposure level, affect the PER relationship (1) --see also Chapter II. The example below illustrates some of these factors.

EXAMPLE 4: In the case of ozone, factors that have been identified as potentially affecting the form of the PER include the duration of the exposure, the exertion level during exposure (exercise, manual labor, etc.), pre-existing disease, the gender and age groups of the receptors (Horvath and McKee, 1994; Rombout and Schwarze, 1995). In practice these factors are specified in advance, and the effect is expressed as a function of the ozone concentration only (Hazucha, 1993). Different health effects have been attributed to short term (1-3 hours), prolonged (6-8 hours or longer), and long-term (months or years) ozone exposures. Respiratory effects have been observed at increased exertion levels (even at low ozone concentrations). Individuals with pre-existing limitations in pulmonary function (e.g., asthmatics) may experience respiratory problems with greater clinical significance than healthy individuals. There is no definitive evidence for sex-based differences in the respiratory susceptibility to ozone. Age plays a role in determining the health effects of ozone (older persons exhibit decreased sensitivity to ozone relative to young adults). □

Environmental policy and health management rely heavily on the rigorous assessment of the damage to receptors caused by air pollution (Krewski *et al.*, 1990; Curtiss and Rabl, 1996). This motivates the introduction of the following population health damage indicators. Consider a geographical region $\upsilon(\mathbf{s})$ centered at point $\mathbf{s}$ within a global domain V [e.g., V=Eastern U.S. and $\upsilon(s)$=city of Raleigh].

Definition 1: *The average local health damage $\Psi_{\upsilon}(\boldsymbol{p})$ to the population of the region $\upsilon(\boldsymbol{s})$ at time t due to the effect $H_P(\boldsymbol{p})$ of Eq. (1) is defined as*

$$\Psi_{\upsilon}(\boldsymbol{p}) = |\upsilon(\boldsymbol{s})|^{-1} \int_{\upsilon(s)} ds' \, \theta(s-s',t) \, \mathcal{F}_{PER}[E(s-s',t)], \tag{2}$$

where $\theta(\boldsymbol{s},t)$ is the density of receptors in the $\upsilon(\boldsymbol{s})$-neighborhood and $\boldsymbol{s}'$ denotes the displacement from location $\boldsymbol{s}$.

The integral in Eq. (2) and in the following equations are over the surface of the area considered. The interpretation of the local damage depends on the definition of the health effect. If $H_P(\boldsymbol{p})$ is a frequency that takes values between 0% and 100%, $\Psi_{\upsilon}(\boldsymbol{p})$ represents the number of people in the population of the region that exhibit the effect.

EXAMPLE 5: If $H_P(\boldsymbol{p})$ is the frequency (%) of physiologic decrements in a population due to ozone concentration level $E(\boldsymbol{p})$, $\Psi_{\upsilon}(\boldsymbol{p})$ will denote the anticipated number of people per unit area with such problems in the region of interest. □

From the exposure-damage relationship (2), the following definition can be obtained.

Definition 2: *The dimensionless local damage index is given by*

$$\psi_{\upsilon}(\boldsymbol{p}) = \Psi_{\upsilon}(\boldsymbol{p})/\Psi_{V}(t), \tag{3}$$

where

$$\Psi_{V}(t) = |V|^{-1}\int_{V} ds\, \theta(s,t)\, \mathcal{F}_{PER}[E(s,t)] \tag{4}$$

is the average (global) damage.

Hence, the $\psi_{\upsilon}(\boldsymbol{p})$ represents the ratio of the anticipated local number of incidents (e.g., people affected by ozone exposure in a specific county) over the anticipated global (e.g., Eastern U.S.) number of incidents.

The surface-averaged exposure within a local region $\upsilon(\boldsymbol{s})$ is (§II.3)

$$E_{\upsilon}(\boldsymbol{p}) = |\upsilon|^{-1}\int_{\upsilon(\boldsymbol{s})} ds'\, E(\boldsymbol{s}-\boldsymbol{s}',t). \tag{5}$$

It would be useful to know how changes in E_{υ} affect the local health damage index ψ_{υ} for a prespecified set of factors that potentially affect the PER relationship. Such an indicator is defined below.

Definition 3: *The elasticity of* ψ_{υ} *with respect to* E_{υ} *is defined as*

$$\varphi_{x}(\psi_{\upsilon}) = (d\psi_{\upsilon}/dE_{\upsilon})(E_{\upsilon}/\psi_{\upsilon}) = d\ell n\psi_{\upsilon}/d\ell nE_{\upsilon}. \tag{6}$$

The elasticity $\varphi_{x}(\psi_{\upsilon})$ is a dimensionless indicator that measures the ratio of the fractional change in ψ_{υ} over the fractional change in E_{υ}. The elasticities of Ψ_{V} and Ψ_{υ} can be defined in a similar fashion. The following relationship is satisfied (Christakos and Vyas, 1998b)

$$\varphi_{x}(\psi_{\upsilon}) = \varphi_{x}(\Psi_{\upsilon}) - \varphi_{x}(\Psi_{V}). \tag{7}$$

While the $\varphi_x(\Psi_V)$ has a single value at each time t that is representative of the global region, $\varphi_x(\Psi_\upsilon)$ and $\varphi_x(\psi_\upsilon)$ vary in both space and in time. The following conclusions can be drawn on the basis of Eqs. (6) and (7). Assume a positive value of $\varphi_x(\Psi_V)$ at time t. If the $\varphi_x(\psi_\upsilon)$ over a region υ is also positive, the elasticity of damage to the population in the local region υ is larger than the elasticity of global damage over V; if, on the other hand, the $\varphi_x(\psi_\upsilon)$ is negative, the elasticity of global damage is larger than the elasticity of local damage. A plausible scenario that leads to negative $\varphi_x(\psi_\upsilon)$ is as follows: Assume that an increase in the point exposure dE leads to an increased impact dH_p to receptors at point $\boldsymbol{s}$. At the same time, it is likely that an increase in the average exposure dE_υ may lead to a decrease in the damage index $d\psi_\upsilon$ over υ and, thus, to a negative $\varphi_x(\psi_\upsilon)$. The reason for the decrease in the damage index is the space-time variation of $\theta(s,t)$ (*population density effect*), as well as the size of the region υ (*observation scale effect*) involved in the calculation of $d\psi_\upsilon$.

Health effect and damage functions may vary with changing scales, partially due to spatiotemporal correlations (Christakos and Lai, 1997). The $H_P(\boldsymbol{p})$, $\Psi_\upsilon(\boldsymbol{p})$ and $\psi_\upsilon(\boldsymbol{p})$ may exhibit different correlation patterns when mapped for counties, state-economic areas, and random aggregates of adjacent counties. The scales of the space-time pollutant concentration maps should be consistent with the domain at which the exposure-response and damage relationships are defined. In other words, the size of local regions υ should be consistent with the size of the exposure mapping grid.

5.2 Linear PER Model

The design of any remedial structure regarding pollution caused by anthropogenic and biogenic sources requires a rigorous analysis of the exposure-response relationship (Clark, 1980; Chechile and Carlisle, 1991). In practice, a useful approximation of the PER function is given by --Eq. (II.4.10)--

$$\mathcal{F}_{PER}[E(\boldsymbol{p})] = \alpha(\boldsymbol{p})\,E(\boldsymbol{p}), \tag{8}$$

where $\alpha(\boldsymbol{p})$ is the local slope of the PER curve. As we mentioned in Chapter II, several recent studies have demonstrated the validity of linear PER relationships. Based on the evidence provided by such studies, Eq. (8) has been used to obtain some numerical results on health damage due to ozone exposure (Christakos and Vyas, 1998a).

EXAMPLE 6: Assume that the slope of Eq. (8) is uniform in space, i.e. $\alpha(\boldsymbol{p}) = \alpha(t)$, and that the population density is time-independent for the time period considered, i.e. $\theta(\boldsymbol{p}) = \theta(\boldsymbol{s})$. Then, the local health damage and health damage index are directly related to the exposure by, respectively,

$$\Psi_{\upsilon}(\boldsymbol{p}) = \alpha(t)\, E_{\theta}^{\upsilon}(\boldsymbol{p}), \tag{9}$$

and

$$\psi_{\upsilon}(\boldsymbol{p}) = E_{\theta}^{\upsilon}(\boldsymbol{p}) / E_{\theta}^{V}(t), \tag{10}$$

where $E_{\theta}^{\upsilon}(\boldsymbol{p}) = |\upsilon(\boldsymbol{s})|^{-1} \int_{\upsilon(s)} ds'\, \theta(\boldsymbol{s}-\boldsymbol{s}')\, E(\boldsymbol{s}-\boldsymbol{s}',t)$ and $E_{\theta}^{V}(t) = |V|^{-1} \int_{V} ds'\, \theta(\boldsymbol{s}')\, E(\boldsymbol{s}',t)$ are the population-weighted average exposures within the regions $\upsilon(\boldsymbol{s})$, and V respectively -- see also §II.3. Note that in this case the $\psi_{\upsilon}(\boldsymbol{s},t)$ is independent of the slope of the PER curve. □

A flowchart summarizing the steps involved in the computation of the health damage indicators from exposure maps is given in Fig. 16. Eqs. (2)-(7) can also be used to

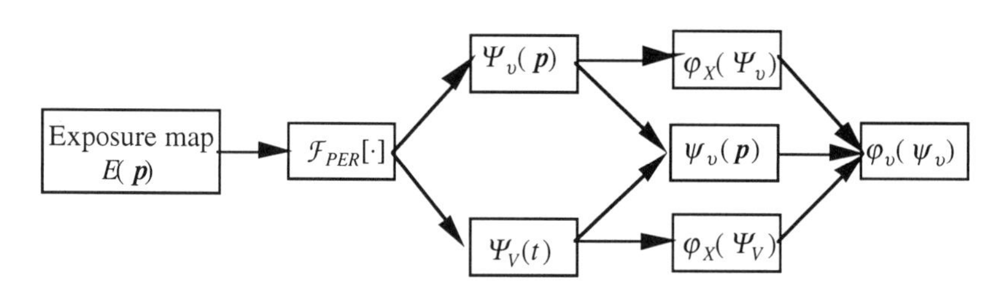

FIGURE 16: Flowchart summarizing the computation of the health damage indicators from exposure map.

evaluate health impacts due to small exposure increments for any PER model, because Eq. (8) can be considered as the 1st-order (linearized) increment of an arbitrary PER model.

5.3 Uniform Population Density

Recent studies have obtained insight into air pollution impact by assuming regions with uniform population density, i.e., $\theta(\boldsymbol{p}) = \theta_0$ (Curtiss and Rabl, 1996). In this case, the health damage indicators reduce to

$$\Psi_{\upsilon,0}(\boldsymbol{p}) = \theta_0\, \alpha(t)\, E_\upsilon(\boldsymbol{p}), \tag{11}$$

$$\psi_{\upsilon,0}(\boldsymbol{p}) = E_\upsilon(\boldsymbol{p})/E_V(t), \tag{12}$$

$$\varphi_{x,0}(\psi_\upsilon) = 1 - (dE_V/dE_\upsilon)(E_\upsilon/E_V). \tag{13}$$

Obviously, in this case, given the population density and the PER slope, the health damage indicators can be obtained directly from the space-time ozone maps.

5.4 Stochastic Trends and Uncertainty Assessment in Exposure-Damage Models

Since the exposure is an S/TRF, so are the health damage indicators. The mean values of these indicators represent stochastic trends in their space-time distribution, which can be expressed in terms of the mean value of the exposure distribution. The stochastic trend of the local health damage $\Psi_\upsilon(\boldsymbol{p})$ is given by

$$\overline{\Psi_\upsilon}(\boldsymbol{p}) = |\upsilon(\boldsymbol{s})|^{-1} \int_{\upsilon(\boldsymbol{s})} d\boldsymbol{s}'\, \theta(\boldsymbol{s}-\boldsymbol{s}',t)\, \overline{\mathcal{F}_{PER}[E]}(\boldsymbol{s}-\boldsymbol{s}',t). \tag{14}$$

The exposure-response mean $\overline{\mathcal{F}_{PER}[E]}(\boldsymbol{p})$ can be calculated using exact or approximate methods.

EXAMPLE 7: If the exposure distribution is Gaussian, the exposure-response mean can be estimated by the approximate expression

$$\overline{\mathcal{F}_{PER}[E]}(\boldsymbol{p}) \cong \tfrac{1}{2}\{\mathcal{F}_{PER}[\overline{E} - \sigma_E] + \mathcal{F}_{PER}[\overline{E} + \sigma_E]\}(\boldsymbol{p}),$$

where $\overline{E}$ denotes the exposure trend and σ_E denotes the exposure standard deviation (this expression can be generalized to non-Gaussian distributions by including the coefficient of skewness, etc.). Similar expressions can be derived for the other health indicators. □

In the special case of a linear PER model, Eq. (14) reduces to

$$\overline{\Psi_\upsilon}(\boldsymbol{p}) = \alpha(t)|\upsilon(\boldsymbol{s})|^{-1} \int_{\upsilon(\boldsymbol{s})} d\boldsymbol{s}'\, \theta(\boldsymbol{s}-\boldsymbol{s}',t)\, \overline{E}(\boldsymbol{s}-\boldsymbol{s}',t). \tag{15}$$

Note that the same result could have been obtained by taking the mean value of Eq. (8). In the case of uniform population density, the health damage trend is given by

$$\overline{\Psi_{\upsilon,0}}(s,t) = \theta_0\, \alpha(t) |\upsilon(s)|^{-1} \int_{\upsilon(s)} ds' \overline{E}(s-s',t). \tag{16}$$

Given the exposure trend map $\overline{E}(s,t)$, the above equations readily provide the corresponding health damage trends.

EXAMPLE 8: Assuming a homogeneous-stationary exposure S/TRF within the region $\upsilon(s)$, i.e., $\overline{E}(p) = \overline{E}$, Eq. (15) becomes $\overline{\Psi_{\upsilon}}(p) = \overline{E}\, \theta_{\upsilon}(s)\, \alpha(t)$. Hence, the health damage trend is directly proportional to the mean exposure, the average population density $\theta_{\upsilon}(s)$ of the region $\upsilon(s)$ and the slope $\alpha(t)$ of the PER curve. Similarly, Eq. (16) yields $\overline{\Psi_{\upsilon,0}}(t) = \overline{E}\, \theta_0\, \alpha(t)$, i.e., the health damage trend is now only a function of time. □

The S/TRF analysis can provide additional indicators expressing the contribution of exposure uncertainty to the uncertainty in population health damage. If we express the degree of uncertainty in $E(p)$ by its standard deviation $\sigma_{E_{\upsilon}}(p)$, its contribution to the space-time uncertainty in $\Psi_{\upsilon}(p)$ may be given by the local health damage uncertainty indicator

$$\zeta_x(\Psi_{\upsilon}) = (d\Psi_{\upsilon}/dX_{\upsilon})\, \sigma_{E_{\upsilon}}. \tag{17}$$

Eq. (17) allows a direct calculation of the uncertainty in population health damage from that in space-time ozone distribution. A more rigorous indicator of the degree of space-time uncertainty in the health damage $\Psi_{\upsilon}(p)$ is offered by its space-time variance

$$\sigma^2_{\Psi_{\upsilon}}(p) = |\upsilon(s)|^{-2} \int_{\upsilon(s)} \int_{\upsilon(s)} ds'ds''\, \theta(s-s',t)\, \theta(s-s'',t)\, c_{H_p}(s-s',s-s'',t), \tag{18}$$

where

$$c_{H_p}(s-s',s-s'',t) = \overline{H_p(s-s',t) H_p(s-s'',t)} - \overline{H_p(s-s',t)}\; \overline{H_p(s-s'',t)}$$

is the space-time covariance of the health effect between two geographical locations at time t. The form of the health effect covariance depends on the exposure covariance and the shape of the PER curve.

EXAMPLE 9: For a linear PER model with $\alpha(p) = \alpha(t)$, the health effect covariance is

$$c_{H_p}(s,s',t) = \alpha(t)^2 c_E(s,s',t), \tag{19}$$

and the space-time health damage uncertainty (18) reduces to

$$\sigma^2_{\Psi_\upsilon}(p) = \alpha(t)^2 |\upsilon(s)|^{-2} \int_{\upsilon(s)} \int_{\upsilon(s)} ds' ds''\, \theta(s-s',t)\, \theta(s-s'',t) c_E(s-s',s-s'',t), \tag{20}$$

which expresses directly the health damage uncertainty in terms of the space-time exposure covariance. □

5.5 Time Delayed PER Relationships

In certain cases, the effect of environmental exposure on the health of the population is not immediate, that is, there is a delayed health damage caused by the exposure. Then, the relationship (1) should be replaced by

$$H_p(p) = \int_{-\infty}^{t} dt'\, \mathcal{F}_{PER}[E(s,t')]\, g(t-t'), \tag{21}$$

where $g(t-t')$ is a normalized kernel describing the delayed response. If $g(t-t') = \delta(t-t')$, Eq. (1) is obtained. If the delayed response appears after time T then $g(t-t') = \delta(T)$. Eq. (21) implies similar modifications to the other health damage indicators.

EXAMPLE 10: The average local health damage $\Psi_\upsilon(p)$ in the case of time-independent population density, i.e. $\theta(s,t) = \theta(s)$, can be expressed as

$$\Psi_\upsilon(p) = |\upsilon(s)|^{-1} \int_{\upsilon(s)} ds' \int_{-\infty}^{t} dt'\, \theta(s-s')\, \mathcal{F}_{PER}[E(s-s',t')]\, g(t-t'), \tag{22}$$

which accounts for time delay health effects □

6. SPATIOTEMPORAL MAPS OF HEALTH DAMAGE INDICATORS - THE OZONE CASE

The wide-ranging flexibility of S/TRF analysis makes it a powerful tool in the study of relationships between environmental exposure and health effect. Mapping of health damage indicators offers a useful description of the data and a quantitative basis for further analysis. Health damage and elasticity maps have significant value in identifying the response of specific regions to variation of exposure levels or to new health management approaches. Or, they are useful for generating hypotheses regarding the etiology of health deterioration (in terms of a specific pollutant, or a whole class of pollutants acting in synergy). In combination with exposure maps, health damage maps can be used for establishing the form of the exposure-response relationship.

The exposure-response relation transfers information about the space-time variation of exposure levels into indicator maps representing expected damage to the population health. Exposure and damage are considered in a space-time continuum. Previous statistical studies allowed only for spatial variability of exposure and risk measures (e.g., Hayes, *et al.*, 1988). As it is shown in Fig. 1 information from a variety of sources is incorporated at each of the three main stages of the approach. Although exposure health impacts can be defined at various scales, they may not be equally important at every scale (King, 1979). Generally, exposure-health damage analysis should be carried out consistently at an appropriate scale for the targeted effect --although the identification of the optimal scale may not always be an easy task. In the following discussion we show how the ozone maps obtained in Chapter VI can be used in the context of health damage assessment.

Space-time ozone analysis and its connection to health issues is considered as one of the great challenges facing ozone science and integrated assessment modelling (e.g., Simpson, 1995). During the last few decades human exposure to high levels of ozone has increased leading to an increased risk of both acute (short-term) and chronic (long-term) health effects (McJilton *et al.*, 1972; Lippmann, 1989; McKee, 1994). The ozone data set considered in this study includes ambient ozone concentration observations from the Aerometric Information Retrieval System (AIRS) maintained by U.S. EPA. Hourly ozone concentration measurements (ppm) have been used from all 1,228 monitoring stations for the years 1994 and 1995. The locations of the monitoring stations are shown in (Christakos and Vyas, 1998a) The daily maxima of the hourly measurements were computed. In order to assess the impact of ozone concentrations on human health, population statistics were obtained from the U.S. Bureau of Census (1992), including population by county, as well as land and surface water area in each county. Hence, the population density $\theta(s)$ can be computed for each county.

In the present study, health impact mapping involves a mapping scale which is consistent with the domain at which the damage indicators are observed. The grid size for the health impact map was selected on the basis of the area statistics of the counties. The minimum county area in Eastern U.S. is 4.744 km^2, the maximum is 17,688 km^2, and the median is 1,305 km^2. A total of 170 out of the 1,693 counties considered have an area of less than 625 km^2. In light of these data, a spatial grid size of $\Delta s_1 \times \Delta s_2 = 25km \times 25km$ was chosen, which represents the 10 percentile area for the entire region. In order to obtain numerical results regarding ozone's health impacts, the linear PER model of Eq. (5.7) is assumed. For the short time periods considered, a time-independent population density is a realistic assumption. The PER slope is assumed spatially uniform. This simplification was necessary due to the unavailability of detailed space-time data for the PER slope values over large geographical regions, like the entire Eastern U.S.. Nevertheless, it allows a damage indicator analysis that illustrates the main steps of the S/TRF approach and offers a useful means for obtaining at least some preliminary insight regarding possible health situations. When a complete set of space-time PER models become available --covering Eastern U.S. in considerable detail-- more realistic results could be obtained by following exactly the same S/TRF approach outlined here.

The ozone map of Fig. VI.5 provides the input for an analysis of the population health state on July 16, 1995. If $E(\boldsymbol{p})$ is the ozone concentration in *ppm* and the effect $H_p(\boldsymbol{p})$ is the frequency of people with respiratory symptoms, the damage indicator $\Psi_\upsilon(\boldsymbol{p})$ represents the density of people with symptoms within each region υ. The corresponding damage index $\psi_\upsilon(\boldsymbol{p})$ --which is the ratio of the damage for each grid block over the mean damage for Eastern U.S.-- is plotted in Fig. 17a.

REMARK 1: This map should be studied in the light of the three main factors involved in its construction: the non-homogeneous ozone variation over Eastern U.S., the non-uniformly distributed population density, and the assumption of a geographically uniform PER slope. The latter implies that people's response to ozone exposure is uniform over Eastern U.S. and, thus, population density plays a predominant role in the calculation of the health damage.

The damage index map permits to consider health impact as an integrated spatiotemporal system, by looking at the whole picture --not just certain isolated parts. It can help to detect health risk heterogeneities across areas and to identify spatiotemporal patterns (in the form of local damage trends and clustering). For the specific data set of Fig. 17a certain areas where ozone pollution should be of concern are clearly distinguishable: One such area is located along the Eastern sea coast, between N.Y. and D.C., where the highest damage is expected to occur. Other hot spots are present over heavily populated areas. Relationships between hot spots in space-time may be established

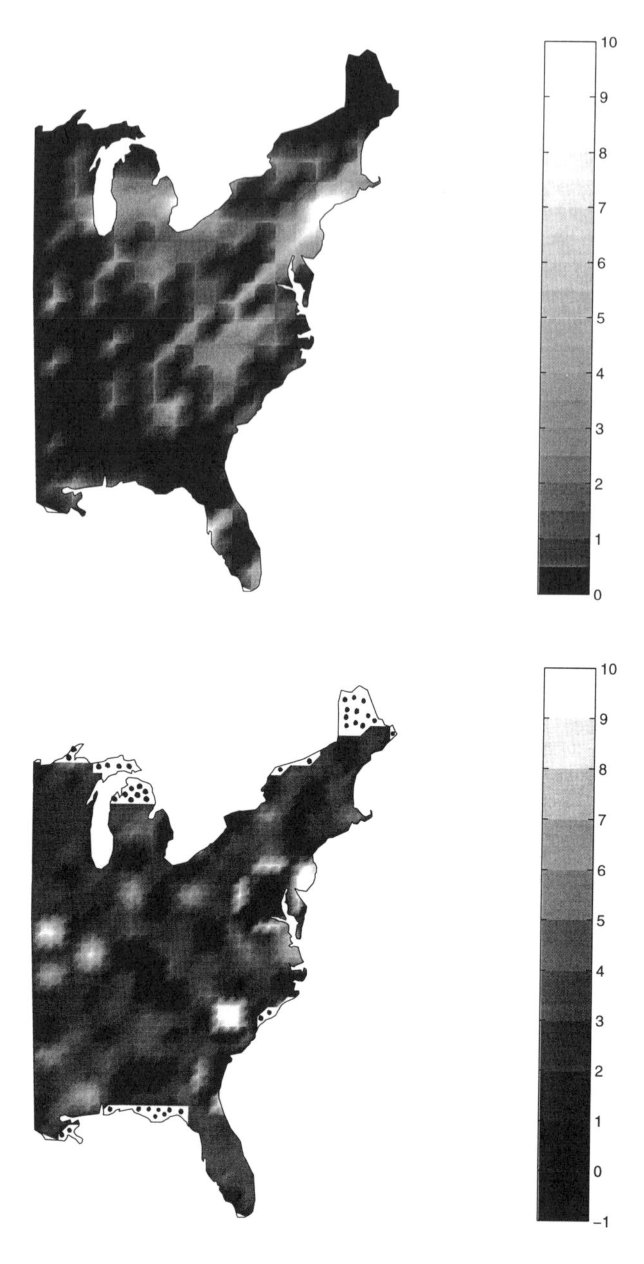

FIGURE 17: (a) Map of the dimensionless health damage indicator $\psi_{\upsilon}(\boldsymbol{p})$ for the ozone map of Fig. VI.5. (b) Map of the dimensionless health damage elasticity indicator $\varphi_{x}(\psi_{\upsilon})$, July 15th-16th, 1995 (the dotted areas indicate regions where estimation was not possible due to insufficient data).

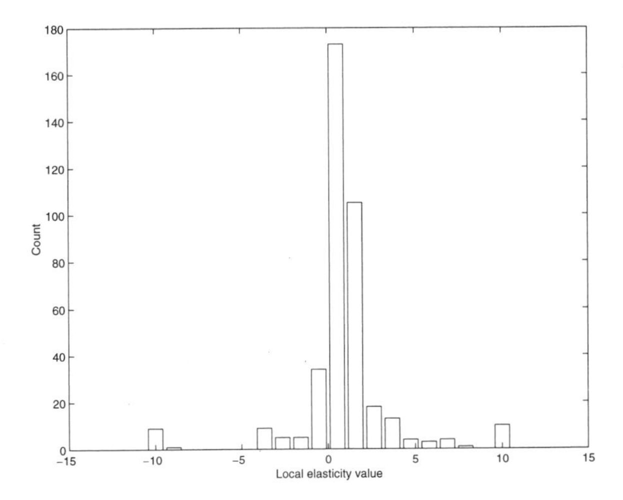

FIGURE 17: (c) Histogram of the $\varphi_x(\psi_\upsilon)$-values.

and several sequential maps of ozone distribution at regional or global scales can be generated by means of an automatic procedure (SANLIB, 1995).

Considering various possibilities for ozone concentration and the resulting population damage maps, regulatory standards and criteria may be analyzed in a probabilistic context. A key parameter in this analysis would be the expected number of people within a region that will develop specific health problems (e.g., decrements in lung function in the range of 10-20%), as revealed by the relevant health damage map. Such health impact criteria may be considered in combination with conventional air quality criteria (e.g., that the number of days/year with maximum ozone concentration above 0.12 *ppm* be equal to or less than one) to obtain new insight regarding the implications of ozone standards for regional health management and administration.

In addition to information regarding the current conditions of health damage provided by Fig. 17a, the health damage elasticity $\varphi_x(\psi_\upsilon)$ of Fig. 17b describes the potential for health damage change during the time period July 15-16, 1995 (i.e., change of the health damage index ψ_υ with respect to the average ozone concentration change E_υ). As it is shown in the histogram of Fig. 17c, most $\varphi_x(\psi_\upsilon)$-values are positive. Elasticity $\varphi_x(\psi_\upsilon)$ is positive when a positive (negative) change in the E_υ causes a positive (negative) change in average health damage over a region. The $\varphi_x(\psi_\upsilon)$ is negative whenever a positive (negative) change in E_υ causes a negative (positive) change in ψ_υ. This can happen if large negative changes in ozone concentration dE occurred at points within $\upsilon(s)$ with large population densities, thus leading to a negative ψ_υ-change, even if the average change in ozone concentration dE_{υ} is positive. In addition to this population density effect, the size of the region $\upsilon(s)$ can play an important role (observation scale effect). So, while for a certain size the $\varphi_x(\psi_{\upsilon})$ is positive, it may be negative for a different $\upsilon(s)$-size.

Unlike the $\varphi_x(\psi_{\upsilon})$, which is independent of the exposure-response slope $\alpha(t)$, the elasticities $\varphi_x(\Psi_V)$ and $\varphi_x(\Psi_{\upsilon})$ depend on the specific form for $\alpha(t)$. For simplicity, let $\alpha(t)=1$ (in suitable units). Then $\varphi_x(\Psi_V)\approx 0.73$ for Eastern U.S. at any time, but the $\varphi_x(\Psi_{\upsilon})$ varies significantly in space-time. In regions with positive $\varphi_x(\psi_{\upsilon})$-values the elasticity of the local health damage with respect to ozone concentration is larger than the elasticity of the global damage over Eastern U.S. This result suggests that measures may need to be taken in certain regions during time periods when the ozone concentration exceeds the permissible levels. On the other hand, in regions with negative $\varphi_x(\psi_{\upsilon})$-values the elasticity of the mean damage is larger than the elasticity of the local damage, and health management measures may not be necessary.

S/TRF methods for determining the space-time ozone trends over Eastern U.S. were

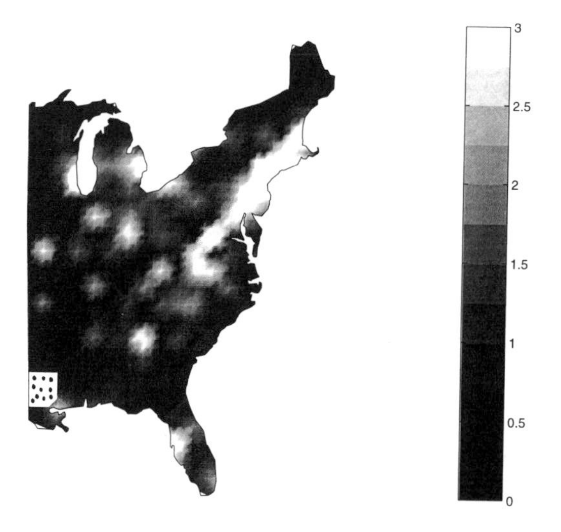

FIGURE 18: Spatiotemporal trend of the health damage normalized by the slope of the PER model, i.e. $\overline{\Psi_{\upsilon}}(s,t)/\alpha(t)$, on July 16, 1995. The dotted areas indicate regions where trend estimation was not possible due to numerical problems (ill conditioned matrices associated with the space-time arrangement of the data set available, etc.).

discussed in Chapter VI. Fig. VI.6 is a map of the ozone trend for July 16. The corresponding spatiotemporal trend of the health damage normalized by the exposure-response slope, i.e. $\overline{\Psi_{\upsilon}}(p)/\alpha(t)$, is plotted in Fig. 18. Figs. 17a and 18 have strong similarities because of the predominant role of the population density on the computation of the health damage indicator.

7. ANALYSIS OF DISEASE - EXPOSURE ASSOCIATION

In Chapter II, we discussed individual and population exposure-response relationships. These relationships represent the effects of the environment on health. Health effects are *multi-stage* (i.e., a series of changes occur in the cells, tissues or organs, etc.) and *multi-causal* (i.e., several pollutants may be involved).

Epidemiologic studies of the determinants of human disease involve a variety of techniques, including visual comparisons of the patterns exhibited by environmental and disease factors, and statistical analysis of population exposure-disease occurrence across a geographic area (Blot and McLaughlin, 1995). Problems with the use of these techniques for identifying disease-exposure relationships are discussed in Mayer (1983) and Hoel and Landrigan (1987). An useful approach to disease-exposure association is in some cases provided by the *scalar vs. vector prediction (SVP)* model (Christakos, 1998c) established on the basis of the random field theory.

SVP Model: *An exposure* $E(\boldsymbol{p})$ *should be identified as a causal factor for a disease* $H(\boldsymbol{p})$, *if the following three criteria are satisfied:*

(i) $E(\boldsymbol{p})$ *precedes* $H(\boldsymbol{p})$.

(ii) $E(\boldsymbol{p})$ *and* $H(\boldsymbol{p})$ *are contiguous in time and space.*

(iii) Disease rate predictions $\hat{H}_a(\boldsymbol{p})$ *obtained from vector estimation using both rate and exposure data are superior to the rate predictions* [$\hat{H}_b(\boldsymbol{p})$] *obtained from scalar estimation using only rate data.*

An illustration of the SVP model is given in Fig. 19. Before the criteria of the SVP model are applied, the effects of confounding factors should first be removed. Criterion (i) is satisfied if there is a history of precedence in previous events, or if there is a biological cause for precedence in light of existing knowledge about disease etiology. Thus, this criterion establishes whether exposures preceding a disease are causally connected with the disease. The term "causes" is usually employed to denote both necessary conditions (exposures) and sufficient conditions. Necessary conditions are present if the effect is observed, but they do not produce the effect (disease) by themselves. However, if a necessary condition is eliminated, the effect will not occur. In this sense we can speak of a necessary condition as a cause of disease. Sufficient conditions, on the other hand, will produce the effect if they are realized. If the sufficient conditions of a disease are known and can be effected the disease can be caused at will.

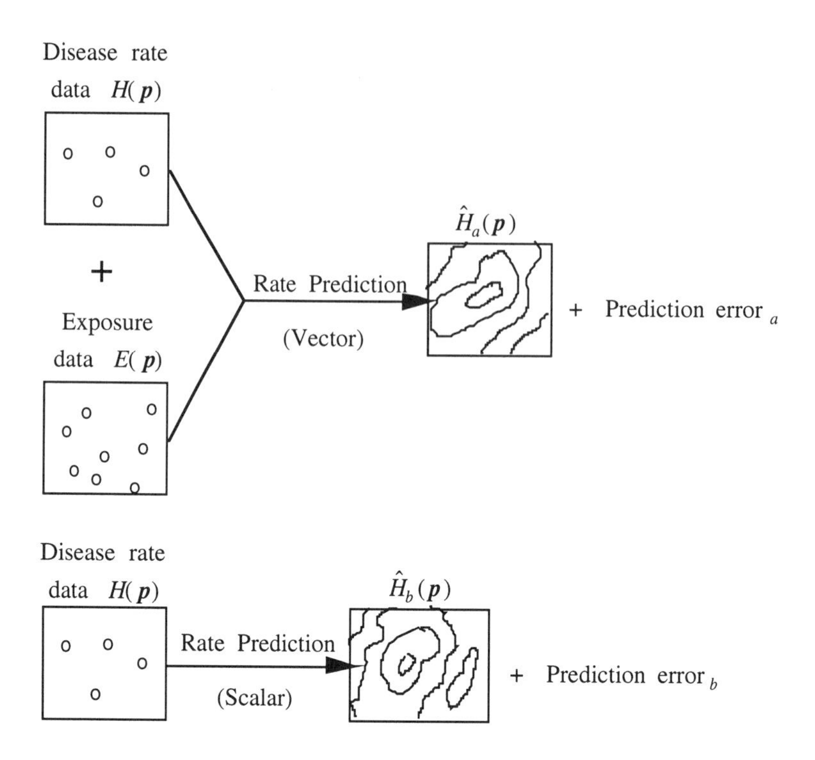

FIGURE 19: An illustration of the SVP model. Prediction error$_a$ < Prediction error$_b$ supports a disease-exposure association; Prediction error$_a$ ≈ Prediction error$_b$ does not support a disease-exposure association

Criterion (ii) requires the existence of a spatiotemporal connection between exposure and disease. An air pollutant can cause illness to a group of receptors, provided that both the pollutant and the receptors are in the same geographical area. Similarly, germs can cause illness to a receptor if they enter the receptor's body. In many cases this contiguity is not a trivial issue, for biological systems are in a constant state of interaction with the surrounding environment.

Criterion (iii) provides a falsifiability test for hypotheses of exposure-disease associations based on the SVP model. The adequacy of the SVP model is judged on the basis of the successful predictions it leads to. The assessment of the prediction accuracy is made either in terms of the prediction error variance; or by means of a set of "control points," i.e., points where the disease rates are known and can be compared with the rate predictions obtained from scalar estimation or vector estimation. The quality of disease prediction is a central component of the SVP model, which distinguishes it from other techniques for detecting environmental causes of disease. The better the vector rate predictions $\hat{H}_\alpha(\boldsymbol{p})$ are compared to the scalar rate predictions $\hat{H}_b(\boldsymbol{p})$, the stronger is the disease/exposure association, thus offering a measure of the strength and consistency of the association. The $E(\boldsymbol{p})$ cannot be considered as a causal factor, if $\hat{H}_\alpha(\boldsymbol{p})$ does not improve the rate prediction compared to $\hat{H}_b(\boldsymbol{p})$. Vector and scalar estimation may be performed using, respectively, co-kriging and kriging techniques (both these estimation techniques were discussed in Chapter VI) Extension of the SVP model to multiple environmental exposures $H_i(\boldsymbol{p})$, $i = 1,...,m$, is straightforward.

EXAMPLE 1: For illustration, consider the case $H(\boldsymbol{p})$ = lung cancer incidence rate and $E(\boldsymbol{p})$ = cigarette smoking rate. According to the SVP model, the latter should be a legitimate cause of lung cancer if the criteria (i), (ii) and (iii) are satisfied, i.e., if cigarette smoking precedes lung cancer (there is spatial connection between the receptor and cigarette smoke), and the prediction of lung cancer rate on the basis of cigarette smoking and lung cancer data is more accurate than prediction based on lung cancer rate data alone. In the following section an application of the SVP criterion in a numerical simulation study is discussed. □

Below we consider examples of logical fallacies in causal reasoning that are due to a violation of the criteria in the SVP model. Common fallacies include: The confusion of necessary and sufficient conditions [see criterion (i); e.g., cyanide is a sufficient but not necessary condition for death]; the assumption that one of several necessary conditions was the sole and exclusive cause of a disease. Other logical errors are due to the assumption that two events which show a high incidence of correlation are causally connected. Consider, e.g., the case $H(\boldsymbol{p})$ = human birth rate and $E(\boldsymbol{p})$ = brooding storks in Germany during the years 1965-80. Although the statistical correlation between these two data sets has been found to be high, brooding storks are not an acceptable cause of change in human birth rate, because (a) there is no clear indication of precedence or etiological information that supports the precedence, and (b) there is no spatial connection between these events [criteria (i) and (ii)].

The reasoning of the SVP model does not violate neither Hume's nor Mill's fundamental criteria of cause and effect (Harris, 1996). Also, the SVP model satisfies the criteria of testability and predictability, in agreement with the Popperian concept of scientific reasoning (Popper, 1934). In addition to suggesting possible environmental causes of disease, the SVP model could prove helpful in testing and confirming causation hypotheses developed within the framework of other investigations (medical, biological, toxicological, ecological, etc.). In some cases sufficient information may be obtained on the relation between exposure and effect to set realistic exposure standards.

Simulation is a powerful tool for environmental health studies, for it allows to perform controlled tests and sensitivity analysis that would not be possible with a real data set .

EXAMPLE 2: The environmental variable (e.g., exposure to sulfur dioxide) and the disease rate (e.g., lung cancer incidence) and is modelled as a spatial random field of order v_1, $H(\boldsymbol{s})$; and the environmental variable (e.g., exposure to sulfur dioxide) by the spatial random field of order v_2, $E(\boldsymbol{s})$. For the purposes of the simulation study, the exposure $E(\boldsymbol{s})$ is assumed to be related to the disease rate $H(\boldsymbol{s})$ by the simple linear regression model

$$E(\boldsymbol{s}) = a\,H(\boldsymbol{s}) + \varepsilon(\boldsymbol{s}), \tag{1}$$

where $\varepsilon(\boldsymbol{s})$ is a zero mean Gaussian white noise with variance σ_ε^2 uncorrelated to $E(\boldsymbol{s})$ and $H(\boldsymbol{s})$ and a is a constant with appropriate disease rate/exposure units (for simplicity, let $a = 1\,unit$). The variance σ_ε^2 provides a measure of the strength of this association: small values of σ_ε^2 implying strong association, large values of σ_ε^2 implying weak or nonexistent association (for illustration, below we will consider four values for the variance, i.e., 0.001, 0.01, 0.1 and 1.0).

Determination of the rate-exposure variations involves the orders of continuity and the generalized covariances. All generalized auto- and cross-covariances of the $E(\boldsymbol{s})$ and $H(\boldsymbol{s})$ are assumed isotropic, so that the matrix of the generalized covariances and cross-covariances is

$$\boldsymbol{K}_{EH}(r) = \begin{bmatrix} k_E(r) & k_{EH}(r) \\ k_{HE}(r) & k_H(r) \end{bmatrix}. \tag{2}$$

Polynomial models of the form of Eq. (IV.8.32) are used for the generalized covariances and cross-covariances. The continuity orders of the exposure and the disease rate were found to be $v_E = v_H = 2$. A disease rate field is obtained using the turning bands simulation technique (Cassiani and Christakos, 1998), which provides the actual rate values $H(s_q)$. Vector MMSE (co-kriging) predictions $\hat{H}_a(s_q)$, as well as scalar MMSE (kriging) predictions $\hat{H}_b(s_q)$ of the disease rate are made at the same points, using the

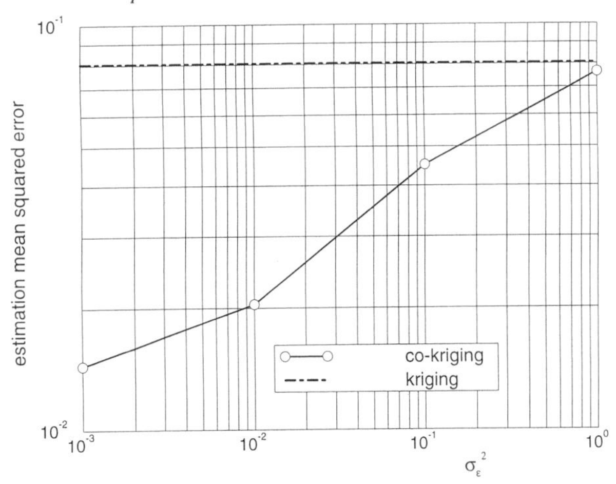

FIGURE 20: Estimation MMSE of kriging and co-kriging disease rate predictions as a function of the data error variance (σ_ε^2).

methods of Chapter VI. Then, the MMSE prediction error $\overline{\sigma_{H_a}^2}$ is estimated by the sample average of the squared differences between the actual rates $H(s_q)$ and the vector (co-kriging) predicted rates $\hat{H}_a(s_q)$, i.e.

$$\overline{\sigma_{H_a}^2} = \tfrac{1}{75} \sum_{q=1}^{75} [\hat{H}_a(s_q) - H(s_q)]^2 ; \tag{3}$$

similar calculations in terms of the (kriging) predicted rates $\hat{H}_b(s_q)$ lead to

$$\overline{\sigma_{H_b}^2} = \tfrac{1}{75} \sum_{q=1}^{75} [\hat{H}_b(s_q) - H(s_q)]^2 . \tag{4}$$

The results are plotted in Fig. 20 for various σ_ε^2 values. Clearly, $\overline{\sigma_{H_a}^2} < \overline{\sigma_{H_b}^2}$ for all σ_ε^2. The prediction error $\overline{\sigma_{H_a}^2}$ increases almost linearly with σ_ε^2 on a log-log plot. In other words, stronger disease-exposure associations imply that the accuracy of rate prediction when the exposure data are used is increased. Only at larger σ_ε^2 values --reflecting a very weak or inexistent disease/exposure association-- the $\overline{\sigma_{H_a}^2}$ approaches the $\overline{\sigma_{H_b}^2}$ value. □

Bibliography

Abramowitz M. and I. A. Stegun, (eds.) (1970). *Handbook of Mathematical Functions,* Dover Publ., New York, NY.

Abrikosov A. A., L. P. Gorkov, and I. E. Dzyaloshinski (1975). *Methods of Quantum Field Theory in Statistical Physics,* Dover Publ., New York, NY.

Aczel, J. and Z. Daroczy (1975). *On Measures of Information and Their Characterization.* Academic Press, New York, NY.

Adler, P. M. (1992). *Porous Media, Geometry and Transports.* Butterworth and Heinemann, Stoneham.

Adler, P. M., C. G. Jacquin, and J. A. Quiblier (1990). "Flow in simulated porous media," *Int. J. Multiphase Flow*, **16**(4), 691-712.

Adler, R. J. (1981). *The Geometry of Random Fields.* Wiley, New York, NY.

Allaby, M. (1996). *Basics of Environmental Science.* Routledge, New York, NY.

Amit, D. J. (1984). *Field Theory, the Renormalization Group and Critical Phenomena.* World Scientific, Singapore.

Anderson, P. W. (1984). *Basic Notions of Condensed Matter Theory.* Benjamin & Cummings Publ. Comp., Inc., Menlo Park, CA.

Anderson, P. W., B. S. Shastry, D. Hristopulos (1989). "A class of variational singlet wavefunctions for the Hubbard model away from half-filling," *Phys. Rev. B* **40**(13), 8939-8944.

Andricevic, R., and E. Foufoula-Georgiou (1991). "A transfer function approach to sampling network design for groundwater contamination," *Water Resour. Res.*, **12**(10), 2759-2769.

Aoki, M. (1967).*Optimization of Stochastic Systems.* Academic Press, New York, NY.

Arlinghaus, S.L. (ed.) (1996). *Practical Book of Spatial Statistics.* CRC Press, New York, NY.

Arneodo, A., E. Bacry, P. V. Graves and J. F. Muzy (1994). "Characterizing long-range correlations in DNA sequences from wavelet analysis," *Phys. Rev. Let.,* **74**(16), 3293-3296.

Aronovitz, J.A. and D.R. Nelson (1984). "Anomalous dispersion in steady fluid flow through a porous medium," *Phys. Rev. A* **30**, 1948-1954.

Asante-Duah, D.K. (1996). *Management of Contaminated Site Problems.* CRC Lewis Publ., Boca Raton, FL.

Auriault, J. L. (1991). "Heterogeneous medium, Is an equivalent macroscopic description possible?" *Intern. Jour. of Eng. Sci.*, **29**(7), 785-795.

Austin, R. H., J. P. Brody, E. C. Cox, T. Duke and W. Volkmuth (1997). "Stretch Genes," *Phys. Tod.,* **50**(2), 32-38.

Avellaneda M. and A. J. Majda (1992a). "Approximate and exact renormalization theories for a model for turbulent transport," *Phys. Fluids A* **4**(1), 41-57.

Avellaneda M. and A. J. Majda (1992b). "Mathematical models with exact renormalization for turbulent transport. II Non-gaussian statistics, fractal interfaces and the sweeping effect," *Comm. Math. Phys.,* **146**, 381-429.

Bachelier, L. (1900). *Theorie de la speculation.* Thesis for the Doctorate in Mathematical Sciences. Annales Scientifiques de l' Ecole Normale Superieure **III-17**, 21-86.

Bailey, C. (1928). *The Greek Atomists and Epicurus.* Clarendon Press, London, U.K.

Bailey, R.G. (1996). *Ecosystem Geography.* Springer-Verlag, New York, NY.

Bak, P. and K. Chen (1989). "The physics of fractals," *Physica D* **38**, 5-12.

Bak, P., C. Tang and K. Wiesenfeld (1988). "Self-organized criticality," *Phys. Rev. A* **38**(1), 364-374.

Barthelemy M. and H. Orland (1993). "Replica field theory for composite media," *J. Phys. I* **3**, 2171-2177.

Barthelemy, M., H. Orland and G. Zerah (1995). "Propagation in random media, calculation of the effective dispersive permittivity by use of the replica method," *Phys. Rev. E* **52**(1), 1123-1127.

Bataille, G. (1988-90). *The Accursed Share.* 3 vols. Translated by R. Hurley. Zone, New York, NY.

Baveye, P. and G. Sposito (1984). "The operational significance of the continuum hypothesis in the theory of water movement through soils and aquifers," *Water Resour. Res.*, **20**(5), 521-530.

Bear, J. (1972). *Dynamics of fluids in porous media.* Dover Publ., New York, NY.

Bender C. M., and S. A. Orszag (1978). *Advanced Methods for Scientists and Engineers.* McGraw-Hill, New York, NY.

Berrino, F., S. Panico and P. Muti (1989). "Dietary fat, nutritional status and endocrine associated cancers," In *Diet and the Etiology of Cancer* (Miller *et al.*, eds.) Springer-Verlag, Berlin, Germany, pp. 3-12.

Berryman, G. B. (1987). "Relationship between specific surface area and spatial correlation functions for anisotropic porous media," *J. Math. Phys.*, **28**(1), 244-245.

Bilonick, R.A. (1985). "The space-time distribution of sulfate deposition in the northeastern U.S," *Atm. Env.*, **19**(11), 2513-2524.

Blot, W.J. and J.K. McLaughlin (1995). "Geographic patterns of breast cancer among American women," *Jour. Nat. Cancer Inst.* **87**(24), 1819-1820.

Bochner, S. (1933). "Monotone funktionen Stieltjessche integrale und harmonische analyse," *Math. Ann.*, **108**, 378-410.

Bochner, S. (1959). *Lectures on Fourier Integrals.* Princeton Univ. Press, Princeton, N.J.

Bogaert, P. and G. Christakos (1997a). "Stochastic analysis of spatiotemporal solute content measurements using a regressive model," *Stoch. Hydrol. and Hydraul.*, **11**(4), 267-295.

Bogaert, P. and G. Christakos (1997b). "Spatiotemporal analysis and processing of thermometric data over Belgium," *Jour. Geophys. Res.-Atmospheres*, **102**(D22), 25831-25846.

Bohr, N. (1958). *Atomic Physics and Human Knowledge.* Wiley, New York, NY.

Bonham-Carter, G.F. (1994). *Geographic Information Systems for Geoscientists.* Pergamon, Kidlington, UK.

Boorstin, D.J. (1994). *Cleopatra's Nose.* Random House, New York, NY.

Botkin, D.B. and E.A. Keller (1998). *Environmental Science.* J. Wiley and Sons, New York, NY.

Bouchaud, J.-P. and A. Georges (1990). "Anomalous diffusion in disordered media, Statistical mechanics, models and physical applications," *Phys. Rep.*, **195**, 127-293.

Boulding, J.R. (1995). *Practical Handbook of Soil, Vadose Zone, and Groundwater Contamination.* CRC Lewis Publ., Boca Raton, FL.

Boyle, P. (1989). "Relative value of incidence and mortality data in cancer research," In *Recent Results in Cancer Research-Cancer Mapping*, **114** (Boyle, P. *et al.*, eds.) Springer Verlag, New York, pp. 41-63.

Briggs, D.J. and P. Elliott (1995). "The use of geographical information systems in studies on environment and health," *Wld Health Statist. Quart.*, **48**, 85-94.

Brown, R. (1828). "On the existence of active molecules in organic and inorganic bodies," *Phil. Mag.* **4**, 162-173.

Buck, B. and V. A. Macaulay (eds.) (1991). *Maximum Entropy in Action.* Oxford Science Publ., Oxford.

Buckingham, E. (1914). "On physically similar systems; illustrations of the use of dimensional equations," *Phys. Rev. IV*, **4**, p. 345.

Bunde, A. and S. Havlin (eds.) (1996). *Fractals and Disordered Systems,* Springer Verlag, New York, NY.

Burbank, F.A. (1972). "A sequential space-time cluster analysis of cancer mortality in the US, Etiologic implications," *Am. J. Epid.*, **95**, p. 393.

Burkitt, D.P. (1962). "A 'tumor safari' in east and central Africa," *Br. J. Cancer*, **16**, p. 379.

Burks, A. (1977). *Chance, Cause, Reason.* The University of Chicago Press, Chicago, IL

Byron, F. W. and R. W. Fuller (1992). *Mathematics of Classical and Quantum Physics,* Dover Publ., New York, NY.

Carrat, F. and A-J Valleron (1992). "Epidemiologic mapping using the "kriging" method, Application to an influenza-like illness epidemic in France," *Amer. Jour. of Epidem.*, **135**(11), 1293-1300.

Casado, L.S., S. Rouhani, C.A. Cardelino and A.J. Ferrier (1994). "Geostatistical analysis and visualization of hourly ozone data," *Atm. Env.*, **28**(12), 2105-2118.

Cassiani, G. and G. Christakos (1998). "Analysis and estimation of spatial non-homogeneous natural processes using secondary information," *Math. Geol.*, **30**(1), 57-76.

CEQ-Council on Environmental Quality (1986). "National Environmental Policy Act regulations, incomplete or unavailable information, Final Rule," *Fed. Reg.*, **51**, 15618-15626.

Char B. W., K. O. Geddes, G. H. Gonnet, B. L. Leong, M. B. Monagan and S. N. Watt (1992a). *First Leaves, A Tutorial Introduction to Maple V.* Springer Verlag, New York, NY.

Char B. W., K. O. Geddes, G. H. Gonnet, B. L. Leong, M. B. Monagan and S. N. Watt (1992b). *Maple V, Library Reference Manual.* Springer Verlag, New York, NY.

Chechile, R.A., and Carlisle S. (eds.) (1991). *Environmental Decision Making.* Van Nostrand Reinhold, New York, NY.

Checkoway, H., N.E. Pearce and D.J. Crawford-Brown (1989). *Research Methods in Occupational Epidemiology.* Oxford Univ. Press, New York, NY.

Christakos, G. (1984a). "On the problem of permissible covariance and variogram models," *Water Res. Res.*, **20**(2), 251-265.

Christakos, G. (1984b). "The space transformations and their applications in systems modelling and simulation," *Proc. 12th Intern. Confer. on Modelling and Simulation (AMSE)*, **1**(3), 49-68, Athens, Greece.

Christakos, G. (1986). "Space transformations in the study of multidimensional functions in the hydrologic sciences," *Adv. in Water Res.*, **9**(1), 42-48.

Christakos, G. (1987a). "The space transformations in the simulation of multidimensional random fields," *Jour. of Math. and Comp. in Simul.* **29**, 313-319.

Christakos, G. (1987b). "Stochastic simulation of spatially correlated geoprocesses," *Jour. Math. Geol.*, **19**(8), 803-827.

Christakos, G. (1990a). "A Bayesian/maximum-entropy view to the spatial estimation problem," *Jour. of Math. Geol.*, **22**(7), 763-776.

Christakos, G. (1990b). "Some applications of the Bayesian, maximum entropy concept in Geostatistics," *Fund.Theor. of Phys.*, Kluwer Acad. Publ., 215-229.

Christakos, G. (1991). "A theory of spatiotemporal random fields and its application to space-time data processing," *IEEE Trans. Systems, Man, and Cybernetics*, **21**(4), 861-875.

Christakos, G. (1992). *Random Field Models in Earth Sciences.* Academic Press, San Diego, CA.

Christakos, G. (1998a). "Spatiotemporal information systems in soil and environmental sciences," *Geoderma*, in press.

Christakos, G. (1998b). "Modern Geostatistics in the analysis of spatiotemporal environmental data, the BME approach," Course Notes. *Intern. Assoc. of Math. Geol. Confer.*, Ischia Island, Naples, Italy.

Christakos, G. (1998c). "An analysis of the disease/exposure association using a vector random field model," *Jour. Expos. Anal. and Env. Epidem.*, in review.

Christakos, G. and C. Panagopoulos (1992). "Space transformation methods in the representation of geophysical random fields," *IEEE Trans. Geosc. and Rem. Sensing*, **30**(1), 55-70.

Christakos, G., and B. R. Killam (1993). "Sampling design for classifying contaminant level using annealing search algorithms,"*Water Res. Res.*, **29**(12), 4063-4076.

Christakos, G., and G. A. Thesing (1993). "The intrinsic random field model and its application in the study of sulfate deposition data," *Atm. Env.*, **27A**(10), 1521-1540.

Christakos, G., C. T. Miller, and D. Oliver (1993a). "The development of stochastic space transformation and diagrammatic perturbation techniques in subsurface hydrology," *Stoch. Hydrol. and Hydraul.*, **7**(1), 14-32.

Christakos, G., C. T. Miller and D. Oliver (1993b). "Stochastic perturbation analysis of groundwater flow. Spatially variable soils, semi-infinite domains and large fluctuations," *Stoch. Hydrol. and Hydraul.*, **7**(3), 213-239.

Christakos, G., C. T. Miller, and D. Oliver (1993c). "Cleopatra's nose and the diagrammatic approach to flow modelling in random porous media," In *Geostatistics for the Next Century*, Kluwer Acad. Publ., Amsterdam, the Netherlands, pp. 341-358.

Christakos, G., C. T. Miller, and D. Oliver (1993d). "Stochastic flow modelling in terms of interactive perturbation, Feynman diagrams and graph theory," *IEEE*, **BT-93**, 77-84.

Christakos, G. and D. T. Hristopulos (1994). "Stochastic space transformations in subsurface hydrology-part 2, generalized spectral decomposition and Plancherel representations," *Journal of Stoch. Hydr. and Hydraul.*, **8**(2), 117-138.

Christakos, G., D. T. Hristopulos and C. T. Miller (1995). "Stochastic diagrammatic analysis of groundwater flow in heterogeneous porous media," *Water Resour. Res.*, **31**(7), 1687-1703.

Christakos, G. and P. Bogaert (1996). "Spatiotemporal analysis of springwater ion processes derived from measurements at the dyle basin in Belgium," *IEEE Trans. Geosc. & Rem. Sens.*, **34**(3), 626-642.

Christakos, G. and D. T. Hristopulos (1996a). "Characterization of atmospheric pollution by means of stochastic indicator parameters," *Atm. Env.* **30**(22), 3811-3823.

Christakos, G. and D. T. Hristopulos (1996b). "Stochastic indicators for waste site characterization," *Water Resour. Res.*, **32**(8), 2563-2578.

Christakos, G., and V. R. Raghu (1996). "Dynamic stochastic estimation of physical variables," *Math. Geol.*, **28**(3), 341-365.

Christakos, G. and D. T. Hristopulos (1997). "Stochastic Radon operators in porous media hydrodynamics," *Quarterly of Appl. Math.*, **LV**(1), 89-112.

Christakos, G. and J. Lai (1997). "A study of the breast cancer dynamics in North Carolina," *Soc. Sci. Medicine*, **45**(10), 1503-1517.

Christakos, G. and D. T. Hristopulos (1998). "Stochastic indicator analysis of contaminated sites," *Jour. of Appl. Prob.*, **34**(4), 988-1008.

Christakos, G., D. T. Hristopulos, and X. Li (1998). "Multiphase Flow in Heterogeneous Porous Media from a Stochastic Differential Geometry Viewpoint," *Water Resour. Res.*, **34**(1), 93-102.

Christakos, G., K. Choi and M. Serre (1998). "Recent development in a vectorial multi-point spatiotemporal prediction," *Intern. Assoc. of Math. Geol. Confer.*, Ischia Island, Naples, Italy.

Christakos, G. and X. Li (1998). "Bayesian maximum entropy analysis and mapping: A farewell to kriging estimators?" *Math. Geol.*, **30**(4), forthcoming.

Christakos, G. and V. Vyas (1998a). "A composite spatiotemporal study of ozone distribution over eastern United States" *Atm. Env.* , in press.

Christakos, G. and V. Vyas (1998b). "A novel method for studying population health impacts of spatiotemporal ozone distribution," *Soc. Sci. Medicine*, in press.

Clark, T.L. (1980). "Annual anthropogenic pollutant emissions in the US and the southern Canada east of the Rocky Mountains," *Atm. Env.*, **14**, 961-970.

Cleek, R.K. (1979). "Cancers and the environment, the effect of scale," *Soc. Sci. & Med.*, **13D**, 241-247.

Covello, V.T. and M.W. Merkhofer (1993). *Risk AssessmentMethods.* Plenum Press, New York, NY.

Cramer, H. and M.R. Leadbetter (1967). *Stationary and Related Stochastic Processes.* J. Wiley & Sons, Inc., New York, NY.

Crawford, F.G., J. Mayer, R.M. Santella, T. Cooper, R. Ottman, W.Y. Tsai, G. Simon-Cereijido, M. Wang, D. Tang and F.P. Perera (1994). "Biomarkers of environmental tobacco smoke in preschool children and their mothers," *J. Natl. Cancer Inst.*, **86**, 1398-1402.

Crawford-Brown, D.J. (1997). *Theoretical and Mathematical Foundations of Human Health Risk Analysis.* Kluwer Acad. Publ., Boston, Massachusetts.

Cressie N. (1991). *Statistics for Spatial Data.* J. Wiley, New York, NY., p. 900

Cressie, N. and T.R.C. Read (1989). "Spatial data analysis of regional counts," *Biom. Jour.*, **6**, 699-719.

Creswick, R.J., H.A. Farach and C.P. Poole, Jr. (1992). *Introduction to Renormalization Group Methods in Physics.* J. Wiley & Sons, New York, NY.

Curtiss, P.S. and Rabl, A. (1996). "Impacts of air pollution, general relationships and site dependence," *Atmospheric Environment*, **30A**(19), 3331-3347.

Cushman, J. H. (1984). "On unifying the concepts of scale, instrumentation and stochastics in the development of multiphase transport theory," *Water Resour. Res.*, **20**(11), 1668-1676.

Cushman J. H. (1986). "On measurement, scale and scaling," *Water Resour. Res.* **22**(2), 129-134.

Cushman, J. H., B. X. Hu, and T. R. Ginn (1994). "Nonequilibrium statistical mechanics of preasymptotic dispersion," *J. Stat. Phys.* **75**, 859-878.

Dagan, G. (1984). "Solute transport in heterogeneous porous formations," *J. Fluid Mech.*, **145**, 151-177.

Dagan, G. (1988). "Time-dependent macrodispersion for solute transport in anisotropic heterogeneous aquifers," *Water Res. Res.* **18**(4), 835-848.

Dagan, G. (1989). *Flow and Transport in Porous Formations*, Springer Verlag, Berlin.

Dagan, G. (1992). "Higher-order correction of effective conductivity of heterogeneous isotropic formations of lognormal conductivity distribution," *Trans. in Por. Media*, **12**, 279-290.

Davies, S., K. J. Packer, A. Baruya & A. I. Grant (1991). "Enhanced information recovery in spectroscopy," in *Maximum Entropy in Action.* (B. Buck and V. A. Macaulay, eds.) Oxford Science Publ., Oxford.

De Waard, F. and D. Trichopoulos (1988). "A unifying concept of the aetiology of breast cancer," *Int. Jour. Cancer*, **41**, 666-669.

Dean, D. S., I. T. Drummond and R. R. Horgan (1994). "Perturbation schemes for flow in random media," *J. Phys. A: Math. Gen.*, **15**, 5135-5144.

Deans, S.R. (1993). *The Radon Transform and Some of Its Applications*. Krieger Publ. Corp., Malabar, FL.

Deem, M. W. and D. Chandler (1994). "Classical diffusion in strong random media," *J. Stat. Phys.*, **76**(3/4), 911-927.

Deng, F. W., and J. H. Cushman (1995). "On higer-order corrections to the flow velocity covariance," *Water Resour. Res.*, **31**(7), 1659-1672.

Deutsch C. V., and A. G. Journel (1992). *Geostatistical Software Library and User's Guide*. Oxford Univ. Press, Oxford.

Dicke, R.H. and J.P. Wittke (1960). *Introduction to Quantum Mechanics*. Addison-Wesley, Reading, MA.

Doi, M. and S. F. Edwards (1989). *The Theory of Polymer Dynamics,* Oxford Univ. Publ., Oxford, England.

Doll, R. (1980). "The epidemiology of cancer," *Cancer*, **45**, 2475-2485.

Dowd, P.A. (1991). "A review of recent developments in geostatistics," *Comp. & Geosc.*, **17**(10), 1481-1500.

Drife, J.Q. (1986). "Breast development in puberty," *Ann. NY. Acad. Sci.*, **464**, 58-65.

Droz, P.O. (1993). "Biologic monitoring and pharmacokinetic modelling for the assessment of exposure." In *Molec. Epidem.,* Schulte (P.A. and F.P. Perera, eds.) pp. 137-157, Academic Press, San Diego, CA.

Duan, N., A. Dobbs and W. Ott (1989). "Comprehensive definitions of exposure and dose to environmental pollution." In *Proceed. of the EPA/A&WMA Specialty Confer. on Total Exposure Assessment Methodology*, Las Vegas, NV, pp. 166-195.

Dyson, F. J. (1949). "The radiation theories of Tomonaga, Schwinger and Feynman," *Phys. Rev.* **75**, 486-502.

Dzombak, D,A., Labienec P.A. and R.L. Siegrist (1993). "The need for uniform soil cleanup goals," *Environ. Sci. Technol.*, **27**, 765-766.

Earnshaw, R.A. and N. Wiseman (1992). *An Introductory Guide to Scientific Visualization.* Springer-Verlag. New York, NY.

Eder, B.K., J.M. Davis and P. Bloomfield (1993). "A characterization of the spatiotemporal variability of non-urban ozone concentrations over the eastern US," *Atm. Env.*, **27A**(16), 2645-2668.

Edwards S. F. and P. W. Anderson (1975). "Theory of spin glasses," *J. Phys. F* **5**, 965-975.

Eerens, H., C.J. Sliggers and K.D. van den Hout (1993). "The CAR model, the Dutch method to determine city street air quality," *Atm. Env.*, **27B**(4), 389-399.

Einstein, A. (1905). "Über die von der molekularkinetischen theorie der wärme gefordete bewegung von in ruhenden flüssigkeiten suspendierten teilchen," *Annalen der Physik,* **17**, 549-560.

Emmelot P. and E. Kriek (eds.) (1979). *Environmental Carcinogenesis-Occurence, Risks Evaluation and Mechanisms.* Elsevier, Amsterdam, the Netherlands, p. 401

EPA (Environmental Protection Agency), *Remediation Technologies-Screening Matrix and Reference Guide.* EPA 542-B-93-005, July 1993.

EPA-Environmental Protection Agency (1983). *Health Assessment Document for Tetrachloroethylene (Perchloroethylene).* External review draft, EPA-600/8-82-005B.

EPA-Environmental Protection Agency (1987). *Unfinished Business, A Comparative Assessment of Environmental Problems, Appendix II, Non-Cancer Risk Work Group.* Office of Policy Analysis, Office of Policy, Planning and Evaluation, Washington, DC.

Esteve, J., E. Benhamou and L. Raymond (1994). *Descriptive Epidemiology.* International Agency for Research on Cancer Sci. Publ. no. 128, Lyon, France.

Farrar, D., B. Allen, K. Crump and A. Shipp (1989). "Evaluation of uncertainty in input parameters to pharmacokinetic models and the resulting uncertainty in output," *Toxicol. Lett.*, **49**, 371-385.

Feder, J. (1988). *Fractals.* Plenum Press, New York, NY.

Feller, W. (1966). *An Introduction to Probability Theory and its Applications.* Vol.**II**, J. Wiley & Sons, New York, NY.

Fetter, A. L. and J. D. Walecka (1971). *Quantum Theory of Many-Particle Systems.* McGraw Hill, New York, NY.

Feynman, R. P. (1962). *Quantum Electrodynamics.* Benjamin, New York, NY.

Fisher R., M. Mayer, W. von der Linden, & V. Dose (1997). "Enhancement of the energy resolution in ion-beam experiments with the maximum entropy method," *Phys. Rev. E*, **55**(6), 6667-6673.

Fisher, D. (1984). "Random walks in random environments," *Phys. Rev. A* **30**, 960-964.

Flaum, J.B., S.T. Rao and I.G. Zurbenko (1996). "Moderating the influence of meteorological conditions on ambient ozone concentrations," *J. Air & Waste Manage. Assoc.* **46**, 35-46.

Fleiss, J.L. (1981). *Statistical Methods for Rates and Proportions.* J. Wiley & Sons, New York, NY.

Fougere, P. F. (ed.) (1990). *Maximum Entropy and Bayesian Methods,* Kluwer Acad. Publ., Dordrecht.

Frisch U. and S. A. Orszag (1990). "Turbulence, challenges for theory and experiment," *Phys. Tod.*, **43**(1), 24-32.

Furuberg, L., J. Feder, A. Aharony and T. Jossang (1988). "Dynamics of Invasion Percolation," *Phys. Rev. Let.*, **61**(18), 2117-2120.

Gel'fand, I. M. and G. E. Shilov (1964). *Generalized Functions,* Vol.**I**. Academic Press, New York, NY.

Gel'fand, I. M. and N. Ya. Vilenkin (1964). *Generalized Random Functions*, Vol.**IV**. Academic Press, New York, NY.

Gel'fand, I. M. (1955). "Generalized Random Processes," *Dok. Akad. Nauk. SSSR,* **100**, 853-856.

Gelhar, L.W. and C. L. Axness (1983). "Three-dimensional stochastic analysis of macrodispersion in aquifers," *Water Resour. Res.*, **19**(1), 161-180.

Gelhar, L.W. (1993). *Stochastic Subsurface Hydrology.* Prentice Hall, Englewood Cliffs, NJ.

Gell-Mann, M. (1995). *The Quark and the Jaguar.* W. H. Freeman and Company, New York, NY.

Georgopoulos, P.G. and P.J. Lioy (1994). "Conceptual and theoretical aspects of human exposure and dose assessment," *Jour. of Expos. Anal. and Env. Epidem.*, **4**(3), 253-285.

Gnedenko, B. V. and A.N. Kolmogorov (1954). *Limit Distributions for Sums of Independent Random Variables.* Addison Wesley, Reading, MA.

Golberg A. M. (ed.) (1979). *Solution Methods for Integral Equations.* Plenum Press, New York, NY .

Graham, J.D. and J.B. Wiener (eds.) (1997). *Risk vs. Risk, Tradeoffs in Protecting Health and the Environment.* Harvard Univ. Pres, Cambridge, MA.

Grasso, D. (1993). *Hazardous Waste Site Remediation* , Lewis Publ., Boca Raton, FL.

Gull S. F. and G. J. Daniel (1978). "Image reconstrunction from incomplete and noisy data," *Nature* **272**, 686-690.

Gupta, R.S. (1997). *Environmental Engineering and Science.* Goverment Institutes, Rockville, MD.

Gutjahr, A. L. , L. W. Gelhar, A. A. Bakr, and J. R. Macmillan (1978). "Stochastic analysis of spatial variability in subsurface flows. 2. Evaluation and applications," *Water Resour. Res.*, **14**(5), 953-959.

Haack, S. (1996). *Deviant Logic Fuzzy Logic.* The Univ. of Chicago Press, Chicago, IL.

Haas, T.C. (1995). "Local prediction of a spatiotemporal process with an application to wet sulfate deposition," *J. Amer. Statistical Assoc.*, **90**(432), 1189-1199.

Haining, R. (1990). *Spatial Data Analysis in the Social and Environmental Sciences.* Cambridge Univ. Press, Cambridge, UK, p. 409

Haken, H. (1983). *Synergetics*. Springer Verlag, New York, NY.

Haltiner, G.J. and Williams, R.T. (1980). *Numerical Prediction and Dynamic Meteorology*. J. Wiley, NY., p. 477.

Harley, R., G. Cass, J.H. Seinfeld, L. McNair, A. Russell and G. McRae (1992). "Application of the CIT photochemical airshed model to SCAQS data base.1" *Southern California Air Quality Study Data Conference*, Los Angeles, CA.

Harris, E.E. (1996). *Hypothesis and Perception-The Roots of Scientific Method*. Humanities Press International, Atlantic Highlands, NJ.

Hayes, S.R., B.S. Austin and A.S. Rosenbaum (1988). *A Technique for Assessing the Effects of ROG and NO_x Reductions on Acute Ozone Exposure and Health Risk in the South Coast Air Basin*. Systema Applications, Inc. San Rafael, CA.

Hazucha, M.J. (1993). "Meta-analysis and "effective dose" revisited," *Advances in Controlled Clinical Inhalation Studies*. (U. Mohr *et al.*, eds.). Springer-Verlag, Berlin, Germany, pp. 247-256.

Helgason, S. (1980). *The Radon Transform*. Birkhauser, Boston, MA.

Herfindahl, O.C. and A.V. Kneese (1974). *Economic Theory of Natural Resources*. Charles E. Merrill Publ. Co., Columbus, OH.

Hernández, D. B. (1995). *Lectures on Probability and Second Order Random Fields*. World Scientific, River Edge, NJ.

Higginson, J. (1979). "Environmental carcinogenesis, a global perspective," *Environmental Carcinogenesis-Occurence, Risks Evaluation and Mechanisms* (Emmelot P. and E. Kriek, eds.) Elsevier, Amsterdam, The Netherlands, pp. 9-24.

Highfill, J.W. and D.L. Costa (1995). "Statistical response models for ozone exposure. Their generality when applied to human spirometric and animal permeability functions of the lung," *Jour. Air & Waste Assoc.*, **45**, 95-102.

Hillel, D. and Elrick, D.E. (1990). *Scaling in Soil Physics, Principles and Applications*. Soil Science Society of America, **25**, p. 122

Hinkfuss, I. (1975). *The Existence of Space and Time*. Oxford Univ. Press, Oxford, U.K.

Hoel, D.G. and P.J. Landrigan (1987). "Comprehensive evaluation of humans' data," In*Toxic Substances and Human Risk, Principles of Data Interpretation* (Tardiff R.G. and J.V. Rodricks, eds.) Plenum Press, New York, NY, pp.121-130.

Hori M. (1977). "Statistical theory of effective, electrical, thermal and magnetic properties of random heterogeneous materials. VII. Comparison of different approaches," *J. Math. Phys.* **18**(3), 487-501.

Horvath, S.M. and D.J. McKee (1994). "Health risk assessment of ozone," In *Tropospheric Ozone, Human Health & Agricultural Impacts* (D. J. Mckee, ed.) Lewis Publ., Boca Raton, FL, pp. 39-83.

Horvath, S.M., J.A. Gliner and L.J. Folinsbee (1981). "Adaption of ozone, duration of effect," *Am. Rev. Respir. Dis.*, **123**, 496-499.

Hristopulos D. T. and G. Christakos (1997a). "A Variational calculation of the effective fluid permeability of heterogeneous media," *Phys. Rev. E*, **55**(6), 7288-7298.

Hristopulos D. T. and G. Christakos (1997b). "Diagrammatic theory of nonlocal effective hydraulic conductivity," *Jour.of Stoch. Hydr. & Hydraul.*, **11**(5), 369-395.

Hristopulos, D. T. and G. Christakos (1997c). "An analysis of hydraulic conductivity upscaling," *Proceedings of the Second World Congress of Nonlinear Analysts*, In *Nonl. Anal.*, **30**(8), 4979-4984.

Hristopulos, D. T., G. Christakos and M. L. Serre (1998). "Numerical implementation of a space transformation approach for solving the three-dimensional flow equation," *SIAM Journal on Scientific Computing*, in press.

Hu, X. and J.H. Cushman (1994). "Nonequilibrium statistical mechanical derivation of a nonlocal Darcy's law for unsaturated/saturated flow," *Stoch. Hydr. and Hydraul.*, **8**(2), 109-116.

Hulka, B.S., J.D. Griffith and T.C. Wilcosky (1990). *Biological Markers in Epidemiology*. Oxford Univ. Press, Oxford, UK.

Hurst, H. E. (1951). "Long-term storage capacity of reservoirs," *Trans. Am. Soc. Civ. Eng.*, **116**, 770-808.

Hurst, H. E., R. P. Black and Y. M. Simaika (1965). *Long-term storage, an experimental study.* Constable, London, UK.

Hutt, M.S.R. and D.P. Burkitt (1986). *The Geography of Non-Infectious Disease.* Oxford Univ. Press, New York, NY, p. 164.

Indelman, P. and B. Abramovich (1994). "Nonlocal properties of nonuniform averaged flows in heterogeneous media," *Water Resour. Res.*, **30**(12), 3385-3393.

IPCC-Intergovermental Panel on Climate Change (1990). *Climate Change, The IPCC Scientific Assessment.* World Meteorological Organization/U.N. Environment Program. Cambridge Univ. Press, UK.

Isichenko, M. B. (1992). "Percolation, statistical topography, and transport in porous media," *Rev. Mod. Phys.,* **64**(4), 961-1043.

Itô, K. (1954). "Stationary random distributions," *Mem. College Sci., Univ. Kyoto,* **A**28, 209-223.

Jackson, J. D. (1975). *Classical Electrodynamics.* John Wiley & Sons, New York, NY.

Jaekel, U. and H. Vereecken (1997). "Renormalization group analysis of macrodispersion in a directed random flow," *Water Resour. Res.,* **33**(10), 2287-2299.

Jain, A. K. (1989). *Fundamentals of Digital Image Processing.* Prentice Hall, Englewood Cliffs, NJ.

Jaynes, E.T. (1983). *Papers on Probability, Statistics and Statistical Physics.* Synthese Library, V.**158** (Rosenkrantz, R.D., ed.) Reidel, Dordrecht.

Jeans, J. (1981). *Physics and Philosophy.* Dover Publ., New York, NY.

John, F. (1955). *Plane Waves and Spherical Means.* Springer-Verlag, New York, NY.

Johnson, D.H. and Dudgeon, D.E. (1993). *Array Signal Processing.* Prentice Hall, Englewood Cliffs, NJ.

Journel, A.G. and C. Huijbregts (1978). *Mining Geostatistics.* Academic Press, New York, NY.

Journel, A.G. (1983). "Non-parametric estimation of spatial distributions," *Math. Geol.,* **15**(3), 445-468.

Journel, A.G. (1987). *Geostatistics for the Environmental Sciences.* EPA Project No. CR 811893, Tech. Report, US EPA, EMS Lab, Las Vegas, NV.

Journel, A.G. (1989). *Foundamentals of Geostatistics in Five Lessons.* American Geophysical Union, Washington, DC.

Kendall,M. and A. Stuart (1977). *The advanced theory of statistics,* Vol.**I**. Griffin, London, UK.

King, P.E. (1979). "Problems of spatial analysis in geographical epidemiology," *Soc. Sci. & Med.*, **13D**, 249-252.

King, P. R. (1987). "The use of field theoretic methods for the study of flow in a heterogeneous porous medium," *J. Phys. A* **20**, 3935-3947.

King, P. R. (1989). "The use of renormalization for calculating effective permeability," *Trans. in Por. Media,* **4**, 37-58.

Kitanidis. P.K. (1983). "Statistical estimation of polynomial generalized covariance functions and hydrologic applications," *Water Resour. Res.,* **19**(4), 909-921.

Kitanidis P. K. (1988). "Prediction by the method of moments of transport in a heterogeneous formation," *J. Hydrol.,* **102**(1-4), 453-473.

Koch, D. L. and J. F. Brady (1989). "Anomalous diffusion due to long-range velocity fluctuations in the absence of a mean flow," *Phys. Fluids A* **1**(1), p. 47.

Koch, D. L., and E. S. G. Shaqfeh (1992). "Averaged-equation and diagrammatic approximations to the average concentration of a tracer dispersed by a Gaussian random velocity field," *Phys. Fl. A* **4**(5), 887-894.

Kolmogorov, A. N. (1941). 'The local structure of turbulence in an incompressible fluid at very large Reynolds numbers," *Dok. Akad. Nauk. SSSR,* **30**, 229-303.

Kondo J. (1991). *Integral Equations.* Oxford University Press, Oxford, UK.

Kosko, B. (1993). *Fuzzy Thinking.* Hyperion, New York, NY.

Kraichnan, R. H. (1961). "Dynamics of nonlinear stochastic systems," *Jour. Math. Phys.,* **2**(1), 124-148.

Krewski, D., D. Wigle, D.B. Clayson and G.R. Howe (1990). "Role of epidemiology in health risk assessment," In *Recent Results in Cancer Research-Occupational Cancer Epidemiology,* **120** (Band, P. *et al.*, eds.) Springer Verlag, New York, NY, pp. 1-24.

Kubo, R., H. Ichimura, T. Usui and N. Hashitsume (1981). *Statistical Mechanics.* North Holland Publ. Comp., New York, NY.

Kubo, R., M. Toda and N. Hashitsume (1991). *Statistical Physics II: Nonequilibrium Statistical Mechanics.* Springer Verlag, New York, NY.

La, D. K. and J. A. Swenberg (1996). "DNA adducts, biological markers of exposure and potential applications to risk assessment," *Mutation Research,* **365**, 129-146.

Lake, W. L. (1989). *Enhanced Oil Recovery.* Prentice Hall, Englewood Cliffs, NJ.

Landau, L. D. and E. M. Lifshitz (1958). *Statistical Physics.* Addison-Wesley Publishing Co., Inc. Reading, MA.

Landis, W.G. and M-H Yu (1995). *Introduction to Environmental Toxicology.* Lewis Publ., Boca Raton, FL.

Langran, G. (1992). *Time in Geographic Information Systems.* Taylor & Francis, London, UK.

Last, J.M. (ed.) (1995). *A Dictionary of Epidemiology.* Oxford Univ. Press, New York, NY.

Lee, Y-M and J.H. Ellis (1997). "On the equivalence of kriging and maximum entropy estimators," *Math.Geology*, **29**(1), 131-151.

Lefohn, A.S. and A.A. Lucier (1991). "Spatial and temporal variability of ozone exposure in forested areas of the U.S. and Canada, 1978-1988," *Jour. Air Waste Manag. Assoc.*, **41**, 694-701.

Lefohn, A.S., Knudsen, H.P., Logan, J.A., Simpson J. and C. Bhumralkar (1987). "An evaluation of the kriging method to predict 7-h seasonal mean ozone concentrations for estimating crop losses," *Jour. Air Pollut. Control Ass.*, **37**, 595-602.

Leung, H-W and D.J. Paustenbach (1988). "Application of pharmacokinetics to derive biological exposure indexes from threshold limit values," *Am. Ind. Hyg. Assoc. Jour.*, **49**(9), 445-450.

Leung, H-W (1991). "Development and formulation of physiologically-based pharmacokinetic models for toxicological applications," *Jour. Toxicol. Environ. Health*, **32**, 247-267.

Leung, H-W (1992). "Use of physiologically based pharmacokinetic models to establish biological exposure indexes," *Am. Ind. Hyg. Assoc. Jour.*, **53**(6), 369-374.

Levin, S. A., B. Grenfell, A. Hastings and A. S. Perelson (1997). "Mathematical and computational challenges in population biology and ecosystems science," *Science,* **275**, 334-343.

Lippmann, M. and R. B. Schlesinger (1979). *Chemical Contamination in the Human Environment.* Oxford Univ. Press, New York, NY.

Lippmann, M. (1989). "Health effects of ozone, a critical review," *Jour. Air Pollut. Control Ass.*, **39**, 672-695.

Loaiciga, H. (1989). "An optimization approach for groundwater quality monitoring network design," *Water Resour. Res.*, **25**, 1771-1782.

Loeve, M. (1953). *Probability Theory.* Van Nostrand, Princeton, Englewood Cliffs, NJ.

Logan, J.A. (1989). "Ozone in rural areas of the US," *J. Geophys. Res.* **94**(D6), 8511-8532.

Longair, M.S. (1984). *Theoretical Concepts in Physics.* Cambridge Univ. Press, Cambridge, UK.

Loomis, T.A. (1978). *Essentials of Toxicology.* Lea & Febiger, Philadelphia, PA.

Louis, A. (1986). "Incomplete data problems in X-Ray computerized tomography," *Numerische Mathematik,* **48**, 251-262.

Lucas, J.R. (1973). *A Treatise on Time and Space.* Methuen Publ., London, UK.

MacDonald, J.A. and M.C. Kavanaugh (1995). "Superfund, The cleanup standard debate," *Policy & Planning*, 55-61.

MacMahon, B. and T.F. Pugh (1970). *Epidemiology-Principles and Methods.* Little, Brown and Co., Boston, MA.

Makse, H. A., G. W. Davies, S. Havlin, P. Ch. Ivanov, P. R. King and H. E. Stanley (1996a). "Long-range correlations in permeability fluctuations in porous rock," *Phys. Rev. E* **54**(4), 3129-3134.

Makse, H. A., S. Havlin, M. Schwartz and H. E. Stanley (1996b). "Method for generating long-range correlations for large systems," *Phys. Rev. E* **53**(45), 5445-5449.

Mandelbrot, B. B. and J. W. Van Ness (1968). "Fractional Brownian motions, fractional noises and applications," *SIAM Review,* **10**(4), 422-437.

Mandelbrot, B. B. (1982). *The Fractal Geometry of Naure.* Freeman & Company, New York, NY.

Manton, K.G., M.A. Woodbury, E. Stallard, W.B. Riggan, J.P. Creason, and A.C. Pellon (1989). "Empirical Bayes procedures for stabilizing maps of U.S. cancer mortality rates," *Jour. of the Amer. Stat. Assoc.*, **84**, 637-650.

Marcus M. and H. Minc (1992). *A survey of matrix theory and matrix inequalities.* Dover Publ., NY.

Marshall, R.J. (1991). "Mapping disease and mortality rates using empirical Bayes estimators," *Appl. Stat.*, **40**(2), 282-294.

Massmann, L., and R.A. Freeze (1987). "Groundwater contamination from waste management sites, the interaction between risk-based engineering design and regulatory policy,"*Water Resour. Res.*, **23**(2), 351-367.

Matheron, G. (1967). *Elements pour une théorie des milieux poreux.* Masson, Paris, France.

Matheron, G. (1973). "The intrinsic random functions and their applications," *Adv. Appl. Prob.*, **5**, 439-468.

Matheron, G. and G. de Marsily (1980). "Is transport in porous media always diffusive? A counterexample,"*Water Resour. Res.*, **16**(5), 901-917.

Mattuck, R. D. (1976). *A Guide to Feynman Diagrams in the Many-Body Problem.* McGraw Hill, New York, NY.

Mausner, J.S. and A.K. Bahn (1974). *Epidemiology, An Introductory Text.* W. B. Saunders, Philadelphia, PA.

Mayer, J.D. (1983). "The role of spatial analysis and geographic data in the detection of disease causation," *Soc. Sci. Med.*, **17**(16), 1213-1221.

Mayo, D.G. and R.D. Hollander (1991). *Acceptable Evidence, Science and Values in Risk Management.* Oxford Univ. Press, New York, NY.

McComb, W. D. (1990). *The Physics of Turbulence,* Oxford Univ. Press, New York, NY.

McCurdy, T.R. (1994). "Human exposure to ambient ozone," In *Tropospheric Ozone, Human Health and Agricultural Impacts* (D. J. Mckee, ed.) Lewis Publishers, Boca Raton, FL, pp. 85-127.

McDonnell, W.F.III, R.S. Champan, M.W. Leigh, G.L. Strope and A.M. Collier (1985). "Respiratory responses of vigorously exercising children to 0.12 ppm ozone exposure," *Am. Rev. Respir. Dis.*, **132**, 875-879.

McJilton, C., J. Thielke and R. Frank (1972). "Ozone uptake model for the respiratory system,," *Abstr. Technical Papers, Amer. Industrial Hygiene Confer.*, San Francisco, CA, *Am. Ind. Hyg. Assoc. Jour.*, **22**, Paper No. 45.

McKee, D.J. (ed.) (1994). *Tropospheric Ozone, Human Health and Agricultural Impacts.* Lewis Publishers, Boca Raton, FL.

Mei, C.C. and J.-L. Auriault (1989). "Mechanics of heterogeneous porous media with several spatial scales," *Proc. R. Soc. Lond.* A **426**, 391-423.

Miller, C.T., G. Christakos, P.T. Imhoff, J.F. McBride, J.A. Pedit, and J.A. Trangenstein (1998). "Multiphase flow and transport modelling in heterogeneous porous media, challenges and approaches," *Adv. Water Res.*, **21**(2), 77-120.

Miller, F.J., J.H. Overton, Jr., R.H. Jaskot and D.B. Menzel (1985). "A model of the regional uptake of gaseous pollutants in the lung. I. The sensitivity of the uptake of ozone in the human lung to lower respiratory tract secretions and to exercise," *Toxic. Appl. Pharmac.*, **79**, 11-27.

Molz F. J., O. Guven, J. G. Melville and C. Cardone (1990). "Hydraulic conductivity measurement at different scales and contaminant transport modeling," In *Dynamics of Fluids in Hierarchical Porous Media* (J. Cushman, ed.) pp. 37-59.

Moore, K.L. and A.M.R. Agur (1995). *Essential Clinical Anatomy.* Williams & Wilkins, Baltimore, MD.

Morris, S.C. (1990). *Cancer Risk Assessment.* Marcel Dekker, Inc., New York, NY.

Morrison, M.A. (1990). *Understanding Quantum Physics.* Prentice-Hall, Inc., Englewood Cliffs, NJ.

Muir, C.S. (1975). "International variation in high-risk populations," In *Persons at High Risk of Cancer. An Approach to Cancer Etiology and Control.* (Fraumeni J.F. Jr., ed.) Acad. Press, New York, NY, pp. 293-305,.

Murray, J. D. (1993). *Mathematical Biology.* Springer Verlag, New York, NY.

Nabholz, J.V. (1991). "Environmental hazard and risk assessment under the U.S. Toxic Substances Control Act," *Science of the Total Env.*, 109/110, 649-665.

Nayfield, S.G., J.E. Karp, L.G. Ford, A. Dorr and B.S. Kramer (1992). "Potential role of tamoxifen in prevention of breast cancer," *Jour. Nat. Cancer Inst.*, **83**, 1450-1459.

Neely, W.B. (1994). *Introduction to Chemical Exposure and Risk Assessment.* CRC Lewis Publ., Boca Raton, FL.

Neuman, S.P., C. L. Winter, C.M. Newman (1987). "Stochastic theory of field-scale Fickian dispersion in anisotropic porous media," *Water Res. Res.* **23**(3), 453-466.

Neuman, S. P. (1990). "Universal scaling of hydraulic conductivities and dispersivities in geologic media," *Water Resour. Res.*, **26**(8), 1749-1758.

Neuman, S. P., S. Orr, O. Levin and E. Paleologos (1992). "Theory and high-resolution finite element analysis of 2-D and 3-D effective permeabilities in strongly heterogeneous porous media," In *Mathematical Modeling in Water Resources,* Vol.**2** (T. F. Russell, R. E. Ewing, C. A. Brebbia, W. C. Gray and G. F. Pinder, eds.) Elsevier Appl. Sci., New York, NY, pp. 118-13.

Neuman, S. P. and S. Orr (1993). "Prediction of steady state flow in nonuniform geologic media by conditional moments, exact nonlocal formalism, effective conductivities and weak approximation," *Water Resour. Res.*, **29**(2), 341-364.

Neuman, S. P. (1994). "Generalized scaling of permeabilities, validation and effect of support scale," *Geoph. Rev. Let.*, **21**(5), 349-352.

Noettinger, B. (1994). "The effective conductivity of a heterogeneous porous medium," *Trans. in Por. Media,* **15**, 99-127.

NRC -National Research Council (1991). *Rethinking the Ozone Problem in Urban and Regional Air Pollution.* National Academy Press, Washington, DC.

Nyer, E.K. (1992). *Practical Techniques for Groundwater and Soil Remediation.* CRC Lewis Publ., Boca Raton, FL.

Olea, R. A. (1997). *Understanding Geostatistics.* Class Notes. Civil Engineering Dept., Univ. of Kansas, Lawrence, KS., p. 290.

Oliver, L.D., and G. Christakos (1995). "Diagrammatic solutions for hydraulic head moments in 1-D and 2-D bounded domains," *Stoch. Hydrol. and Hydraul.*, **9**, 269-296.

Oliver, L.D., and G. Christakos (1996). "Boundary condition sensitivity analysis of the stochastic flow equation," *Adv. Water Res.*, **19**(2), 109-120.

Oliver, M.A., K.R. Muir, R. Webster, S.E. Parkes, A.H. Cameron, M.C.G. Stevens and J.R. Mann (1992). "A geostatistical approach to the analysis of pattern in rare disease," *Jour. of Public Health Med.,* **14**(3), 280-289.

Pais, A. (1982). *Subtle is the Lord, The Science and the Life of Albert Einstein.* Oxford Univ. Press, New York, NY.

Papameletiou, D. (1995). "Clean Production Technologies, Figuring out key issues for the future," In *IPTS Report,* Dec.95 (D. Kyriakou, ed.). IPTS, Joint Research Centre, European Commission, pp. 20-24 .

Parkin, D.M. (1989). "Cancers of the breast, endometrium and ovary, geographic correlations," *Eur. Jour. Cancer Clin. Oncol.*, **25**, 1917-1925.

Patterson, D.W. (1990). *Introduction to Artificial Intelligence and Expert Systems.* Prentice Hall, Englewood Cliffs, NJ.

Pearson, K. (1951). *The Grammar of Science.* J.M. Dent & Sons Ltd., London, UK.

Peng, C.-K., S. V. Budyrev, S. Havlin, H. E. Stanley and A. L. Goldberger (1994). "Mosaic organization of DNA nucleotides," *Phys. Rev. E* **49**(2), 1685-1689.

Perera, F. P. (1996). "Molecular epidemiology in cancer prevention," In *Cancer Epidemiology and Prevention* (D. Schottenfeld and J.F. Fraumeni, Jr., eds.) Oxford Univ. Press, New York, NY.

Pickle, L.W., T.J. Mason, and J.F. Fraumeni, Jr. (1989). In *The New U.S. Cancer Atlas: Recent Results in Cancer Research-Cancer Mapping*, **114** (Boyle, P. *et al.*, eds.) Springer Verlag, New York, NY, pp. 196-207.

Pielke, R.A., W.R. Cotton, R.L. Walko, C.J. Tremback, L.D. Lyons, L.D. Grasso, M.E. Nicholls, M.D. Moran, D.A. Wesley, T.J. Lee and J.H. Coperland (1992). "A comprehensive meteorological modeling system-RAMS," *Met. Atmos. Phys.* **49**, 69-91.

Pietronero, L. (1995). "Theoretical concepts for fractal growth and self-organized criticality," *Fractals*, **3**(3), 405-414.

Pilinis, C., P. Kassomenos and G. Kallos (1993). "Modeling of photochemical pollution in Athens, Greece. Application of the RAMS-CALGRID modeling system," *Atm. Env.* **27B**(4), 353-370.

Piotrowski, J. (1971). *The Application of Metabolic and Excretion Kinetics to Problems of Industrial Toxicology.* National Instit. of Health, Washington DC.

Pipes, L. A. and L. R. Harvill (1970). *Applied Mathematics for Physicists and Engineers.* McGraw-Hill, New York, NY.

Pisani, P. (1992). "Breast cancer, geographic variation and risk factors," *J. of Env. Pathol. & Oncol.*, **11**(5/6), 313-316.

Poincare, H. (1929). *La Science et l'Hypothese.* Flammarion, Paris, France.

Polanyi, M. (1958). *Personal Knowledge, Towards a Post-Critical Philosophy.* Univ. of Chicago Press, Chicago, IL.

Polyak, I., G.R. North and J.B. Valdes (1994). "Multivariate space-time analysis of PRE-STORM precipitation," *J. Appl. Meteor.*, **33**, 1079-1087.

Popper, K.R. (1934). *Logik der Forschung.* Springer, Vienna, Austria.

Powell, D.M. (1994). "Selecting innovative cleanup technologies, EPA resources," *Chem. Engin. Progr.*, 33-41.

Prigogine, I. (1980). *From Being to Becoming.* W. H. Freeman and Company, New York, NY.

Pukkala, E. (1989). "Cancer maps in Finland, an example of small area-based mapping," In *Recent Results in Cancer Research-Cancer Mapping*, **114** (Boyle, P. *et al.*, eds.) Springer Verlag, New York, NY, pp. 208-215.

Quintard, M. and S. Whitaker (1987). "Ecoulement monophasique en milieu poreux, effet des heterogeneites locales," *Jour. de Mecha. Theor. et Appl.*, **6**(5), 691-726.

Raiffa, H., and R. Schlaifer (1961). *Applied Statistical Decision Theory.* Harvard Business School, Boston, MA.

Ramsey, J.C. and M.E. Andersen (1984). "A physiologically-based description of the inhalation pharmacokinetics of styrene in rats and humans," *Toxicol. Appl. Pharmacol.*, **73**, 159-175.

Rao, S.T. and I.G. Zurbenko (1994). "Detecting and tracking changes in ozone air quality," *Jour. Air and Waste Manag. Assoc.* **44**, 1089-1092.

Rao, S.T., E. Zalewsky and I.G. Zurbenko (1995). "Determining temporal and spatial patterns in ozone air quality," *Jour. Air and Waste Manag. Assoc.* **45**, 57-61.

Rappaport, S.M. (1991). "Exposure assessment strategies," *Expos. Asses. for Epidem. and Haz. Control* (S.M. Rappaport and T.J. Smith, eds.) Lewis Publ., Chelsea, MI, pp. 219-249.

Ravenscroft, P.J. (1992). "Recoverable reserve estimation by conditional simulation," In *Case histories and methods in mineral resource evaluation* (Annels, A.E., ed.) *Spec. Publ. Geol. Soc. London*, **63**, 289-298.

Ray Smith C., R. Inguva & R. L. Morgan (1984). "Maximum entropy inverses in physics," *SIAM-AMS Proceedings*, **14**, 127-137.

Reiss, R.D. and M. Thomas (1997). *Statistical Analysis of Extreme Values.* Birkhauser Verlag, Basel, Switzerland.

Rhodes, C. J. and R. M. Anderson (1996). "Power laws governing epidemics in isolated populations," *Nature*, **381**, 600-602.

Riemann Hershey, R. (1997). "Using geostatistical techniques to map the distribution of tree species from ground inventory data," In *Model. Longitud. and Spat. Correl. Data* (T. G. Gregoire, D.R. Brillinger, P.J. Figgle, E. Russek-Cohen, W.G.Warren, R.D. Wolfinger, eds.) Springer Verlag, New York, NY.

Riesz, F. and B. Sz.-Nagy (1990). *Functional Analysis.* Dover Publ., New York, NY.

Rivoirard, J. (1994). *Introduction to Disjunctive Kriging and Non-linear Geostatistics.* Clarendon Press, Oxford, UK.

Rombout, P.J.A. and P.E. Schwarze (1995). "Quantitative exposure-response relationships for ozone," *Proceed. of the Nordic Expert Meet. on the Estimation of Potential Health Effects from Air Pollut. Exposure on a Regional Scale*, Oslo, Norway.

Rossi, H.H. and M. Zaimer (1996). *Microdosimetry and its Applications.* Spinger-Verlag, New York, NY.

Rouhani, S. and T.J. Hall (1989). "Space-time kriging of groundwater data," In *Geostatistics* (Armstrong, M., ed.) Kluwer Acad. Publ., **2**, pp. 639-651.

Ruelle, D. (1991). *Chance and Chaos.* Princeton Univ. Press, Princeton, NJ.

Sadowski, F.G. and S.J. Covington (1987). *Processing and Analysis of Commercial Satellite Image Data of the Nuclear Accident near Chernobyl, U.S.S.R.* U.S. Geol. Survey, Bulletin 1785, U.S. Gov. Printing Off., Washington, DC.

Salles, J., J. F. Thovert and P. M. Adler (1993). "Reconstructed porous media and their application to fluid flow and solute transport," *Jour. Contam. Hydr.*, **13**, 3-22.

SANLIB-Stochastic Analysis Software Library and User's Guide (1995). *Stochastic Research Group*, Research Rept. n.SM/1.95, Dept. of Environmental Sci. and Engin., Univ. of North Carolina, Chapel Hill, NC.

Schmidt, R. and K. Housen (1995). "Problem solving with dimensional analysis," *Ind. Phys.* **1**(1), 16-24.

Schnoor, J.L. (ed.) (1984). *Modeling of Total Acid Precipitation Impacts.* Butterworth, Boston, MA.

Schulte, P.A. and F.P. Perera (eds.) (1993). *Molecular Biology.* Acad. Press, San Diego, CA.

Schwartz, L. (1950; 1951). *Théorie des distributions,* Vols.**I-II**. Actualités Scientifiques et Industrielles, Hermann & Cie, Paris, France.

Searle, S. R. (1971). *Linear Models.* Wiley, New York, NY.

Seinfeld, J.H. (1986). *Atmospheric Chemistry and Physics of Air Pollution.* J. Wiley, New York, NY.

Serre, M., P. Bogaert, G. Christakos (1998). "Latest computational results in spatiotemporal prediction using the Bayesian maximum entropy method," *Intern. Assoc. for Math. Geol.*, Ischia Island, Italy.

Shafer, S.Q. (1978). "Mapping mortality, the search peaks of bone cancer death rates in Pennsylvania counties," M.A. thesis, *Dept. of Geography, Columbia University*, NY.

Shannon. C. E. (1948). "A mathematical theory of communication," *Bell System Tech. Jour.*, **27**, 379-423 and 623-656.

Sheldrake, R. (1995). *The Presence of the Past.* Park Street Press, Rochester, VE.

Shields, J. (1990). *Environmental Health, New Directions.* Princeton Sci. Publ. Co., Inc., Princeton, NJ.

Shlesinger, M. F., G. M. Zaslavsky and U. Frisch (eds.) (1994). "Lévy Flights and Related Topics in Physics," In *Proceed. of the International Workshop, Nice, France,* Springer Verlag, NY.

Shvidler, M. I. (1964). *Filtration Flows in Heterogeneous Media.* Consultants Bureau, New York, NY.

Simpson, D. (1995). "Modelled ozone concentrations in relation to health issues," Proceed. of the *Nordic Expert Meeting on Estimation of Potential Health Effects from Air Pollut. Exposure on a Regional Scale*, Oslo, October 15-17, Norway.

Sinopoli, N.T., G.E. Trivers, C. Ficorella, S. Tomao, M. Martelli, P. Cagnazzo, N. Sama, C.C. Harris, and I. Frati (1990). "Immunoassay detection of carcinogen-DNA adducts in tumor and noninvolved lung tissue from lung cancer patients," *Proc. Am. Assoc. Cancer Res.*, **31**, p. 97.

Skilling, J. (ed.) (1989). *Maximum Entropy and Bayesian Methods,* Kluwer Acad. Publ., Dordrecht.

Smans, M., Muir, C.S. and P. Boyle (1992). *Atlas of cancer mortality in the European Community.* IARC Sci. Publ., No. 107, IARC, Lyon, France.

Smart, J.J.C. (ed.) (1964). *Problems of Space and Time.* Macmillan, New York, NY.

Smith, A.E. and D.J. Fingleton (1982). *Hazardous Air Pollutant Prioritization System (HAPPS).* EPA 450/5-82-008. Prepared for the Environmental Protection Agency by Argonne National Laboratory, Argonne, IL.

Smith, L. and R.A. Freeze (1979). "Stochastic analysis of steady-state groundwater flow in a bounded domain, 2. Two-dimensional simulations," *Water Resour. Res.*, **15**(6), 1543-1559.

Stanley, H. E. and N. Ostrowsky (eds.) (1986). *On Growth and Form*, in NATO ASI Series, Kluwer Acad. Publ., Hingham, MA.

Stauffer, D. and A. Aharony (1992). *Introduction to Percolation Theory.* Taylor and Francis, London, UK.

Stratton, J.A. (1941). *Electromagnetic Theory.* McGraw Hill, New York, NY.

Sturgeon S.R., C. Schairer, M. Gail, M. McAdams, L.A. Brinton, R.N. Hoover (1995). "Geographic variation in mortality from breast cancer among white women in the US," *Jour. of the Nat. Cancer Inst.*, **87**(24), 1846-1853.

Swenberg, J. A., N. Fedtke, T. R. Fennell and V. E. Walker (1990). "Relationships between carcinogen exposure, DNA adducts and carcinogenesis," *Progress in Predictive Toxicology* (D. B. Clayson, I. C Munro, P. Shubik and J. A. Swenberg, eds.) Elsevier, Amsterdam, pp. 161-184.

Swerling, P. (1962). "Statistical properties of the contours of random surfaces," *IRE Trans. Infor. Th.*, **IT-8**, 315-321.

Tatarski, V. I. (1964). "Propagation of waves in a turbulent atmosphere," *Sov. Phys. JETP*, **19**, 946-953.

Thompson, A. M., J. A. Chappellaz, I. Y. Fung and T. L. Kucsera (1993). "The atmospheric CH_4 increase since the Last Glacial Maximum, 2, Interaction with oxidants," *Tellus*, **45B**, 242-257.

Thurston, G.D., K. Ito, P.L. Kinney and M. Lippmann (1992). "A multi-year study of air pollution and respiratory hospital admissions in thre New York State metropolitan areas, results for 1988 and 1989 Summers," *Jour. of Exp. Anal. and Env. Epidem.*, **2**(4), 429-450.

Tikhonov, A.N. and A.V. Goncharsky (1987). *Ill-Posed Problems in the Natural Sciences.* Mir Publishers, Moscow.

Tompson, A.F.B., R. Ababou, and L.W. Gelhar (1989). "Implementation of the three-dimensional turning bands random field generator," *Water Resour. Res.*, **25**(10), 2227-2243.

Trichopoulos, D., S. Yen, J. Brown, P. Cole and B. MacMahon (1984). "The effect of Westernization on urine estrogens, frequency of ovulation, and breast cancer risk. A study of ethnic Chinese in the Orient and the USA," *Cancer*, **53**, 187-192.

U.S. Bureau of the Census (1992). "1990 Census of Population and Housing, Summary Tape File 1-C," *US Department of Commerce*, Bureau of Census, Washington, DC.

U.S. Office of Technology Assessment, U.S. Congress-OTA (1981). "Assessment of technologies for determining cancer risks from the environment," *US Governemnt Printing Office*, Washington, DC.

Vander, A., J. Sherman and D. Luciano (1998). *Human Physiology.* WCB McGraw-Hill, Boston, MA.

Viecelli, J. A. and E. H. Canfield (1991). "Functional representation of power-law random fields and time series," *Jour. Comp. Phys.*, **95**, 29-39.

Vinegar, A., D.W. Winsett, M.E. Andersen and R.B. Conolly (1990). "Use of physiologically-based pharmacokinetic model and computer simulation for retrospective assessment of exposure to volatile toxicants," *Inhal. Toxicol.*, **2**, 119-128.

Von der Linden, W. (1995). "Maximum entropy data analysis," *Appl. Phys. A* **60**, 155-165.

Voss, R. F. (1985). "Random fractal forgeries" In *Fundamental Algorithms in Computer Graphics* (R. A. Earnhshaw, ed.) Springer Verlag, Berlin, pp.805-835,.

Vukovich, F.M. (1994). "Boundary layer ozone variations in the Eastern U.S. and their association with meteorological variations, long-term variations," *Jour. of Geophys. Res.*, **99**(D3), 16839-16850 .

Vyas, V. and G. Christakos (1997). "Spatiotemporal analysis and mapping of sulfate deposition data over the conterminous U.S.A," *Atm. Env.*, **31**(12), 3623-3633.

Vyas, V.M. (1997). *Random Field Analysis of Space/Time Natural Processes.* PhD Thesis Report, Dept. of Environmental Sci. and Engin., Univ. of North Carolina, Chapel Hill, NC.

Vyas, V.M. and G. Christakos (1998). "Spatiotemporal analysis and mapping of sulfate deposition data over the conterminous USA," *Atm. Env.*, in press.

Wagner, B.J., and S.M. Gorelick (1987). "Optimal groundwater quality management under parameter uncertainty," *Water Resour. Res.*, **23**(7), 1162-1174.

Wassertheil-Smoller, S. (1996). *Biostatistics and Epidemiology.* Springer-Verlag, New York, NY.

Weinberg, S. (1972). *Gravitation and Cosmology, Principles and Applications of the General Theory of Relativity.* Wiley, New York, NY.

Wentz, C.A. (1989). *Hazardous Waste Management*, McGraw-Hill, New York, NY.

Weyl, H. (1952). *Space Time Matter.* Dover Publ. New York, NY.

Weyl, H. (1987). *The Continuum. A Critical Examination of the Foundation of Analysis.* Dover Publ., Inc., New York, NY.

Whitfield, R.G., H. M. Richmond, S. R. Hayes, A. S. Rosenbaum, T. S. Wallsten, R. S. Winkler, M. J. G. Absil, and P. Narducci (1994). "Health risk assessment of ozone," *Trop. Ozone* (D. J. Mckee, ed.) Lewis Publ., Boca Raton, FL, pp. 129-173.

Whitfield, R.G., W.F. Biller, M.J. Jusko and J.M. Kaisler (1995). "A probabilistic assessment of health risks associated with short-term exposure to tropospheric ozone," ANL/DIS-3, Decision and Information Sci. Div., Argonne National Lab, Argonne, IL, p.219.

Whittaker, E. T. and G. N. Watson (1927). *A Course of Modern Analysis.* Cambridge Univ. Press, Cambridge, UK.

Wielicki, B.A., R.D. Cess, M.D. King, D.A. Randall and E.F. Harrison (1995). "Mission to planet earth, role of clouds and radiation in climate," *Bull. of the Amer. Meteorological Soc.*, **76**(11), 2125-2153.

Wiener, N. (1930). "Generalized harmonic analysis," *Acta Mathema.* **55**(2-3), 117-258.

Wilkinson, D. and J. F. Willemsen (1983). "Invasion percolation: A new form of percolation theory," *J. Phys. A* **16**, 3365-3376.

Wilson, K.G. and J. Kogut (1974). "The renormalization group and the ε expansion," *Phys. Rep. C* **12**(2), 75-200.

Wolfram, S. (1991). *Mathematica, A System for Doing Mathematics by Computer.* Addison-Wesley Comp., Reading, MA.

Yaglom, A. M. (1957). "Some classes of random fields in n-dimensional space related to stationary random processes," *Theory of Prob. and its Appl.*, **II**(3), 273-320.

Yaglom, A. M. (1987). *Correlation Theory of Stationary and Related Random Functions I.* Springer Series in Statistics, Springer Verlag, New York, NY.

Yakhot, V. and Orszag, S.A. (1986). "Renormalization group analysis of turbulence,"*Phys. Rev. Lett.*, **57**(14), 1722-1724.

Yamartino, R.J., J.S. Scire, G.R. Carmichael and Y.S. Chang (1992). "The CALGRID mesoscale photochemical grid model-I. Model formulation," *Atm. Env.*, **26A**, 1493-1512.

Yow, Jr., J.L., Aines R.D. and R.L. Newmark (1995). "Demolishing NAPLs," *Civil Eng.*, 57-59.

Zartarian, V. G., W. R. Ott and N. Duan (1997). "A quantitative definition of exposure and related concepts" *Jour. of Exp. Anal. and Env. Epidem.*, **7**(4), 411-437.

Zhang, D. and S. P. Neuman (1995). "Eulerian-Lagrangian analysis of transport conditioned on hydraulic data, 1. Analytical-numerical approach," *Water Resour. Res.*, **31**(1), 39-51.

Zhang, Q. (1995). "Transient behavior of mixing induced by a random velocity field," *Water Resour. Res.*, **31**(3), 577-591.

Ziman, J. (1978). *Reliable Knowledge.* Cambridge Univ. Press, Cambridge, UK.

Zwart, P.J. (1976). *About Time.* North Holland, Amsterdam, the Netherlands.

Index